it the year before. Last January when I was "billed" for it, I had an attendance of over 250, so that poultry-keepers must enjoy it and the lessons therefrom. The one advantage is that I demonstrate on different birds, which varies the lecture each time.

I am not unmindful of the fact that the interest shown in the internal workings of the laying hen at these lectures has done much to show me that the desire for knowledge on the subject is greater than it was. Such interest has in fact tempted me to alter my views of years ago, and to attempt this volume, hoping that it will supply the need.

Another aim is to teach observation on the part of every poultry-keeper, and to prevent him from being automatic. Being connected officially with so many societies, I know fully well that observation and a sound understanding of the little hen are other very weak links. It is to teach observation and to encourage the personal side in poultry-keeping that I practise my score-card system. For the same reasons I have been pioneer in Utility Shows, which are of a great educational value to backyarders. They bring out the personal side of the man, making him keen and a student of detail. As a result the hens receive better treatment, and poultry-keeping becomes profitable and pleasurable instead of a nuisance.

In conclusion, this volume has been made possible by the support of my friends—especially those who have taken advertising space herein. If you enjoy the book, you may show appreciation by supporting them. My thanks, too, to those who have ordered copies in advance, and have generously promised to recommend the volume to their friends. Two editions are being published—one for breeders, and a less costly issue for backyarders and small poultry-keepers.

W. POWELL-OWEN.

CONTENTS.

SELECTING THE LAYERS.

CHAPTER I.

THE HEN FROM WITHIN.

IN an interesting way I will go through the photographic Plates given in the book of the internal organs of a hen, keeping to simple explanations and names, so that what has too often been regarded as a dry and laborious subject will not be too hurriedly studied.

Plate 1 is of an ordinary hen, which is to be the medium for demonstrating the internal organs. The body is nailed to a board in an upright position, ready for cutting up, and no poultry-keeper without inside knowledge would be able to guess the wonderful mechanism within and the masterful placings of the organs. To me the plump breast would in, say, February point the bird out as not a very good winter layer from the previous October. I take my key from her plump condition, which is altogether foreign to a hen in February or March that has been laying heavily throughout the five or six winter months. Before we have completed the Plates I am hoping to have shown you that fat condition and heavy production are well apart, and that in certain seasons you cannot have eggs and excessive internal fat; but must choose between the two.

The rounded breast is so marked that if the carcase was on the poulterer's slab, and you were buying a bird for table, you would have no hesitation in picking this one. Yet were I to ask you if your hens were fat, or your table cockerels fit for killing, you might be in difficulties. The only difference lies in the fact that, when living, the parts are covered with feathers, and out of sight. That being so, your first lesson in handling will commence, and if you are broad-minded you will admit that to judge the condition of the flock, whether layers, youngsters or table cockerels, you MUST take them in the hand. What the eye alone sees may prove deceiving, but the practised hand will keep you on the right road. The thighs are well fleshed also, and coming to the abdomen we feel plenty of fat inside, the gizzard being difficult to find. In February, if the bird had been a heavy layer during the winter, I should have expected to have felt the gizzard like a potato at the end of the breast-bone. Again, the pelvis bones on either side of the vent are thickly covered with fat, whereas after five or six months of good production they should be pliable and just nicely coated with soft workable flesh.

You will better understand the demonstration if you take that hen which died yesterday and cut the body up as I show you. You ought really to open every bird that dies, and endeavour to locate the cause of death. It will save you much worry. Often a hen dies and the owner becomes nervous. He changes his feeding and puts the whole flock out of lay and tone. This lack of understanding the hen from within is responsible for many false steps which eat away the profits. Why should not a hen die? If you have first-grade laying stock, you will, by the time I have finished this volume, understand how every good layer risks her life with each egg laid, and more so if she is not in fit condition internally. That is your part of routine to keep her fit, and therein lies her insurance policy. Always open every body, whether chick or adult, and in poultry-keeping be sure and call a spade a spade. Many owners of poultry keep hens while they are unproductive, and then sell them just as they are going to start to lay for the other fellow to get the profits, just for lack of a simple method of handling the birds for condition as to laying. Not a few keepers of poultry declare daily for months that their hens will lay any day now, whereas they have been producing, and, sad to say, *eating* the eggs. I have known a duck mated to a pen of other ducks because the owner thought she was a drake. One of our largest breeders betted me £500 that a hen in male plumage was a *cockerel*; but my knowledge of handling stood me in good stead, as the alleged cockerel laid eggs and sat on them, and they proved fertile. Remember my advice, and call a spade a spade every time.

You will need for post-mortem work a few simple tools. First comes a sharp pair of scissors; secondly, a pair of pliers for breaking the bones; thirdly, a small saw for getting through bones; and finally, a hammer and nails. An apron, india-rubber or other gloves, a box to put the feathers in as plucked off, a towel to wipe on, and a bowl of water containing disinfectant to wash in after the cutting up, will complete the accessories. The tools will, of course, be washed in disinfectant after use, and immediately dried.

The first operation is to pluck the feathers from the breast, abdomen, thighs and neck, and the body can be nailed flat to a table or upright to a board, with breast facing the operator.

Drive a nail through the neck near the head, and then stretch out the wings and secure each with a nail or two; the legs can also be stretched out and nailed down.

Plate 2 confirms our suspicions of excessive internal fat, and having removed some skin from the abdomen we receive the first introduction to excessive internal fat which I contend is the great enemy of egg-production. In B we have the first fat between the skins so to speak. Next we have at A the second fat commonly known as the " bladder of lard " fat. It is of interest to bear these in mind because we are going to see what organs are beneath them, and to glean what harm the fats do. If the bird had to be cut open at once we would run the scissors along each side of the breast-bone as shown by the dotted lines. The best plan is to pinch up the skin at the base of the breast-bone and with the scissors to cut it through, afterwards continuing the cut along the dotted lines, but taking care not to cut into any organs below.

Plate 3 gives a clear illustration of the " bladder of lard " fat (A) which is encircling the internal organs. The end of the breast-bone is marked C, and the dotted lines will be the first cut when removing the breast-bone. The lobe of the liver is marked L. It is noticeable from this Plate that the abdomen is very full or heavy and considerably out of proportion in length to the breast-bone. Such a bird when walking would have a dropped abdomen, which I am very much against. I vote every time for the breast-bone that supports the abdomen and prevents it from sagging. It must be straight, of course, and not the other extreme in being too long. Of that later.

Plate 4 supplies us with the solution to the problem of why overfat hens die of heart trouble. The photograph is a side-view showing the breast-bone, the end of which is marked C. Running from the abdomen (A) is a membrane or skin (two really) which passes between the two lobes of the liver, no doubt to keep them apart. Whatever Nature's reason, we are to face facts and the build of the hen; nothing else can count. This membrane (M) runs up to the heart seen at H, and carries with it particles of fat to deposit eventually on the heart. Every overfat hen then is " two inches " away from a fatty heart. What this latter means to the laying hen called upon to turn out eggs in plenty, to prove a useful breeder, and to lay on in the second season at a good and profitable rate may well be imagined. In the second year the hen with a fat heart can have a considerably reduced total yield to offer. She may fail as a breeder through becoming bronchial. Worse than that she may die suddenly in the nest-box during warm weather and prove a loss entirely. Lacking tone in the breeding season she may give trouble from infertility and dead-in-shell, or be responsible for weakly stock. She may lay soft eggs which lead up to egg-eating. Whichever way I look at over-fat stock I cannot see any good point in their favour, although we do not want half-starved hens. The two extremes are very undesirable.

When removing the breast-bone be careful of the bones at the junction S; break them with the pliers and do the same to the corresponding bones on the other side of breast-bone. Be careful not to prick your finger or tear the rubber gloves with these sharp bones and then, with scissors and saw, cut through flesh and bones from S to T which will see the breast-bone removed showing the internal organs as in Plate 5. A simple plan is first to cut along the dotted lines on either side of the breast-bone with the scissors. Next break the bones at the junctions S and T with the pliers, and cut deep through the junctions with the scissors. Then, taking hold of the end of breast-bone, bend the latter right back towards the bird's head until its whole gives way at the broken junctions. Finally, with the scissors cut through the flesh from S to T at top of breast-bone, using pliers or saw to get through the small bones. The whole of the breast-bone is then removed. In Plate 5 the two lobes of the liver are marked L, and the dotted line between illustrates where the membrane extends from the abdomen (A) to the heart (H).

If we decided to get rid of some of the "bladder of lard" fat we reveal in Plate 6 the gizzard (G), and no wonder this organ cannot be felt prominently when it is buried in fat as in overfat stock. The intestines (I) now become noticeable neatly folded round the end of the gizzard.

We have not, however, dispensed with the internal fat, and in Plate 7, after cutting away the second or "bladder of lard" fat we come to fat No. 3 on the organs themselves. We notice the stomach (S) is covered with fat and there cannot be much chance of that little organ doing the needful in softening the food passed down to the gizzard (G) for grinding up. The action of the gizzard must be considerably interfered with, seeing what a lot of fat coats its walls. The heart (H) has also a plentiful supply of fat (F) around it showing that where there is fat there really is FAT! We are in this Plate also introduced to the spleen (SP) which so many when "drawing" a fowl take for the kidney. It is a kidney-shaped article perhaps to some, but, hiding behind the stomach, it is a blood-forming organ and an important one. It will be noticed in Plate 7 that the liver has been removed.

Before disturbing the fat on the gizzard we see that organ and also the stomach drawn aside to show the organs below.

Here we notice the ovary or yolks therein at O, and the oviduct at Z. We thus arrive at the only organs that seem to count with nine poultry-keepers out of ten. In Plate 9 with the fat removed (melted especially in the sun for you while I kept away the cats and the flies—here I apologise for any flies which did get " snapped " on the subjects) from stomach and gizzard we notice the large yolks (E) in the ovary beneath these two organs.

I wish you particularly to bear in mind the placings of the ovary and oviduct because I contend they are not well planned for their work by Nature. We have, for instance, the stomach over (under when the hen walks) the ovary and the gizzard over the oviduct and, more important, the shelling department. Had I been asked to build the hen for laying 300 eggs per annum, I think I should have placed the laying organs on one side of the bird and the digestive organs on the other, and I might have dropped a strip of asbestos between to keep everything in order permanently no matter whether the hen laid 200 or 300 eggs. Built to lay a clutch of eggs every season the planning may be sound, but I submit, and am attempting to prove, that built as she is the hen requires every care in conditioning that the owner can give her. Readers will understand as we proceed why one man gets plenty of eggs and another does not, and how the best of hens on breeding or quality can be compelled to be low-grade producers on unsound feeding and bad conditioning. Condition holds the key to heavy and continuous laying.

CHAPTER II.

THE LAYING ORGANS.

NATURE must have been nervous of what would happen to the little hen when man domesticated her and turned her from a 20-egg layer into a 300-egg MACHINE. Throughout the bird there seem to be safety curtains as will be seen in Plates 10 to 12.

Plate 10 discloses a membrane or thin skin (M) protecting the laying organs, and maybe you have opened many hens and never discovered it. Plate 11 gives a better idea as we have removed the stomach and gizzard and only left the spleen (SP). Under the membrane (M) are to be seen the large yolks (Y) in the ovary showing through as if under ice. One notices in this Plate as well as in Plate 12 the excessive depth of fat ("bladder of lard" brand), marked F, which is taking up some of the abdominal capacity that could be plainly utilised for the many organs to be revealed in Plate 12.

Having removed very carefully the membrane or skin we see the five large yolks (Y) in the ovary, the oviduct (O), and the intestines (I). When we consider that the latter can, when fully functioning, measure six feet and the oviduct two feet or so, I think you will agree that a hen has plenty of organs in her body. Incidentally I ask if we can really afford any room for excessive internal fat.

I forgot to mention, too, that when an egg is in the shelling-department room has to be found for that shelled product. We see a shelled egg (E) in Plate 13, also the oviduct (O), the yolks at Y, and the smaller yolks awaiting development at R. I have removed the intestines except for an inch, marked I, at the end where it joins the finish of the oviduct in the same channel.

So many know that I do not like overfat hens, and they will therefore pardon me for again referring to internal fat. But we have fat No. 4 to contend with, and this is the layer which coats the kidneys (K) as in Plate 14, where I have removed the ovary and oviduct. One cannot see the kidneys for fat and, coming on top of the laying organs with the other fats underneath, you will follow the sandwich they make of our poor ovary and oviduct—the two organs that count as far as laying goes. The end of the intestine (I) is noticed, and as that represents the vent the fat from the bottom of the Plate marked by arrows is quite surplus. It is there to be carried about by the hen like a football; one can pardon her for thinking it might be an egg that needs to be passed out, and, in straining the muscles of the oviduct to expel it, for laying soft eggs ever after.

With the fat removed we arrive at the kidneys (K) which are naturally diseased and raised from their sockets. And we also discover the cloaca full of droppings (Z) showing a stoppage and sluggish digestion. No wonder with the fat interfering. Any such bird could be poisoned internally under those conditions from the gases. A blockage of the intestines is often the cause of hens going off their legs, but I have never seen a man syringe out the cloaca as a first attempt at a cure, and to be on the safe side. I have seen poultry-keepers hold the hen's legs in hot water and rub them with embrocation, and others have changed the feeding and omitted fishmeal, thinking the mash to be too forcing. Calling a spade a spade in all poultry matters will ever be our best slogan after all. And a better knowledge of the little hen will help you in that direction. I am sincerely hoping that a close study of the present book will supply the needful. In Plate 15 I have marked the lungs L.

Plates 16 to 18 interest me very much because I think they are excellent photographs of the ovary and oviduct when removed from the hen. It is no easy matter to remove these organs and to fix them in position again. Plate 16 demonstrates the ovary at Y, the oviduct at O, and the shelling department at S. The oviduct would appear to be tied up anyhow from Plate 16, and the average poultry-keeper might just be aware that the membrane at the top attaches the oviduct to the back of the bird. When I have made a few judicious cuts the fold of the oviduct appears quite graceful and curtain-like in its folds as seen in Plate 17. Here the membrane which attaches itself to the back of the bird near the kidneys is marked M, the ovary Y, oviduct O, and shelling department S. At P we notice the casing from which the egg seen in the shelling department dropped.

A few more cuts and the oviduct channel is unfolded to the fullest extent. Plate 18 gives us the idea, and the egg will be seen shelled in the shelling department. C is of great interest to me because it demonstrates the opening to the oviduct into which the ripe yolk drops.

The process of egg-making is a very unique one and needs simple explanation for you to follow why it is that so many heavy layers have mishaps in ovary and oviduct. An ovary in full working order represents many hundreds of yolks or prospective eggs, about five of which are very large, while the rest behind them, and waiting their turn to develop, are small. With a needle one can remove each small yolk or prospective egg and count into hundreds. The marvel to me is that we have not made the best use of these yolks seeing that we are content with two years' production only from a hen, and many try to argue that we should get out all the eggs possible in twelve months and then market the hen for table. I should like to see some longevity put into the laying hen by selective breeding, seeing that there are these hosts of potential eggs in the ovary and we get out but a small part of them.

The working of the ovary is wonderful, to be sure, and always reminds me of two things, namely, the mushroom where you pluck one and another comes up next day, and the acorn. A matured oak tree must shed thousands of acorns every year but—Nature's way—one never seems able to find many young trees as a result. Along comes the nurseryman and gathers in all the acorns and he can by artificial methods obtain a tree from almost every acorn. In my opinion we have the same thing in the hen's ovary. We have never put more yolks into the ovary by special breeding but we have by artificial means, whether in breeding, feeding, housing, or management, got the yolks already there to mature quickly and become eggs. We never seem to sap the ovary because when a hen dies at the end of two or three seasons there are still plenty of yolks within.

My view of the ovary is that Nature gave the hen sufficient yolks to last twenty or thirty years in her wild state. Knowing she would lay around twenty eggs per annum and live on for years till old age and enemies took her off Nature provided a good egg store. I have seen hens over ten years of age and still in lay. Is it at all possible that Nature allowed the hen an abundance for her to live upon in the winter? She could thus be both producer and consumer, and later I will give some very interesting photographs. I have always argued that a hen has at maturity in her ovary every egg she will make and lay in her lifetime; no more are added, and it is for every poultry-keeper to get them out in the shelled state, preferably as many as he can yearly over a long spell. In the natural state the hen would go short of food in the winter months and may have wintered in the trees consuming internally some yolks from her ovary! Now that we provide her with the necessary food and protection in housing she delivers the yolks instead of consuming them! But we must connect clever feeding and housing, also management, with the increased egg-production in addition to breeding. We must not neglect the personal side of getting eggs out by artificial methods. Even if the

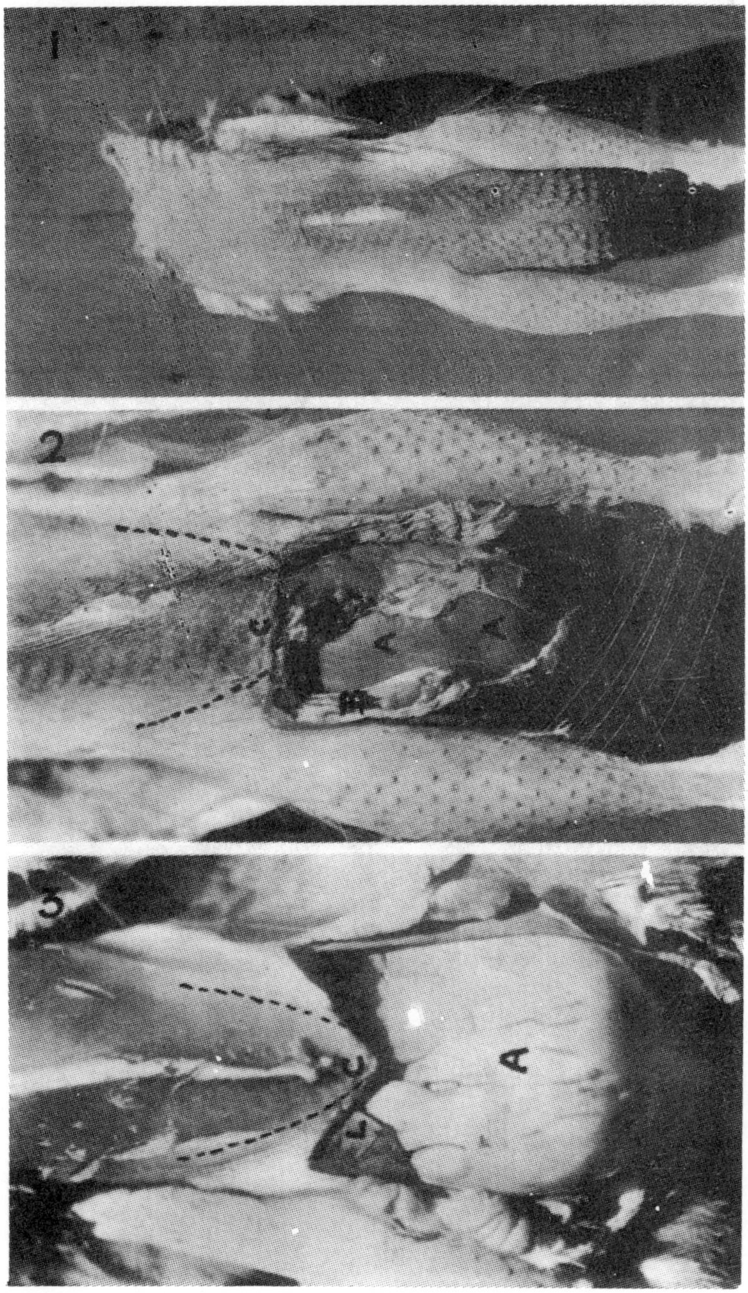
1
2
3

farmer had the best-bred stock his antiquated methods in the winter would fail to get the eggs, which is proof in itself that to obtain eggs in plenty there must be the combination of breed, strain, type, and personal management.

Egg-making is quite mechanical, and I am of the opinion that a hen in laying condition, with the ovary and oviduct free of fat, must continue delivering the eggs. The more she lays the more eggs there are to ripen and to follow those gone before. Taking the five large yolks, the first is quite large and ripe. It is in the membrane or skin as we find it in the shelled egg at breakfast-time, and the whole is again enclosed in a case which is attached to the ovary by a stalk. This outer casing shows a white streak where the wall will rupture when the yolk is full or ripe and ready to drop into the oviduct. Having released the yolk, the casing seen at P in Plate 17 passes along the ovary like a little snowdrop and wastes away. You can count the number of eggs the hen has laid within, say, a week or ten days by the number of "snowdrop" casings found in the ovary when examined.

But it will come as a surprise to you perhaps when you endeavour to find the cup or funnel shaped opening into the oviduct. To make the aperture clear most writers in the past have drawn a proper cup-like opening so that you could follow the direction of the egg. Each has copied the other. I remember an American photograph showing the oviduct with one egg being shelled and another yolk going down, and this has been generally copied by all. You will cut up a few hundred hens and never see two yolks in the ovary at the same time.

It has become popularly known that the cup-like opening to the oviduct resembles a large basin into which the yolk as large as a walnut cannot help falling. If you had never opened a hen before I would willingly offer you a good sum if you could even locate the opening. I have often set this puzzle at lectures without finding any person in the audience able to locate it. In Plate 18 the opening to the oviduct is marked at C, and this channel ends in two very fine skins which come up and take the yolk when it is ripe. The parting in the skins is invisible, and it will be an educational lesson for you to cut out the oviduct whole from along the back of the bird, getting the scissors well under, and starting by removing the ovary as well. You will understand why heavy layers go wrong and why ripe yolks fail to find their way into the oviduct channel. Every time a yolk drops it is like threading a needle and accidents are bound to happen.

With the yolk once in the oviduct the albumen layers are added, also the membranes, and the whole passes into the shelling department at the end to be shelled, the muscles of the oviduct passing the article along. When the ripe yolk No. 1 is going down the oviduct, yolk No. 2 ripens and takes the place of the former in the ovary, and a fifth yolk (small) comes along to the ripening queue in the ovary from the main growth. As the egg is shelled and laid, the ripe yolk drops in the oviduct, and the fifth small yolk steps in to fill the ranks. It is clockwork laying to be sure. The snowdrop casings pass along to waste away. The key is simple then, namely, that you must get the shelled egg out quickly to let another yolk down. Again, it is up to the individual owner, because if the shelling department is clogged with fat then there will be a hold-up of the shelled product which may mean two eggs instead of four weekly. While waiting its turn, too, the second yolk may become overripe and burst its case, dropping into the abdomen and causing the loss of the bird after some time. Or the yolk may decide to become absorbed or "sucked" back, which is happening in Plate 17. The egg in the shelling department is being held up, judging by the inflamed blood vessels on the walls of the oviduct, which cannot pass out the shelled product for the internal fat blocking the way. The yolk to the right of Y is glossy and "podgy," showing that the contents are being sucked back. There is not the fulness about this yolk which we find in a healthy developing one.

CHAPTER III.

OVARIAN DISORDERS.

WHEN a pullet competing in a Laying Test dies, the chance of the pen goes with her—one reason why I consider Tests a proper lottery. Viewed as sporting affairs they are all right, but seeing that the public are asked to consider that the winners of 1, 2 and 3 have good strains of layers and the rest have not, Tests have their disadvantages. If the poultry public had copies of the Test reports in their hands every month, that would be different, because they would be able to follow the performances of each pen. But you have to pay for these reports, and John Citizen is not sufficiently interested to buy.

Some Tests allow a reserve bird to be sent with the pen, and if a pullet dies the reserve steps into her place. This is as it should be, because surely it is a lottery if the wins go only to the pens that stand up at the end. Certain Tests will not allow reserve birds, contending that it is proof of stamina when all the pullets in a pen last the trial. Well I remember a fox having a few hens from one Test when they were in broody-coops in the alley-way and not in enclosed runs. Again, in another Test, when a bird dies, her score ceases to count; hence your pen can lay 600 eggs in *eight* months, and then two die, making the pen have a total record of but 380 for the *nine* months.

In this chapter I wish to draw attention to the point that it is very simple for the best layers to die. We take Plate 19, and here the hen died because she was too short in the back, and when a double-yolked egg found its way into the shelling department it could not get out. You will see the egg at E with the gizzard (G) turned aside and the ovary at Y, with the smaller yolks or yolk cases at R. At C we have a cyst. This large egg has touched the ripe or large yolks in the ovary and caused them to stop functioning, so that the yolk matter inside each is being pumped back by the same organs which pump it in. The yolk matter will then be absorbed into the digestive system through the blood. Even the small yolk cases, which appear to be as large as peas in most cases, and full of fluid, are beginning to lose their form and to dry up into one mass like bloater's roe, and we shall see later the exact appearance of the dormant ovary. Watery cysts are odd and yet very common. Some declare that they are undeveloped yolks, but I am not sure that they are not associated with local disorders and irritations. This cyst would appear to be caused by the irritation set up by the abnormally large egg and the straining to expel it.

Plate 19 shows clearly how badly the laying organs have been planned when the gizzard is directly under the large egg in the shelling department. Any fat pushing up the gizzard on to the egg that is being shelled must interfere with its expulsion. In my score-card system I aim at a long back, preferably five inches or so from the wide part of the back to the parson's nose, in order to keep the shelled egg away from the ovary, which, if touched, must cease to function. A long back is an insurance policy against mishaps when abnormally large eggs find their way into the shelling department.

Plate 19 appeals to me too, because I often set a problem in examination papers asking if a new-laid egg can be a stale one. The average poultry-keeper who lacks observation writes down the date of laying on the egg when it is collected, whereas it may have been in the shelling department for many weeks or months. When I removed the large egg in question the contents were putrid. The moral is to test every egg for its contents when incubated, and more so when sold for eating or incubation. I have settled very many arguments between poultry-keeper and consumer, the latter arguing that the former sold addled eggs as new-laids. The poultry-keeper has been so sure that every egg sold was laid by his own hens and collected daily, and therefore *must* have been fresh. The consumer has equally been sure the contents of the egg in dispute were rotten or addled. I have actually found a developed germ in such eggs that have been held up. When opening a hen with Mr. W. G. Arkell on his farm at Cricklade, Wilts., I found an embryonic chick in a similarly held-up egg. The sale of rotten eggs may cause the loss of customers, and a little testing will prevent any such mishap. Often a hen will be mopish as a result of an egg held up in the oviduct, and the owner never tests for it. Gently insert the finger in the vent and the egg will be felt easily enough on the left side.

If you want to tell without trap-nesting how many eggs each of your birds lays in the twelve months, and assuming you are a critic and do not believe in my system of handling, I can help you over the difficulty, but you MUST rely upon HANDLING, which, after all, holds the key to very many problems not explained or noticed by other methods. But I hope I shall have converted you to my handling methods before you have finished this volume. Take each hen in the morning before any has laid and examine as advised. Take the bird under the left arm, with head down and tail up and the open left hand high up between the legs or thighs near the breast-bone. The hen's head will be resting between the arm and the body, so that the bird can see the ground behind the operator and will remain quiet. Gently insert the finger in the vent and the egg will be felt in the shelling department. If it is not felt just inside, do not go further with the finger, as you may rest assured that such a hen will not lay that day. I am all against inserting the finger deeply, as many do, to find out if a hen is a heavy layer or fertile breeder; you may injure many birds that way. I have taken small pens of birds and tested them out for long periods for many years, and have had them trapped to verify my testing, and I have never been wrong in number of eggs laid daily. Naturally, one can have an egg held up and go wrong, but that is the exception that proves every rule.

My system of handling for the egg within is very helpful too in the matter of broodies. I contend that to cure hens of broody spells you must have them out quickly and isolated in airy broody-coops. Thus promptly isolated, they should be fed up for egg-making and be given a good mash with increased fish-meal therein, while grit, shell, and drinking water must be provided just as if they were in the laying flock. To starve them, as many do, or to dip them under the pump, as I was instructed when a lad, merely ensures a proper rest. When a hen goes broody she dries up her ovary, and then stops on the nest without troubling about her rations. As a result, the ovary dries up completely, the oviduct also, and the intestines are lost, so that the abdomen feels empty. Now it takes many weeks to get the laying organs going again and the abdomen full of intestines, and that means fifty eggs lost. We must, in a nutshell, hold on to the intestines, and to do that we must not let the broody hen lose her abdomen and laying organs. Starvation and frights from dipping in water will be the opposite methods to those needed, and allowing the hen to remain on the nest after she has dropped the last egg in the ovary will encourage the broody-rest, seeing that she will not take full rations. The key, then, is to have the broody out into the isolation coop just before the last egg is laid, or rather while she is still in lay. Delay is dangerous.

Where the owner trap-nests, he might adopt a very simple system of recording. When he has a hen in the trap-nest and no egg, let him mark the record-card with a letter "N," and if a bird is found with two such letters against her for the same day, she goes into the broody-coop. On the other hand, if a hen has laid on one day and visits the trap-nest again on that day without laying another egg, she can go into the coop. Prompt isolation, generous feeding, and an iron tonic daily in the drinking water to give tone, will see the hen in lay again in a week or ten days, whereas a loss of fifty eggs may result from letting her win the fight and in her resting properly. When a hen is found on the nest and there is no egg, the owner can feel for the egg inside the vent as advised, and if no egg is in the shelling department such a hen might be placed at once in the broody-coop. Any hens caught on the nests at night might go into the coops if threatening broodiness. I prefer to place likely broodies in their isolation coops while they are still in lay and before they have emptied their ovary of yolks; and handling for the egg will help, because if a hen shows signs of broodiness when in the nest-box, and if we feel an egg there, she can be taken to the broody-coop at once. At a Laying Test, when walking round the pens, I noticed six hens of different breeds in a broody-coop, and several eggs. In Tests I think that each pen and hen should have its own broody-coop and section; all eggs can then be individually recorded, and one man's weakly birds kept from another's vigorous ones. Always record on the egg-chart each visit and no egg by a letter "N." Feel for the egg; if there, put "Y" on chart. One can then locate hens with held-up eggs and the "never" layers.

When selecting the location for the coops there is no need to try and get roup among the broodies by placing the structures on the damp ground. Have them off the ground under cover or protected from the driving rain, so that the interior does not get wet. The back and sides, also roof, can be boarded in, while the front and bottom can be sparred to let in the air and make sitting uncomfortable. A perch can be placed within to good purpose, and the broodies can have a male in charge to tread them when they sit down. Have a trough at the bottom of the front to take the mash, water-pot, grit, shell and any grain that may be given. A hinged, sparred front or door will be useful and convenient for getting the birds out or putting them in. A good coop consists of a slatted box arrangement suspended to a tree by a rope, which is secured to a strong spring attached to the top of the coop. When the hens move, the box swings up and down. Such is often used to effect by small poultry-keepers.

While we cannot afford to let hens be constantly broody, I always advise letting a hen go broody late in the season if she is a splendid layer, because it rests her and brings her up very fit for breeding the next season. We have to take care of our best layers if they are to make good breeding hens in their second year, and I never hesitate to let a hen go broody when

I know the rest will do her good, being always prepared to let her record go somewhat. If I know she is a gem of a layer, what does it matter if by my rest idea she lays 235 eggs, and becomes a good breeding hen for fertility and rearability; as against 260 eggs without the rest, and in return she blesses me with weaklings and unrearable progeny? I commend my methods to those breeders who judge a hen's quality solely by the number of eggs she lays without a thought to her usefulness or otherwise as a breeder, or even the size of egg and its influence for good or bad on the strain. I like all good layers and detest all low-grade ones, and I shall never like a hen that lays 240 better than her sister which produces 239 solely because there is one egg difference. Two years ago I graded out a hen that laid 298 eggs because they were small and she was undersized, but the owner would not listen. This year, when grading out, he asked me to pass out all this hen's progeny, because they had failed badly for size of egg and rearability. There are other things which make the good utility bird besides mere numbers of eggs laid individually.

I am reminded that eggs are, in the main, shelled at night, seeing that nearly every egg to be laid during the day can be felt in the morning inside the vent as advised. Many argue that it takes so many hours for the yolk to get half-way down the oviduct, and, all in all, about thirty-six hours to be laid; but I am thinking that "Score-card Marvel" speeded things up a bit when she was busy with her 315 eggs in 11½ months. The hen will some day be X-rayed, and that will help us; to-day she is thought to be too insignificant for that expense. One of these days perhaps I will fill the gap, but I have but seven days a week and could well do with ten; this leaves little time for experiments. It solves the problem as to whether wet mash or grain should be given at night. Every time I vote for grain for breakfast to ensure activity when the eggs are to be laid, and wet mash to the full when the birds can retire at night and be happy making the eggs. As a change, of course—and gentle changes create appetites—the reverse can be tried.

Passing along to Plate 20, we have a curious mishap. In the shelling department we find a broken egg (BE) with a particle of the shell at S. Such an egg was probably thin-shelled (overfat hens invariably lay abnormal eggs as regards size and shell-texture) and broke before it was expelled. As a result the contents became "hard-boiled" by the heat of the body. Death did not follow from peritonitis, but the broken egg caused a stoppage, and other yolks falling into the oviduct from the ovary were denied a passage and in consequence also became held up and "hard-boiled," or coagulated (see CY).

A yolk at U, in a membrane, has failed to find the passage, and at Y the ovary seems to be drying down. Such is the way of cock-hens, which grow spurs directly the ovarian (female) organs go wrong.

It often happens that there is an escape of egg-matter, or yolks, from the ovary into the abdomen, and the substance illustrated in Plate 21 represents anything up to forty yolks which have dropped from the ovary into the abdominal cavity, there to become "hard-boiled," or coagulated. One will notice that I have cut the substance through to show the "flaky" layers of the yolk. Many take these "balls" of coagulated yolks for tumours. Many years ago the late Revd. George Crawshay, a very keen student of Nature and an admirer of Light Sussex, wrote to say that he had taken a tumour from a hen, and would send it along if I wished to have it for preservation purposes. I replied that from his description I considered the alleged tumour to be a mass or ball of yolks. He then offered me, in a sporting way, a sitting of eggs, individually marked, from his best Dairy and Palace winners if I were right and he were wrong. The sitting of eggs duly came along when he had taken my advice and cut open the "ball" to discover the layers of yolk and the yolk in the centre, which was the first to go wrong and cause the trouble. And pullets hatched out of those eggs which would have won the Dairy, but I had agreed with him not to exhibit the progeny.

It is easy to understand how the egg-matter gets into the abdomen now that I have dealt with the internal and laying organs. It is not so easy as most suppose for the yolk when ripe to find its way into the oviduct, and dropping into the abdomen it is joined by other "dropped" yolks and becomes a hard mass. When it has occupied the full capacity of the abdomen, it gets very often a pretty skin or coat, which makes so many consider the substance a tumour. When cut through, the flaky yolk gives the key, and sometimes one finds both yolk and white in the substance. In such cases it is obvious that the mishap has happened in the oviduct, or the albumen or white would not be present. When, for instance, we have a blood-spot in the yolk, we know that the trouble is in the ovary, but if the spot is in the albumen then the disorder is in the oviduct.

As a rule, these mishaps occur with overfat hens. For instance, the fat on the stomach and surrounding the ovary can disconnect the ovary and the opening in the oviduct, so that when a yolk is ripe it falls into the abdomen. In like manner, the excessive internal fat can clog the oviduct to such an extent that the passage of the egg in the oviduct is prevented; the yolk-matter with albumen forces its way through or right up the oviduct into the abdomen. A twisted oviduct, due to the perch being too high, or to rough handling of the birds, or in flighty temperamental "flying" hens, will cause the yolks to go through the walls, in their big "push," into the abdomen. I have known a cyst with its cord-like attachment tie itself round the opening of the oviduct and cause the ripe yolks to drop into the abdomen. Plate 22 illustrates the kind of watery cyst that would be likely to tie up the opening to the oviduct. It is suspended from the ovary by means of a string-like substance, which is the usual placing for ovarian cysts. Maybe this is due to an undeveloped yolk or a disorder in the ovary, the irritation causing the cyst to develop. This Plate is of extraordinary help to us, because the hen had not laid for months, and yet, when opened, her ovary and oviduct had every appearance of functioning. The opening to the oviduct was wrapped round by the attachment to the cyst, proving, as I have always stated, a hen can make and consume her yolks internally instead of laying and shelling them. The section of fat lining the abdomen is excessive, and the Plate furnishes us with a good idea of a well-developed ovary. We see the large yolks at Y, five in number, and of various sizes up to the largest or ripest. At R we have the smaller yolks in the ovary, at I the end of the intestine, at O the oviduct, and at C the ovarian cyst.

Having removed the large yolks, we have a good view in Plate 23 of the smaller yolks at R, each of which is separate, showing full functioning of the ovary. The oviduct is also full, and shows the appearance of functioning.

Plate 24 is of a hen which had all the appearance of laying, but which was producing eggs within. The shelling department is full of coagulated yolk (BE) from a broken egg, the oviduct (O) is full of hard yolk and albumen, and at U we notice a yolk in a membrane at the junction of the oviduct. My friends who watch with interest and "sympathy" the controversy over the Westminster cock-hen will be interested to see this hen has grown spurs (S).

Plate 25 is of particular help, too, because the yolks dropping from the ovary have kept separate. We see three such "hard-boiled" yolks (Y), and each has had formed around it a skin or membrane, and this seems Nature's way when a small foreign matter gets into the abdomen. At East London Show, recently, a member brought in a chicken that was on its last legs, and, after handling this two-month-old chick, I suggested that it would not be reared, as it had not absorbed the yolk from the first. I killed the chick, and the yolk was "hard-boiled" and quite as large as is usual at the day-old stage, but it was encased in a protecting membrane which had ended with the usual constipation and stoppage of the vent; and the chick had lived for the two months, although, as I have always said, such must die sooner or later if the yolk does not get clear.

The hen photographed in Plate 25 was graded out by me when hand-grading a student's flock, and was placed in a house with other "doubtfuls" to be trap-nested. The hen continued to lay, but shortly afterwards she died in the nest-box. Handling revealed the hard yolks in the abdomen, and one could have graded out such a bird prior to sale to prevent disappointment on the part of the buyer, and put the trap-nest on to verify her laying. She might have been a passenger, but proved the exception, although death claimed her eventually. Even after dying, the unobservant owner, if a critic of handling, might have wasted five shillings on a post-mortem examination all for the want of finding the yolks in the abdomen by a little handling. Another owner might have accused the next door neighbour of poisoning the bird, and it is within the realms of possibility that a third might have suspected an irritant in the dry mash and changed all his flocks on to wet mash at a moment's notice. Handling will help you to call a spade a spade.

CHAPTER IV.

WHEN EGG-MAKING IS CHECKED.

THE development and cessation of the functioning of ovary and oviduct is worth a special chapter. As mentioned, the ovary in a hen in full lay consists of the five large yolks and hundreds of small undeveloped pea-like cases ready in a queue to be filled with yolk-matter when the time comes for each to step up into the line of the ripening five. Every hen in full lay then should have the large yolks present, but the bird has a means of withdrawing the yolk from the large cases and absorbing it into the system. Proof of this can be had when a hen in full lay suddenly has a fright or becomes ill and stops laying. One can understand Nature in this planning, because if the yolks remained in the ovary they would become "hard-boiled" with the heat of the body, and would not be laid. It often happens that a yolk waiting in its place to drop into the oviduct becomes interfered with, and does get "hard," preventing any more yolks from moving up to the opening of the oviduct, and such a bird will never lay again. Internal poisoning may ensue.

When the yolk-contents of the case are pumped back, the case becomes baggy, as noticed at Y in Plate 26, and also in Plate 27. In the former one notices the surplus fat and flesh marked by the arrow from the end of the intestine. The oviduct (O) is still rather full, but in Plate 27 the oviduct (O) is much reduced, and several of the large yolks (Y) in the ovary have been sucked back, while the small yolks at R are beginning to form up into one mass like the bloater's roe of the dormant ovary. The membrane (M) on each side of the ovary is another of Nature's safety-curtains to protect the laying organs.

Plate 28 shows the yolks almost extinct except for the last "baggy" one (Y) under which is placed a match-stick. The oviduct (O) and intestines have decreased in size considerably, while the smaller yolks at R have become quite unrecognisable as separate pea-like cases.

Coming to Plate 29 we have what I consider to be the best photograph yet taken of a dormant ovary (R) and oviduct (O). The kidneys are seen at K. The yolks have now gone into one mass, and the oviduct, instead of being full, resembles a Chinese "grasshopper" so popular on November the 5th.

It has so shortened that the opening is far removed from the ovary. Plate 29 would be true of every hen in moult or broody with chicks, and young pullets before they had laid.

I wish every reader to note especially the coming in and going back of the ovary, oviduct, and the intestines, because they will best understand my handling system by such knowledge. You may hand me a pullet which is well out of lay, and handling alone will reveal the state of the internal organs mentioned. If everything is dormant the abdomen will be empty and the vent small. Before an egg can be laid the pullet must make her ovary by getting all the yolks from the massed or dormant stage to the separated stage, and then must make the five large yolks which are to ripen and take their turns in dropping into the oviduct, each to have the albumen or white added and to be shelled. While the pullet is making the ovary the vent is increasing in size ready for the first egg and receiving Nature's oil to help with the expulsion of it. The abdomen fills because for ordinary purposes of growing and keeping alive the intestines take up little room in the body, and do not come into the abdomen at all. But directly extra digestion takes place to make the yolks, the intestines swell into the abdomen. An empty abdomen, then, will foretell dormant laying organs, and a full one will mean that the egg-organs are active. How useful that is for pullets and hens, telling us when each pullet will commence to lay, whether hens are coming into production or going out, may be understood when there is no other method as accurate as this part of my handling system. Often a poultry-keeper is not getting eggs, and is tempted to sell off the birds for the other man to gain the harvest, seeing that the hens are just on the point of starting production. A little handling will show him the condition of the birds and will stop a sale and often much loss. Again, a man getting bad results is often tempted to change the feeding, or even the breed, just as the birds are coming into lay, or to carry out drastic methods in one direction or the other through lack of knowledge in handling for condition. A little patience after verifying by handling the condition of the pullets or hens may prevent losses. You cannot make money from poultry if you keep the hens while they are unproductive and sell them when they are just on the point of production! One is able to separate hens that have stopped and will follow on with a moult, so that they can be sent to table and not be kept for many weeks for no return. One can also feed up and even force those hens that are still in lay for a final batch of eggs before they go to table. Also, the owner can judge his pullets for Laying Tests, judging them to lay by a given date.

While on this subject I would remind readers that you cannot hurry the development of the egg-making organs to any practical extent. If a hen falls ill through bad management, or is allowed to go broody, or if the flock lose tone and stop laying, the whole process of losing the ovary, oviduct, and intestines has to be gone through, also the redevelopment of these organs. Such takes time, and I put it quite simply that every time a hen goes out of lay we lose approaching fifty eggs, because it takes many weeks to lose and get back the organs. To cure broodies you need to have them isolated promptly and well fed, so that they do not actually lose laying condition. There you have the explanation why hens properly treated come into lay again inside a week or ten days, and why others, neglected and allowed to sit on the nest for days and to refuse to fly down for regular meals, have a long rest from production, and lose thirty to fifty eggs off their "trap" records.

When pullets are sent to a Test in lay, the journey generally puts them out of lay, and in losing their organs they take several months to get them back and to get into their stride again. Their chances of winning are lessened, just as they are if they go into a partial moult. From every pen of five pullets "messed up" in this way you can lose up to 200 eggs or more; that is another lottery part of Laying Tests, especially where the eggs are recorded at high winter market prices, as in Tests where egg-values count. My pen gets going at once, and at fourpence per egg, and your pen goes wrong—your birds can lay like smoke at one penny per egg in the summer, but they will not catch mine. A lead in the winter is the way to win all Tests where value is the deciding point. I have on many occasions sent a pullet to single-bird Tests, but, although laying twenty eggs monthly throughout, she has missed the first month, and has been unable to catch the leaders.

I would especially warn poultry-keepers against shifting birds when in lay, and more so pullets, for fear of losing the internal organs and having a loss of up to fifty eggs per bird. Be very careful to have a definite plan for wintering your pullets, placing them in their winter quarters before they start to lay, and on no account be late with your schedule. You will see how easy it is to make good birds put up decreased records by a little mismanagement—one reason why trap-nesting should not be on cast-iron lines. If a pullet lays 220 eggs in twelve months, and I know that she lost two months at the start through my own fault, then I am going to regard her as a 250-egg layer. Again, if a pullet has laid 220 eggs and falls broody when she has two months to go, I may decide to let her have a rest, so that she will be all the better as a breeder the second season. I count her also as on the 250-egg bird for my common-sense breeding purposes. Many good birds are discarded because one does not use discretion in that direction. Some who trap-nest want a pullet to lay 250 eggs in her first year and 249 in her second, with something around 248½ (the half will be a soft egg) in her third, and force her all the time—where does breeding come in? Are hens to be layers all their lives and yet be asked to reproduce?

Take every care not to have the birds delivered before the houses arrive, which will see them moved several times, and may find them cramped for room in outbuildings with all breeds and both sexes mixed. That is the road to ungraded flocks and a loss in eggs *ad lib.* I remember at one Test the houses were not ready, which resulted in two shifts of a certain section. And I watched how long it took to get the laying organs back and to get the flock into tone. Also remember the danger of letting a whole flock get out of tone by colds, or even by the use of an irritating ingredient of the mash; otherwise you will need three months to condition them again, and will be losing eggs all the time. Now that you understand why it takes time to condition flocks of layers, you will follow my ruling that the personal side of poultry is most important, and yet is so often neglected. I say to all readers: learn hand-grading and conditioning, so that you can "read" your flock at any given time. It is the most vital part of successful egg-production whether you keep six or six hundred layers. The poultry-keeper who is observant knows the condition of his flocks to a nicety, and nothing teaches observation in a more thorough way than my handling system.

If I go into a large laying house in, say, October, and find the two hundred pullets listless, I know that the owner has messed them up. If, on the other hand, I find them all alive and "talking," I know that he has managed them well and will have full winter egg baskets. If the pullets are sneezing I know his housing has been wrong and, maybe, there has been overcrowding at maturity. If many pullets are laying in the litter I am sure he has neglected the birds at the settling-down stage. In like manner, after the pullets have been laying heavily from October to February, should I find them pale in face and comb just then, I may be sure he is letting them lose tone and, maybe, he is running them short of raw greenery and mineral salts. He may have been forcing them too much. They will need toning up, and a general tonic and a slight revision in the feeding. Again I may catch and handle a few to find them like lead, showing they need thinning a little. On the other hand, if they handle light they may have had colds, or are being underfed, or are over-laying for the raw material given by way of food and choice of ingredients.

I am not in favour, however, of finding these things out when they have happened. Rather do I advise every poultry-keeper to master my handling system, so that he knows to a nicety the condition of each flock, and can by his personal care, feeding and management keep up condition every week of the year. Instinct will tell him if anything is wrong, and he will have applied the remedy before there has been a severe loss in eggs and condition. A little extra mash or grain may be all that is necessary for a week or two, or the rations may need a slight drop or a little of the fattening food withdrawn for more bulk. Trap-nesting will tell you after the flock is out of tone; handling tells you before it is too late to alter things, and you can then "stop the rot." I am not attacking trap-nesting by any means, but I have to convert my critics who reply in chorus, "trap-nests," whenever I deal with my handling system. No one has done more for trap-nesting than I have, even among backyarders. My writings and my books speak for me over many years, and I was the first to draw attention to the recording of ducks, which will be found in the first duck-keeping booklet I wrote. Trap-nest where you are able, but use my handling system as well—the ideal combination that keeps you clear of cast-iron automatic methods. Where you cannot trap-nest, use my handling system of grading and selection.

CHAPTER V.

DIGESTIVE DISORDERS.

SO far we have seen that digestion means everything in the direction of eggs. The intestines take up little room in the body when the egg-organs are dormant, but when egg-making starts and is at full pressure, the digestive system is also at full steam with the intestines filling the abdomen. I have always stated that eighty per cent. of deaths among layers may be traced to digestive disorders, and yet feeding is another weak link in the poultry-keeper's chain—so many feeding with the hand and not with the head.

The overfat hen is the culprit of digestive disorders, and cannot help herself. She can go wrong internally in a hundred different ways, and I give a warning to backyarders who have to keep their fowls in confined runs that the tendency to get overfat on the part of most hens, and the heavy breeds in particular, is always present. When a flock are overfat, my experience teaches me that deaths will be numerous spread over a given time. Often, when a bird dies to-day and another to-morrow, the owner becomes alarmed and thinks the neighbour has poisoned the birds; it may be just overfat condition throughout the flock.

Internal poisoning is a very common cause of death in fat hens. The food becomes held up, and the poisonous gases go through the system. Often the ovary becomes affected, and the yolks turn putrid, sending out poisons all over the body. The internal organs, as my Plates show, are so packed together that if one goes wrong the other touching it also becomes affected. Plate 30 gives an idea of a hen which has wasted away solely through internal poisoning of the system following overfat condition. She was actually graded out of a breeding pen, showing that in being mated at the time of killing the owner had never handled his stock. The hen was out of lay, as the empty abdomen shows; and the inclusion of such a non-laying hen in any breeding pen would be sufficient to make the male angry and ill-tempered with her for refusing his attentions. Such would upset the pen as a whole, and the observant reader will see that when a hen is bullied by the male she should be caught and handled. Many [illegible]-grade layers or hens that have gone wrong in the laying

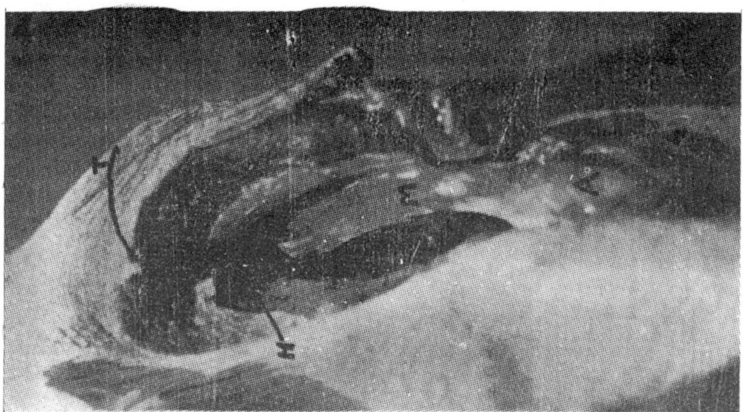

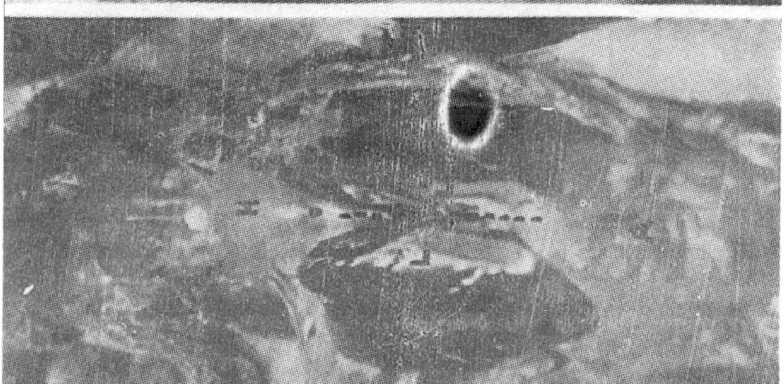

organs can be discovered that way. Certainly infertility will occur where the hen and the male "scrap," and those not in lay at the time should not be mated up unless they are coming into production, when they will accept the male's attentions. Handling at abdomen for "coming in" will advise the owner which hens to mate up and which to discard. When grading at Bulwell I passed through a pen of Reds, and one hen had not laid for months and would not start for another two or three months, by which time the breeding season would have been well over. The male was not on good terms with this hen, and there would have been more peace had she been removed. Catching her, I found that she had severe bronchitis, which accounted for the bad laying condition. Handle before you mate up the pen, and do so now and then while they are mated up, and always be on the look-out for the mopish hen or the one the male bird "scraps." It may happen that she is just a frightened individual, but such high-strung specimens are not desirable in utility. The docile, sweet-tempered bird is an ideal producer.

You can only tell condition by handling, so never be nervous of picking a hen or two up now and then—hens do not bite! In my judge's book I have some notes against exhibits that would interest the critics of handling. Here are a few samples:—"handles light"; "abdominal dropsy"; "Xmas condition"; "half-starved"; "overcrowding"; "been down with colds"; "egg-in-abdomen," etc. At Enfield Show I marked against about twenty exhibits "handles light"; "needs more generous feeding"; "overcrowding"; "no body"; "feather and bone," etc., throughout the Show, and in perusing my notes when I had obtained a catalogue I found that most of the exhibits belonged to one man, although they were in many classes and of different breeds. Some time after I visited the owner's place, and found that colds and catarrh had been troublesome in his pens, confirming my notes. There is nothing like a cold to pull a young bird to pieces, and in a very few days; yet, as is so often the case, it takes but *days* to lose laying condition and MONTHS to get it back. There are diseases like tuberculosis which make a hen go to skin and bone; but often people get alarmed when a bird is found in such a condition, kill and burn the body at once, and fear an epidemic; whereas it may be that the flock have colds and handle light. It takes little to frighten poultry-keepers who are not sure of their ground; hence the importance of a broader knowledge and of calling a spade a spade. I get a weekly letter from a poultry-keeper, and have done for months, who is certain she has had tuberculosis among her hens because one that died had a wasted frame. I have told her that she need not worry, seeing that the others are handling well and turning out the eggs. She notices that I am judging within a hundred miles of her place and thinks I might visit her; hence the weekly letter. From her letters I know that she is nervous and lacks confidence, and places no confidence in those who have been to see the fowls. Should a hen die from egg-binding she will have quite an alarm. I am afraid some hens get severe dosings and medicinal treatment because their owners "think" something awful is the matter with them. Get to know condition by handling and by the eggs the birds turn out.

Coming back to the overfat hen, Plate 31 reminds us that internal fat does not remain solely in the abdomen, as many think. Here we see layers of fat surrounding the crop at C and the gullet at G. Remember that the crop is the container for the food; the more a bird can get into it within reason the better she can go through the long winter evening on the perch, and the more eggs she can make. I grade out pullets with flat chests, because I know that they lack stamina and with the flat chest goes an empty or badly filled abdomen. What the hen omits to get into her crop fails to keep the digestive organs in the abdomen at their fullest extent. The chick with a flat chest and cut-away abdomen (which must follow), and which walks on its toes like a Runner duckling, is a weakling to grade out; and she so often has pins-and-needle shanks and long, thin beak. Such handling properties keep with her throughout life. The rearable chick has a rounded front and a developed abdomen, and walks on the soles of its feet. Such are there for life, and I mention this to show what part the crop plays in egg-making. The weakling pullet that runs too fine in bone, and lacks stamina and digestive power, eats little, is easily bullied, and is not worthy of being graded in. She is the bird the male takes a dislike to, just as he does to the pullet or hen that is nervous or temperamental; and she can be nervous if the comb falls over her eye and obscures vision. The weakling, too, is sat on by the stronger hens, and therefore fails to get her full meals. When she is feeding another hen only has to move and off she runs; when she is in the nest-box laying, a look from another hen sees her fly off the nest without troubling where she falls or what wooden support she hits her head against.

When picking the pullets for the Tests I never include a temperamental bird, and always try and find out the peculiarities of each one. For instance, I must see that every pullet eats well and holds her own at the mash trough. If she bullies others then out she goes, and if she does not eat well she is graded out for that. A pullet must eat well to lay well, and more so to produce large eggs. That being so, I am not going to let the hens get so fat that their crops become "fat-bound."

The result of a sluggish digestive system is shown in Plate 32, where the food is held up in the crop and has turned putrid, sending off poisonous gases throughout the system. Overfat layers will have these troubles, and I know nothing better than milk to drink to shift any stoppage. When the evening mash is given, do not place it in the troughs and go to the next pen; stop with each pen of birds for a short time to notice if any refuse to eat. That is the time to find out the invalids or those with anything wrong, seeing that such birds drink rather than eat, and soon lose appetites. See that any bird which refuses to eat at the trough is caught and handled, and the reason discovered. Often the crop is compacted, and if the bird is placed in a coop and handled again in the morning before feeding, the crop will still be full instead of being empty. Put her on a milk-to-drink-only programme for twenty-four hours or so, and that will generally shift it so long as you have not been careless enough to let her swallow twenty or thirty winkle-shells.

Crop troubles are very easily brought about. Give coarse raw greenery in too large pieces, and a portion will act as an umbrella and hold up the mash. You can get trouble from even a whole maize which happens to get fixed in the opening from crop to stomach or stomach to gizzard, especially in an overfat hen where the organs of digestion are sluggish. Feeding whole large maize is too risky, and I advise the kibbled kind every time. The digestive system is very delicate because the main organ (the gizzard) comes too far down. The crop is the container, and the stomach softens while the gizzard does the grinding. But you will have seen how narrow and short is the stomach, and you have not to take risks. Reduce this down to the day-old chick, and I say that whole grain and large maize means certain trouble in the stomach, as do peas and beans and indigestible dry mashes and meals. It takes a day and a night to boil whole maize, and the stomach cannot soften it in two inches of small tubing. The compacted crop often tells one that the trouble is down lower—in, say, the stomach or the gizzard—so remember the milk.

Grit of too large a size will cause a stoppage, as will husky foods in excess; but remember that the hen with crop, stomach, gizzard and abdomen coated with fat cannot digest much food. Plate 33 has "fogged" most poultry-keepers to whom I have shown the print. I always carry photos about when judging at Shows or lecturing, and then if any destructive critic gets busy arguing over a given point we can check matters with actual photographs. The organ marked GB is the gall-bladder, which is actually larger than the lobe of the liver (L). The bile is stored therein till active digestion takes place, and then the fluid is discharged into the small intestine. If digestion is sluggish too much bile can collect, and the over-loaded gall-bladder is here seen. Jaundice may follow, the symptom being yellow face, headgear and skin.

Plate 34 might be taken for a harp, but it is just the intestines removed to show the fat on them and the connecting membrane; the fat-covered gizzard being marked at G and the stomach at S. A better idea may be had from Plate 35, which illustrates the gizzard (G) and stomach (S) with the surplus fat (marked with dotted lines and arrow) removed. The duodenal tubes are noticed at D, with the pancreas at P. The duodenal tubes represent the first portion of the small intestines, and are divided by the pancreas, which pours digestive secretion into the small intestines. This will dispose of the theory that it is wrong to give mash to fowls for their evening meal because at a certain time in the night the crop becomes empty. Digestion does not end even at the gizzard, seeing that the duodenum follows on.

Liver disorders are common, but very much so to overfat hens. The liver, to my mind, is the most worked organ in the bird, and yet so easily goes wrong, because when another organ gets out of order it always asks the liver to help. Being already overworked, the liver refuses, breaks down, and there is a death. In fact, most birds that die have liver disorders; but one should be prepared for that, and endeavour when conducting a post-mortem examination to locate the direct or original cause. In Plate 36 we see an enlarged liver, which is asking the abdomen to make room for it, as if that part had not already sufficient within. Overfat hens have either enlarged or very small livers, both of which will interfere with health and egg-making.

Internal hemorrhage and rupture of the liver readily affect fat stock, and as a precaution I always advise perches, drop-boards, nest-boxes, fly-up platforms, etc., to be fixed low down—say two feet from the top of the litter. High fly-ups will lead to many disorders, not forgetting ovarian troubles. Prolapse of the oviduct or cloaca can follow high perching, the exposed part becomes pecked by one of the other birds, and death follows. Constipation can of course cause it, as can force-feeding, or straining to lay a large egg. If a hen mopes with tail down, examine for this and oviduct trouble, and isolate at once for treatment before the parts are damaged by other fowls. A common disorder in fat stock is the retention of the excrement as already mentioned. Straining can always affect the ovary, and even make that organ become useless. Yolks may drop before they are ripe, and the walls of the oviduct may be so weakened by straining that they fail to pass a yolk down or a shelled egg out; or, as is common, a hen goes about merely passing albumen in the droppings, and not laying shelled eggs; or the hen may lay soft eggs for good, not even getting right again after the moult. Soft-egg layers need watching, as egg-eating can commence; and it always pays to get such hens to table before they do damage. Again, layers of soft eggs may often be regarded as passengers, and need prompt grading out. Keep both eyes open for egg-eaters, and also layers of soft eggs; do not rest till you have located them. By handling each bird in the morning as advised, by gently passing the finger into the vent you can pass aside all where the hard-shelled egg can be felt. Then you can separate any where the egg inside feels soft and unshelled. Again my handling system will come to your aid even if the flock be composed of hundreds in a single house. If you constantly find soft eggs on the drop-boards, you can locate the culprit this way by ringing every bird which has a shelled or hard egg in the shelling department. When you have gone through the flock (and all birds should wear rings) you can take out the others for close observation. Apart from finding out the best layers, trap-nesting the entire flock discovers all the passengers. Many pullets and hens lay soft eggs for life because the ovary or oviduct is irritated by the presence of a growth or cyst or diseased ovum. Perhaps a yolk fails to enter the oviduct and becomes hard, remaining to cause irritation. Failing trap-nesting, handling is necessary.

Supposing you decided to carry out this practice of feeling for the eggs for a given week twice or so annually. You would on the first morning ring every pullet carrying an egg, and the next morning you would run through those unrung, until in the end you might have some birds which had been carefully handled each morning of the seven without finding an egg. Such could be rung with red rings (always keep to my system of using *red* rings for danger) and be placed together in a small house and be trap-nested to see if they are delivering the "goods." Or you could handle every bird each morning for the week and record the eggs. I know that in every flock sooner or later there will be plenty which will be unproductive. Trap-nesting would very soon locate them, but all breeders seem to trap a few birds and leave the rest untrapped because of the labour and time entailed. Some even cut off the trap-nesting of second-year hens, relying on pullet-year records for their guide. I remember buying a pen of birds for £50 for a beginner, and, although I had the record of each hen when she was competing as a pullet at the Laying Test, I was unable to obtain second-year records because the birds were not trapped. Well, the pen won at the Test and I had reluctantly to tell the buyer (a novice) that breeders did not always trap the second year. It was an eye-opener to me, because if a winning pen is not worth trap-nesting I am sure ordinary pullets are not put through the traps a second and some a third season. The buyer could not understand things, but I assured him that I was honest and the breeder I knew to be.

Supposing, again, trap-nests were not in use, and we desired to know the winter producers. We could feel for the eggs in October and ring all birds in lay with blue leg-bands; pullets carrying eggs in November could be rung white, and those failing until December, say yellow.

Seeing, then, that for every pullet trap-nested there are dozens upon dozens of birds of all ages on the same farm that are not tested for laying, I recommend my handling system for them, knowing that very many hens and pullets will be discovered as mere passengers. They can go to table and not rob the baskets of those which are turning out the eggs. Passengers bring down flock-averages very badly, as can be understood, and the sooner hand-grading finds them, helped by the trap-nest, then it will be the best piece of work the poultry-keeper can do. This in fairness to his other birds, his breeding stock, his employees, together with time, labour, and outlay; not forgetting, too, the man who supplies the foods. You cannot make a non-layer become a paying proposition, and housing is costly, warranting its use for workers only.

I would warn duck-keepers of the importance of having a good landing-place where the ducks use streams, rivers, or even ponds. Insufficient attention is paid to this part of the business. A duck is such a gross feeder that its liver is not always in the best of condition, and slipping down a steep bank is harmful and causes much loss each year with many. Ruptured livers are common. Make a good landing by having a gradual walk to the water and also from it on the other side and put sand down to help matters.

Dropping from a high perch on to hard ground will also cause bumble foot, while hard runs are not very helpful to poultry-breeders. I know that in the dry season the ground gets hard, but many leave sharp stones in the runs without attempting to pick them up; as a result many good birds are lost from leg-troubles. Imagine losing a good stock cock from a poisoned leg because a few sharp stones are left in the grass enclosure day by day! There is plenty of work to be done on a poultry farm, but stone-collecting should not be neglected. A fowl's toes are so constructed on the circular idea that the only natural walk can be had on soft litter or ground. Chicks should never be neglected by being compelled to run when young on hard ground, seeing that such accounts for more leg-deformities than anything else I know. Unable to get a hold on the hard ground with a curved toe, the latter turns round, and the bird when grown up looks unsightly with crooked toes. I have seen whole flocks of birds with such deformities both on farms and in backyards. The heavier the breed or bird the more readily it is affected.

Whenever you are tempted to put down some material to make the ground free of damp, avoid sharp particles, which can cause you endless leg-trouble and bumble foot. If when buying or renting you have choice of farms, select the one cleanest from sharp stones, and you will save much hospital treatment.

CHAPTER VI.

COCK-HENS, AND HENS THAT NEVER LAY.

THAT very little is known of the internal workings of the hen is illustrated by the case of what has become known as the "Westminster" or *Daily Mail* cock-hen. At Westminster Show (the "annual" of the National Utility Poultry Society) in the Horticultural Hall in December, 1921, a class was scheduled for "Graded-out exhibits." Under agreement with the hall management no males were to be shown, and in this class, when I started to judge, I found a cock-hen; strictly speaking, a hen-cock-hen, or a hen in male's plumage. After careful placings I put this bird second and went on with the next class, where I had to sort out the best five of eighty-seven White Wyandotte pullets. Imagine my surprise when Mr. Tom Barron, one of my fellow-judges at the Show, came up and declared that I had given a cockerel a prize. With him was the Editor of one of the poultry papers, and we went to see the "cockerel." I endeavoured to explain that the bird was a cock-hen, but was challenged for £500 that I was wrong, and that the bird was a cockerel. "Mated to a few small hens—not too many—and he would prove fertile and breed some nice chicks," declared the breeder. I replied that I was astounded at how little even the biggest utility breeders knew about the hen from within. This was a case where the ovary had gone wrong, and the spur made its appearance followed by the male's hackles and sickles. I was not out to accept the £500, I declared, but would be happy as an experienced "handler" to prove the bird was a hen, and even in lay. The advertisement for my handling system would be worth the £500 lost to me by not accepting the bet. But explanations were of little use, because you cannot convert anyone who lacks belief, especially when facing him with what looked a male and to say "he" was producing eggs.

The newspaper reporters gathered round, as did many of the usual destructive critics, who hoped perhaps that I had stumbled. The result was that the *Daily Mail*, out for some good copy, made a splash over the bird, declaring that experts disagree—one saying the bird is a cock and the other a cock-hen "in lay." The bird's photo. appeared in the papers, and hundreds not interested in poultry came to the Show to see the bird. The incident thus helped the gate!

After the Show was opened the owner came in and was asked to settle the problem by stating whether or not the bird was in lay, had laid, or was a cockerel; and he replied that she was in lay, and named her trap-nest record. This promptly brought forth a £500 challenge to him by my fellow-judge for daring to argue with a man of experience on utility matters. A free fight was prevented, and we got no further with the main question as to the egg-production idea. We put a steward to guard the pen, and wired down the latter, and on the second day of the Show the bird laid an egg in the pen. Therefore I was prevented further from being the laughing-stock of my critics as the bird delivered the "goods"; and the papers were again full of the performance, "Poy," of the *Evening News*, now coming along with a cartoon of egg-laying males. Luckily for me the bird did not crow at the Show, or I should have lost my reputation, although through ignorance, as some cock-hens not only get the male's plumage but also crow.

After the Show the owner wrote the fellow-judge a letter, challenging him for £50 that he would send the bird to any unbiassed breeder for a week, and if it laid he was to pay £50 to the Eynsford Hospital, and if it did not the owner would pay £50 to a Lancashire hospital. This was not accepted. This bird belonged to the Kent County Council, and by owner I mean poultry manager to the Eynsford Training Centre for Ex-soldiers.

I traced the bird after the Show, and found that she was exhibited in a Maidstone corn and poultry food shop window for a time, and was seen to lay several times, and once by Mr. Venn Carr, the well-known poultry breeder. When the manager died and the training centre was disbanded I lost sight of the bird. Meanwhile I was the object of much criticism on the part of my Lancashire critic, and I have Press cuttings of his speeches in Lancashire where he "slated" me. His explanation to one big audience was that a hen enters a trap-nest by working a wire to make the shutter drop. As was well known, a bird often gets in without working the shutter and escapes after laying her egg. Now this happened apparently on the very many occasions that the cock-hen laid, and the alleged cock got into the nest, worked the shutter, and was credited with the egg. But the hen that escaped had laid each egg.

At the Show he stated that the cockerel would make a fine breeder, and at one of his lectures he said the cockerel was what one called a "bad doer." At another lecture he declared that cocks often got into the trap-nests to call the hens up; but while that is very true you can take it for granted that "bad doers" do not act thus, but spend most of the day on the perches in case they get a good hiding from other birds in the flock, or they go into the nests out of the way—observation will prove that.

Coming to the final stage, in one of my weekly utility articles in *Feathered World*, I gave illustrations of the internal organs of a hen, and explained why she grew spurs and became a cock-hen. This drew a letter from Dr. Crew, of the Edinburgh University Research Station, to say that his experiments confirmed my own views, and that I might be interested to know that the *Daily Mail* cock-hen was in his care and was at the moment of writing sitting on nine of her own eggs, which were fertile. It is fair to say that Dr. Crew had no knowledge of the Westminster affair. Getting down to facts I ascertained that the hen was the same old cock-hen, and that she had been sent on from Kent to the Research Station.

I want you to picture a perfect cock for appearances, because she looked that at the Westminster Show. Now when she arrived at the Research Station the males took no notice of her, and the hens even gave her a good hiding. She stopped laying after a bit and took a summer moult, and in this moulted out her hackles and sickles and became just a spurred hen. The other hens then agreed with her and the male treaded her. As a result her eggs were fertile.

In going over all this ground I do so not for any "swank," but as this volume is my own platform I reply to the speeches that have been made against me over the bird. After all, facts count, and he who laughs —— ! Again, those "Doubting Thomases," who even to-day tell me of the *Daily Mail* cock-hen, and add that such a bird could not lay, may be enlightened by the above and may consider how little they really do know of the little hen "from within."

It did much good; handling succeeded where appearances failed. It explained what "graded-out" meant as a description to the class. Everybody found cock-hens up and down the country, so that poultry-keepers handled their fowls more than ever they did before. I have put this "graded-out" class on at several Shows, but it shows how little is known of grading when poultry-keepers cannot understand what it means. Directly I tell them "duds," they understand, and many have gone so far as to say they have none! At Tottenham Show in 1921, of which Society I have been President for many years, we had the same class, and I gave first prize to a Rhode Island Red pullet which was hatched in February, 1921, and judged as a "dud" by me in the October, 1921. This pullet took fourteen months to shell its first egg, and apart from broodiness was in full moult and out of lay within four months of starting. She belonged to Mr. H. A. Hussey, the Secretary of the Tottenham Branch of the N.U.P.S., and in August, 1922, I was handling his Reds. In one house only two were laying, whereas the rest were in moult. The two in lay were (1) a pullet I had given first to at Westminster in the Branch Team class, and (2) a pullet that was first at Tottenham (same day as the "dud" won, but in the layers' class), first and cup Ilford previously, and second Burton Joyce subsequently in 1922; all under me, and spread over the season; also only times out under me. Taking hold of the birds in turn I came to the "dud" hen, and was asked what I thought of her; and I said that I would give Mr. Hussey 5s. for her as a broiler. He replied that I had better handle her again, as I had given her a "red ticket." To this I replied that if that was so I must have been drunk! Then he added—"as a 'dud'"! The best layers found as pullets in 1921 were, in August, 1922, still hard at production, whereas the "duds" were heavy in moult, showing again that the best layers are the late moulters, while the low-grade performers moult early. Remember this in grading.

Secretaries, fully alive to the drawing power of the cock-hen, wrote me to know where they could get such birds, and to each I replied that they would be successful if they did plenty of grading among members' birds; and I suggested a grading committee to be attached to each Society, as we have at Tottenham. But I am sure that at many Shows cockerels were penned as cock-hens with such labels as "this cock-hen laid 254 eggs in twelve months."

The methods I employed when testing this cock-hen at judging-time, and which failed to convince my fellow-judge, were as follows:—In the first place a pullet placed in the pen was not noticed in the same way that a cockerel would have greeted her. She was a small hen, and for a cockerel should have had, say, a three-finger abdomen, but she measured five. The pelvis bones should have been close in so small a male, but I could get three fingers between the two bones. The abdomen, too, instead of being reasonably empty, as in the male, was quite large and full, and, having first made sure that such increased measurements were not due to abdominal dropsy, I came to the conclusion that her oviduct and ovary were active. The great size of vent when stretched flat, and the moistness of it, confirmed the production or "in lay" side of the question. In abdominal dropsy the abdomen fills with water or fluid, and naturally keeps the bird's measurements very much extended; but here the abdomen was pliable, or workable, as in a bird in lay. To test abdominal dropsy, place the bird on its back, and, if water is present within, the abdomen can be shaken and the movement of the fluid can be noted. The fluid may be clear or straw-coloured, and increases in volume until the bird is unable to walk. When handled, the bird would appear to be in full lay. You can extract the fluid by a medical instrument, but it returns again. In the final stages the abdomen becomes drum tight and swollen. Handle every bird for this disorder if she walks wide and with great difficulty.

Finally I patted the cock-hen on the back and ascertained that she had a pullet's voice.

I can quote one case of another perfect cock-hen, which belonged to Sir G. Cooper, Bart., of Winchester. She was a Light Sussex, which also acquired male's plumage *in toto*, hatched a brood of chicks off, and moulted out into a spurred hen, *i.e.*, losing her saddle and hackle feathers, and continuing to produce eggs.

How then can we utilise my handling system in grading out these cock-hens and those hens which never produce a single egg IN THE NEST-BOX in twelve months? My instance of the cock-hen may set you thinking; and I hope so, as I shall deal with matters in this chapter which are not generally known, and which for fifteen years I have been experimenting with, and studying deeply.

I wish you to bear in mind that there are cock-hens which lay, and others which do not. Also there are hens which do not lay that can be easily found out by handling, and others which appear to lay and even visit the nest-boxes and yet fail to deliver the shelled products. Even these can be discovered by handling, as I will explain.

First of all dismiss from your mind the generally accepted idea that a hen must be in lay when she so handles. A full abdomen and a three-finger space between the pelvis bones, with four or five finger measurement from the end of breast-bone to the pelvis bones, may denote that the ovary and oviduct are active; but they do not prove that the eggs are expelled and deposited in the nest-boxes. Many will think that such is a sweeping statement, but I will prove it by photographs. I am not accustomed to make a definite statement on handling unless I have given the same an exhaustive test. I have always contended, in writings and lectures, that a hen can absorb eggs internally as quickly as she can produce them in the nest-boxes. She can make the yolks and pass them into the system. She visits the nest in the ordinary way every time she "lays an egg inside," and handles in lay, but fails to show you the *shelled article.*

I will deal with a concrete case, taking as my subject a White Leghorn hen which spent a year in the Northern Laying Test at Burnley laying only one egg. This case will be thorough because she was trap-nested officially as a pullet for the full twelve months. At the end of the twelve months she came into my possession, and I wanted to see if after the moult she would commence to lay. When she reached me at close of year she was very fat, and I tried thinning her drastically; but I did not get the shelled product I was aiming at. But from the start she went into the trap-nest several times weekly, but "no egg." She always handled in lay, with full abdomen, large vent, three fingers between the pelvis bones and nearly five fingers at abdomen. After the moult she continued to visit the trap-nest, and after keeping her for another twelve months I decided to have a post-mortem examination. She had one year at an official Test and twelve months with me with but one egg being seen, and yet she was trapped in the nests several times weekly. I soon had my suspicions confirmed of her laying internally. I experimented by removing her from the trap-nest shortly after she had settled down, but on every occasion she would return to the nest. On no occasion when she had cackled and been released from the nest-box would she return. After every visit to nest she would cackle, and acted when in the nest just like a hen at lay.

Upon opening the bird I found that the ovary was full of monster yolks, and I took the opportunity of killing her directly after a visit to lay. The yolk released that day was found among the internal organs, resembling a burst yolk, *i.e.*, broken. Obviously she had for two years been making monster yolks which, when each hăd swollen beyond the casing, would burst and pour the yolk contents into the body, there to be absorbed into the system.

We can best follow her history through by studying the Plates. At Plate 38 we have a photograph of the White Leghorn hen's head, and are struck with the sad look in the face as reflected about the eyes. I wish you to notice that, as I shall refer to the same later. Her comb has thickened and gone partially erect, which please note is a sign of masculine acquirements. She also called the other hens now and then for a tit-bit like a male, and on several occasions I have heard her crow. There were no other male characters.

Plates 39 and 40 show the giant yolks (GY) in the ovary, with oviduct at O. Each of the five larger or developed yolks are monsters, and each almost as large as the other. Plate 41 is of more than general interest, because we see at C the case or sheef from which passed the yolk just before I killed the hen. As will be seen, the casing shows a distinct rupture or tear, so unlike the natural parting as in the "snow-drop" shaped casing previously dealt with. At R we notice the smaller yolks, which are separate, and point to the activity of the ovarian organ.

Plate 42 illustrates the intestines, and throughout the folds were found particles of half-melted yolk-matter. There was a distinctive look about this, so different to the yolk "hard-boiled" by the heat of the body. I can give as an illustration the fat which is hard like suet, and that brand of fat which in a hen seems soft and meltable. I have marked several portions of yolk-matter in Plate 42 by a letter Y, and in some places undoubtedly the remains of a bunch of yolks could be seen suspended in a cluster by strings from the intestines. The string-like connections would probably pass the yolk-matter into the system in the same way as they do with baby chicks. At Z will be noticed the liquid yolk from the last egg laid within, which was running loose. I also found two other soft yolks, which were small and evidently melting away.

When handled, the bird would have been thrown out by my score-card system at any time, because of her low capability, which will later be fully explained. Her abdomen was coarse and handled fat, the gizzard could not be easily felt, while the pelvis bones always carried heaps of flesh and fat. The explanation of this bad handling is followed by the amount of fat that was absorbed into the system by the yolks, which were being constantly made and put back into the system. Also the handling revealed the fact that the shelling department and the oviduct had been out of use for a long time, which matter I will cover later in this chapter.

I have had many hens like this one, and on each occasion the laying within has been confirmed when the birds have been visiting the nests regularly but giving no shelled egg over a long period. It shows that the ovary is automatic, and how sensitive it is.

Wherever the feminine or ovarian organs get out of gear, we may expect to follow at some time or other masculine characters. It is the same even with oysters! The first character to be taken in the hen in the majority of cases is the spur. A spur on a hen warns me to be suspicious of her. I teach you to treat her in the same way. This male character does not mean that the hen has ceased to lay, but that the spur grew when the ovary was out of order. It is for the owner to make sure that every spurred hen is producing, and that is where the trap-nest comes in. So, you see, I pat the trap-nest on the back, as the two are valuable twins, and we cannot do without either. Handling and trap-nesting together make the perfect combination, and I always use my handling system to locate in the flocks on the farm at given times what I think to be wasters or passengers. Then I call in the traps to prove the birds or find them out as drones. I have one rule, then, namely, to put into a trap-nesting house every bird noticed with spurs to see if eggs are coming, and if she is idle after a good spell she gets fattened and sent to market. At the end of the laying season, say in June or July, before the moult sets in, I put all spurred birds together and others I do not like (all of which wear *red* rings for *danger*) and trap them for a time. Any that fail can go to table after being fattened. One can, of course, have a special trapping house where all such suspects can remain, and one can get an idea of what they are as producers by their flock-average, which, after all, is the only test of a farm's tone or quality, and the only key to maximum profits. One can trap them continuously or periodically, or handle for the eggs in the shelling department as advised by inserting the finger in the vent. At any rate, the red rings on their shanks will see that they are never bred from, as a red ring disqualifies a bird from the breeding pens.

It may so happen that the spur is the only male character, or later may follow growth of the comb which, while falling over in, say, a Leghorn hen, will become almost and in some cases erect. The comb will grow thicker also, and the wattles will increase in size. The saddle hackles, etc., may follow, as in the Westminster cock-hen. Crowing may be noticed, and the actions of the male may show themselves up to the point of dropping the wing and running round the other hens, but not to actual treading.

This must not be confused with unmated hens which tread other hens. This is a question of temperament, and may arise through the absence of a male bird and be helped on by the over-use of fattening foods. On the other hand, the bird may just call the other hens when she finds a piece of food, or the comb only may grow larger. There need never be the spurs, for the White Leghorn hen in Plate 38 had no spurs. One cannot be too careful when handling hens to look for male characters, and not to miss them if present.

In grading, one bears in mind one of my simple mottoes: "When in doubt, leave out." The policy should be to like all good layers and to detest all low-grade ones. That being so, we have to grade in and breed from the good ones, and grade out the bad. Poultry are distinctive stock in that they are rapid producers. They may not be like the louse (a grandfather in twenty-four hours), but we can start with a pen of ten hens and breed sixty the next season from them, and six hundred from the originals all in two years. If you start with ten low-grade hens you can have six hundred of the same brand in two seasons, and if you start with five good and five bad hens you can still have three hundred low-grade stock birds. Why not, therefore, commence with ten good hens and have the best results shown up in the progeny? The breeding stock hold the key to quality in the offspring.

I have on three occasions, when lecturing, had members of the audience put up birds for my opinion as to their laying qualities. There are two fields of thought. The first declare for an egg-type, handling a bird and saying it is 220 or 130 egg type. I have always belonged to the other section, which uses handling to grade out the workers from the drones, and to teach conditioning. I have never altered my views, and have not given guesses about the number of eggs a pullet will lay when handled. Those who rely upon egg-types take as their key the thickness or thinness of the pelvis bones and the number of fingers from breast-bone to pelvis bones. Please do not confuse my score-card with those two measurements, and carefully note when critics are busy, and referring generally to abdominal capacity, that my capacity is different, and that my card goes all over the bird. The critic is not going to point out the difference for you, and so often tries to write generally to get his criticism home. And very often letters in reply to the criticism are never inserted, but returned to the writers as "not of sufficient interest to readers," or "the correspondence is closed," or the writer is told that the letter is an advertisement and cannot be inserted. I hold many such letters (sent to me by correspondents) which have been returned to them along with the editorial replies. A reader challenges my handling system under a pseudonym (invariably) in an open correspondence column. A breeder replies, giving records of birds which have won under me, and he is refused publication. On this, my platform, I can mention the facts.

Handling to types has done more than anything to set back the progress of hand-grading along with the interested criticisms. I have never been able to understand why a breeder should be dead set against supplying a graded article when he asks a high price and gets it without arguments. The buyer places himself in the hands of the breeder entirely. I have known a breeder send three birds to a buyer, starting with a two-guinea cockerel. When asked for a better bird he charged £3 3s. and sent an inferior one, and the last one at an even higher price was the worst. It is in the interests of all engaged in poultry-breeding to supply the graded article and then their stock will get a good name and keep it, and the industry will prosper.

But I have worked very hard for handling, and never hesitated to select pullets for Tests to show what they will do, and to hand-grade flocks for breeders to obtain converts. To-day hand-grading is very much alive, and in years to come will be

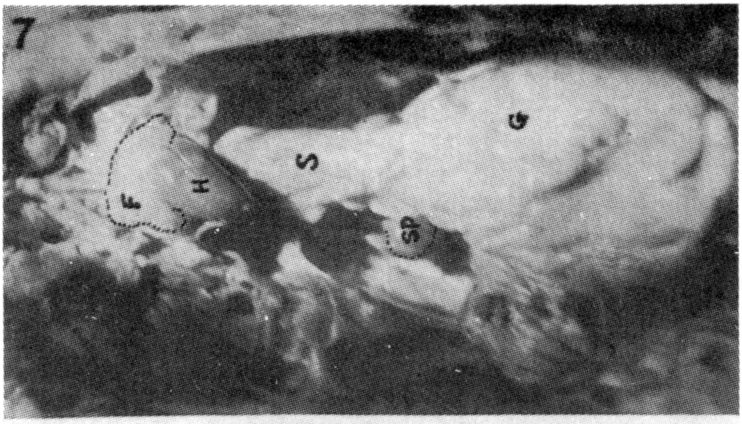
7
F
H
S
G
SP

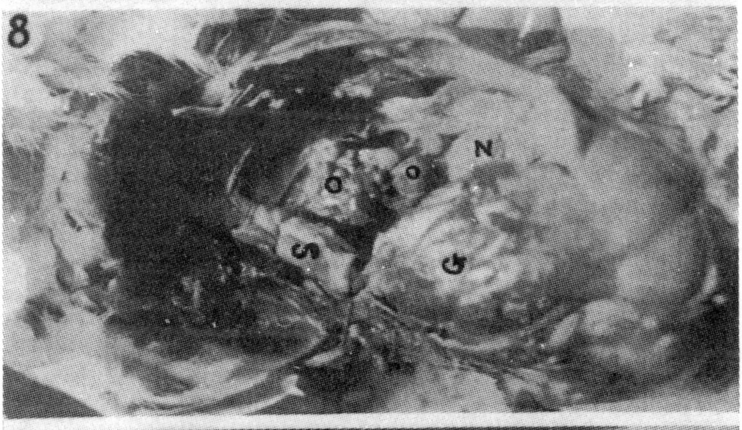
8
N
O
O
S
G

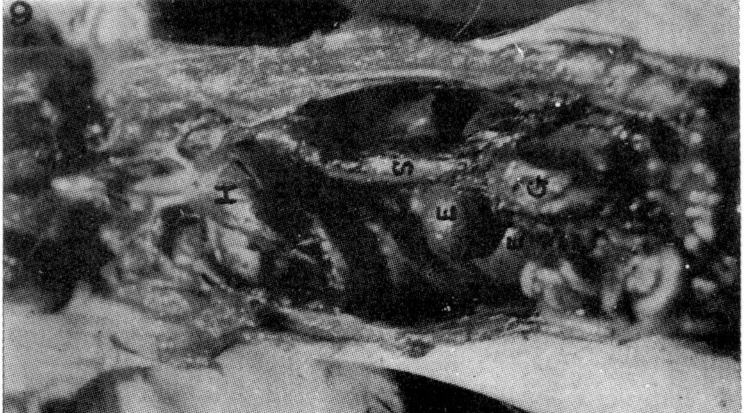
9
H
S
E
G
E

practised extensively by every progressive poultry-keeper, whether backyarder or breeder. How little it is understood may be followed from the incidents at the Harper Adams College Conference, 1921, when three "handlers" were to grade the birds penned, and then would be placed up their egg-records for those present to read and digest. The cards went up, and so did the egg-records, and then picture a tent full of destructive critics, who had a field day. Luckily I had sufficient observation to take the ring number of each bird and the date of hatch. There were just THREE pullets out of over a hundred birds put up for grading, and many were three years old. So the cat was out of the bag. I never have stated that I can handle a three-year-old hen and say what number of eggs she laid three years ago. The three pullets were placed well by me, and two got into the cards with a third prize for one of them. Anyway, it showed that handling is not understood, that hand-grading for types is not possible (or a type should be always there), and, thirdly, that the birds were not trap-nested longer than the first year. Some of the birds were weeds, but up went what they did in the first year, which is no criterion to what they will do in the second or third. Several birds were deformed, and one had cloacitis, which meant that on my handling I had to deduct points which threw them out. The Test seemed to have the following objectives:—Three judges on three different score-cards were to handle the same birds put up in breeds and classes, and award so many cards, as at a Show. Score-carding is useless, because all three did not give the same birds first, second, and third in each class. But while we worked on three different cards, surely even on one and the same card any three men would not agree? I have items on my score-card that are not on the others, and I give points for Show condition, breed characters, and so on. Again I deduct for scaly leg, lice, and the like, and did so here just as if I were at a Show. The bird with cloacitis I located, while the other two judges missed her, which surely does not concern score-carding but individuality in judging. As regards the latter, one judge, when handed his first prize White Wyandotte pullet, said he made a serious error, and she should not have won. Again one of the judges did not on his card place any one of the three pullets, despite their splendid egg-records. I placed the three pullets best, and the rest of the Test was a waste of time and my travelling expenses. I labelled one bird "weed," but the other two "handlers" scored it; the fact that the College officials withdrew the bird after putting up its egg-total of two years back shows that they did not think it advisable for those at the Conference to see it.

No, the experiment was not a *test*, but from it all we learned how little understood is hand-grading, that birds even from the best establishments are valued on their pullet-year records alone, that ailing specimens and weeds can be found in the best of flocks, and that a utility bird has become valued solely by the number of eggs laid in her first year, her value as a breeder being ignored entirely.

If I take a pullet in hand calling her a five-finger bird at abdomen with a pelvis bone thickness of one-eighth of an inch, and set her as a 230-egg type, then deduct or add eggs according to the thickness or fineness of the pelvis bones and the good or bad measurements of the abdomen, that is dangerous. For instance, the beginner takes in hand a pullet too fine in the bone and lacking stamina, and she must be graded on egg-type very high. I want the owner to see that such a pullet proves that he is not breeding well and is losing stamina. Again, he will not send such a pullet to a Laying Test or trap-nest her, because lacking bone she may not keep up laying, but may rest after a batch of eggs while she builds herself up again. Also he will know that she must not be bred from, or "weeds" will result. Going in the other direction, it will be seen that one has but to fatten a good bird and away goes the capability to produce; and, as grading would be too automatic if egg-types were the guide, I should lose my lesson of the value of the personal element in poultry where the owner having the right type of bird must feed and tend it in the correct way. Good feeding will make a bird lay more eggs than bad feeding; bad feeding will make a good hen lay badly! It is conditioning that counts.

I recall an instance where several handlers to types gave their views as to the number of eggs a hen had laid, and whereas all said over two hundred, the trap record said under fifty. On another occasion, when a lecturer said two hundred and fifty, the man in the audience said that he valued the bird so little that the lecturer could kill it. I have gone against these methods, because the next thing handling will have to say is the number of times a hen will go broody, the number of days she will be in the moult, and if her sire died of bumble foot. Let handling stand on its merits, and if it only teaches observation in the strictest sense as against automatic methods, then it will find a high place in poultry-breeding and poultry-keeping.

I have usually said at lectures whether or not I thought well of a bird as a layer, and on three occasions mentioned birds were handed up that had not laid. My reply in one case was that the bird might be worth fourpence, and I graded out the second. At York Show, where I judged, I was handed the third. On the evening before judging I addressed members of the York and District Poultry Society in a marquee on the Show grounds. After dealing with several pullets and explaining my methods of conditioning them for laying, a member said he had a bird whose trap-nest record he knew and he would like me to say what I thought of it. I told the members that I thought little of the hen, but pressed further I got to figures, which is the only occasion I have departed from my rule not to teach numbers. I said that in my opinion the hen had not laid for the last six months, and that I should not be surprised if the owner said she had not laid for a much longer period. The member was full of appreciation, as were all present, when he said that the hen had yet to lay her first egg. Is there nothing in handling?

The next day I judged York Show, and was able to look members in the face and to come home with my reputation unshaken. What if I had said 230-egg type? At this lecture I had referred to spurred hens, the spur denoting that at the time of its appearance the ovary or oviduct had gone wrong. Penned at the Show was a Minorca hen with spurs, and in the abdomen was a shelled egg which had found its way through the walls of the shelling department, never to be laid, but at any time to set up trouble likely to cause death. As to how she was laying I do not know, except that she laid a soft egg that day in the show-pen. Such a bird could lay a soft egg for good, or if a shelled egg came along, the pressure when in the shelling department on the other egg could break it and cause death. Probably this hen laid soft eggs because the irritation of the shelled egg in the abdominal cavity when touching another soft egg in the shelling department might cause the hen to strain and lay the egg without a shell and too previously, so to speak, or constant straining to expel the shelled egg might ruin the muscles of the oviduct and the walls of the shelling department.

It is surprising the number of birds I handle at Shows and on farms where obstacles are found within the abdomen. It may be a cyst, a growth, or a shelled egg or several shelled eggs, including some broken ones or hard balls of yolks or liquid, as in abdominal dropsy. Some such birds have spurs and others have not, while a few show certain characters of the male, if only an enlarged or upright comb, especially in those breeds where the comb falls over to one side in the hens. In hens with upright combs the comb thickens and increases in size. Writing of upright combs in, say, Leghorns, where the comb should fall over in the hen and be upright in the male, it will often be noticed that some hens have upright combs. In breeding exhibition Leghorns it is common to mate a hen with an upright comb to the upright-combed male to breed sound-combed cockerels with wide base to comb, to enable it to keep up well. Such hens are called cock-breeders. In like manner one takes a male whose comb falls over, but not from lack of stamina, to hens with correct falling combs to breed females with soundly-carried combs. Such males are called pullet-breeders. I mention this because I have on many occasions heard utility poultry-keepers say of a hen with an upright comb: "Do not use that hen or she will breed all cocks." Again the need for calling a spade a spade in poultry matters! Naturally pullets starting with upright combs will be so marked, and then difficulties of observation will be removed.

This question of breeding combs should be studied with greater care, even by utility breeders. I always go for a comb that keeps clear of the hen's eye when it drops. Now you want to avoid excessively large combs, in order not to get "beef" there. In like manner it should be remembered that if there is no base to the comb it will fall over like a piece of tissue paper. Often, when one goes for texture in the comb of the male, the bird with a comb like paper from base to tip attracts those without the necessary observation and knowledge, and mated to hens having the same failing will quickly get a thin comb in the progeny with no base to keep it out of the eye. Remember that the base is the medium for throwing a medium comb just clear of the eye, so select both hen and male with a nice base to the comb where it joins the skull. If the hen carries a beefy comb, then mating a male with nice medium comb will reduce size. Never use a male with a large floppy comb, or it will take years to get out of the strain, as will any external failing introduced on the male's side.

When stock are sold to customers, let me advise you very strongly to handle each minutely, not only for deformities, but for any articles in the abdomen. It does not mean pleasure to a client if a hen dies a day or so after delivery, and the man to succeed at poultry-breeding is the one who does give constant satisfaction. I know there are some whom one will never please, but that is no excuse why all customers should be treated haphazardly. Handle each bird at abdomen to make sure there is no egg or the like inside. Next see that there is no deformity and no sign of bumble foot or swollen shank, no watery matter under skin of shank, no abdominal dropsy, and no articles like a large piece of stone in the crop, which may cause a blockage some day. Have the birds placed in a special pen and handle their crops in the morning before giving the breakfast. The crops should be empty, and if full of food then crop-compaction can be suspected and the bird graded out for hospital treatment and not sold. I am always keen on this handling, because we have to look after the beginners if the Poultry Industry is to be strong and sound. Year by year I see beginners fighting shy of well-bred stock because their first purchase has been a fiasco. Imagine a man's feeling when a few days after arrival a bird with a compacted crop begins to waste away without the owner knowing what is the trouble. Imagine the society secretary having to perform the operation by cutting the crop open to remove the food, etc. The deal gets going the rounds and the breeder is black-listed. The small poultry-keepers will talk, and they will advertise your strain to the skies if they are satisfied, but if they are not, well—it cuts the other way. Many may be tempted to say, "What does a beginner want for his money?" To them I give my personal opinion that if they cannot take a little trouble over the supply part of the business, when they have spared no pains or money to get the enquiry and the order, then they should not be stock-breeders, but entertain themselves getting new-laids. They remind me of the Show secretary and his committee who offer 30s. first prize to get the entries, and then take twelve months to pay it out when won. We know what a fiasco his next Show is, and he loses on entries. Once bitten ——!

Again see that every bird has clean, dry nostrils and sweet breath, with no traces of colds, because if any with colds are sent and roup follows, you may get out of it by saying it was caught en route; but think of the buyer who may have risked all his capital, and the sad consequences. Take special care as the time comes round for selling matured stock that they are not overcrowded in housing, and that colds do not come along to spoil your chances of sales through roup, or to cause annoyance if any that are sold get roup, etc. Do not attempt to rear a hundred birds with housing sufficient for but twenty! Birds with colds will have running nostrils or watery eyes, and the feathers on their backs (the hen's handkerchief) may be matted where they wipe their eyes and nostrils.

Handle the birds, too, to see that they are healthy and plump. Often a bird looks well but handles as light as a feather, so rely for condition upon how the bird handles and not what it looks like on the ground or in the run.

Hens may carry substances in the abdomen and yet not go wrong, but the proof is in the testing after locating them. I make it a point about June or July to handle the stock for such articles, especially spurred hens, although these are handled when noticed and kept under check. By testing them up for laying in late summer one can have the birds in table condition if they are passengers, and they need not be kept at a loss over the moult. One needs to sell for table at the most opportune time, and that is before the moult comes along. The birds are then plump, and more so if they have not laid well, whereas if they stop laying, or think of the moult, hens soon lose weight as they lose appetites, and fattening is out of the question unless they are kept on, got over the moult, and fed up into plump condition again. That is where many lose time and profits. Dispose of all birds, then, when they are plump, and even give them a good final period of fattening. Losses are heavy on every farm because handling for condition is not widely practised. By feeling the breast one can get an idea quickly of whether a bird is fit for table. Table condition can also be checked by the flesh on thighs, pelvis bones, etc.

You thus have a chance to "prove" the suspicious hens before the moult comes along, and any can be graded out to time. Any hens retained should again be handled during the moult, when the abdomen is empty of intestines, which will enable one to feel any article that may be difficult to locate with the hen in lay and the abdomen full. Make a practice of this, and any hens found to be wrong can be watched for laying after the moult if kept on. Birds, however, that are to be graded out should be found before the moult, and then there can be a great saving; otherwise they will have to be fed right through the moulting months for no return, and will be sold afterwards, using up valuable housing as well. Hand-graders will be very busy in late summer, because the poultry-keeper will have to decide his programme for the next year, and will be compelled to reduce his adult stock to house his pullets for the winter. As for winter eggs, too, the younger the stock and the better the results with present-day breeding and strains, where longevity has not so far received its full consideration. Do not waste valuable housing, labour and outlay on passengers, but devote the same to the workers. Quality before quantity!

I am a great believer in reading a hen from the eye, just as I do humans and all stock. The eye of the horse will tell his temperament, as it will show the difference between the cow that gives the milk and the one which kicks it over once it is in the bucket. Take my advice, too, and never work for the man who cannot look you full in the face. The hens that have internal disorders and ovarian trouble, and among them the "never" or "seldom-if-ever" layers, can be suspected by the eye. When a man has lost sixpence you can tell it from his eye and worried look. There are exceptions, of course, because one man will argue that he will find a shilling to-morrow. A few have a habit of being thankful for what has happened to-day, and will think of to-morrow when it comes.

At the close of the laying season, about June or July, one begins to locate the loafers, non-layers or low-grade performers. If, too, anything is wrong internally one sees signs of it then. Take the Leghorn hen illustrated in Plate 38, and the two years of internal laying and worry have left her with a sad face, puffed eyelids, and an eye lacking fire and character. We pass to Plate 37, and have a White Wyandotte which has a very worried look and a sad eye. Her history is unique. I was grading out a backyarder's little flock, and came across this pullet, which was sixteen months before she laid her first egg. I bought the bird for observation, and found that she was temperamental, and was afraid to eat and to take her own part. I petted her a bit, and soon she got very friendly and, by special feeding, I got her to start laying. But she provided less than a hundred eggs in twelve months. Always puzzled over the hen I was determined to find out what was worrying her, and found that she had a scar on the skull where the feathers had not grown. This can be seen in the Plate. The cause is not difficult to fathom, and my suggestion is that in the growing stages she got a peck from a young cockerel which lacerated the skull and affected the brain. But the worried expression is there.

As June and July come along the drones begin to mope about, and that is when you can suspect birds and test things by handling. Later still, as the moult starts early with the low-grade birds, moping on their part increases. We need at the close of the laying season to pick out the hens which are active and which have an eye without any worry portrayed in it. The best description to give is the hen which looks as happy as the day is long and as if she had not a care in the world. When you have mastered my handling system instinct will help you to pick them out. With age I prefer all stock to keep their quality—an old hen looking in face like a pullet, and an adult cock like a cockerel. When a cockerel looks like a four-year-old cock, and a pullet like a five-year hen, then each lacks utility quality. If you get your hens too fat, or if they are infested with lice, you can see a worried look, and can have a probe into the likely causes.

There are many hens which go about laying soft eggs; and I am very keen on keeping them under check, never hesitating to send them to table to prevent the habit of egg-eating from catching on. I can see no fun in letting hens eat their eggs *ad lib.* when they lay them; it is enough trouble to get the eggs made, and we cannot afford to have a leakage here.

Where trap-nesting is practised it has the advantage of preventing egg-eating to a great extent, and certainly stops a hen from eating any but her own eggs. But all flocks are not trapped, therefore be very careful in choice of nest-boxes and where you place them. Try all means of prevention. When on your daily round you happen to see a hen with yolk on beak or face, place her in a coop with a few eggs to see if she is egg-eating. A hen that lays soft eggs permanently may be found out by the abdominal fluff being constantly wet and soiled. This may be a sign too of ovarian trouble, and I always suspect hens with soiled fluff at abdomen. Often a yellow stain may mean a cyst inside, although it could be irritation from feeding that is on wrong lines. But whether it be ovarian trouble or an irritant in the food (such as damp Sussex ground oats, middlings over which rats have run, or a bad brand of fish-meal), handling will help in the solution, and the soiled fluff will give the advice to handle.

I remember visiting a well-known White Wyandotte breeder's farm, and noticed quite a goodly number of hens with diarrhœa; and noticing the long grass and the short legs I soon discovered the cause. Asked if he had any low-legged Orpington type of Wyandotte males about, he said that Mr. —— had told him that the short-legged birds made the best layers. He had sold the best and kept the worst types! In the Laying Tests that year he lost many birds from the same cause, and had gone wrong because he had received injudicious instructions in selection. He knows better now and is again back at the head of the Tests.

I am also keen on the feel of the abdomen when handling hens in late summer before the moult. I have already explained that the shelled egg can be felt in the shelling department inside the vent to the left of the bird. Placing the thumb of the right hand on one side of the abdomen, and the first finger on the other side, we need pliability and softness where that egg is located. This will tell us whether or not the oviduct has been in constant action. Keep well up to that part of the abdomen just below the pelvis bones, and go well into the abdomen, not confining the handling to the skin. If the bird has been a poor producer or has laid few eggs, if any, the part mentioned will be hard like leather. This would be particularly so in a hen which laid eggs within, and the abdomen between the pelvis bones at the rear and by the vent would be leathery. In the hens that have laid well one will be able to work the abdomen, and in many birds to touch thumb and finger through near the shelling department, assuming there is no egg being shelled at the time in the latter section. Remember when handling that if an egg is there the abdomen will not be so pliable, so that one must always in handling use the head as well as the hand. If an egg is in the shelling department and the abdomen feels less pliable than expected, you will be able to check up condition by the pliable pelvis bones and the prominence of the gizzard. Hand-grading is a deep matter, but why should it be so simple that, as many suppose, it ought to be mastered in five minutes? Allowances have to be made frequently, and the good grader is the person who balances the bird to a nicety. Take the matured pullet and you forget what is in your hand; but try to picture her after she has laid a hundred eggs. That is where experience in my handling system comes in. You have to start with a knowledge of the hen from within; but I think this volume will give you a better understanding in that direction than you had previously.

I have made a very close study of hens that never lay. If the ovary becomes dormant, the bird does not visit the nest-boxes; if, however, the ovary is active, the bird will visit the nest regularly but never lay an egg. I think there are two types of such birds, the one making giant yolks which burst and send their contents into the abdomen to be absorbed, and the other making the yolks but just sucking them back again. The majority of such birds that have been under my notice for two seasons and more have never laid a shelled egg in the nest-box. But several birds have surprised me by suddenly commencing to lay properly shelled eggs, some after the moult and others during the ordinary season. I have had a few birds lay well for their pullet year, then start visiting the nests the second year but producing no eggs other than within themselves, only to be active and good producers in their third year. This can readily be understood as proper functioning suddenly commences. I have tried exhaustive experiments with fattening tests, and have kept hens out of lay through excessive fat for nine months and longer, yet brought them into lay again by thinning them. In such cases the ovary remains dormant, but with hens that visit the nest and "lay internally" they handle as if in lay owing to the "functioning" of ovary. I am still experimenting in this direction.

Straining to lay any article that happens to be in the abdomen may, as already mentioned, see the hen laying soft eggs permanently through ruining the oviduct muscles or walls.

A few hints on handling chicks at abdomen will here be of help. A chick takes in the yolk about hatching time, and if it does not melt away it hardens with the heat of the body, and causes death. Bad feeding and chills will soon stop the yolk from melting away, and when a brood goes wrong I always grade the chicks out. By feeling the abdomen just below the pelvis bones one will feel the hard yolk, and such a chick can be regarded as lost, and no worry occasioned if it should die. Those with soft and pliable feeling will be rear-able, and can be separated. A little practice in handling chicks for this will soon make one perfect. If a chick ails, and the hard yolk is felt, one need not worry if it dies, because it is almost sure to, sooner or later. The yolk must be kept soft and meltable. Good feeding and attention will do that, but if the digestive system is overtaxed it has enough to do to fight the extra strain, and does not get ahead with the absorption of the yolk, which then hardens and is past the absorption stage. In each case the hard yolk is large.

When handling your first hen do not write and tell me she has a large egg in the abdomen by reason of your mistaking the gizzard for the obstacle; and do not kill a chick because you take the gizzard for the hard yolk. You will not learn my hand-grading system in two minutes, and your first lessons will have to be on such elementary sections as locating the gizzard of a hen and a chick. When handling for the yolk in the chick you keep to the same place where you handle the hen's oviduct for pliability. Take a few healthy chicks at various ages and you will follow, and do not forget to open every chick and hen that dies to perfect your knowledge of the internals.

CHAPTER VII.

CATCHING AND HANDLING FOWLS.

JUST as hand-grading is a weak link, so also is the method of handling. At the Shows I am provided with all kinds of stewards, but I seldom find the man I am after who has a perfect knowledge of the way to get a bird from a pen. As a result I generally get my own birds out. Sometimes when starting with a steward I am amazed at the way he sets to work. If I were the little hen in the pen I should certainly refuse to be caught, getting the impression that the man was after my neck. Sometimes I have a steward schooled in handling exhibition birds who makes a dive for the hen's legs and certainly succeeds in securing a quick catch. But you cannot do " catching-by-leg " tricks with the best layers unless you want ovarian disorders and to turn a hen into perhaps a cock-hen.

At Shows I get a good idea of ways of handling adopted. I see very frequently exhibits at small country Shows brought to the event under the arm, with their legs tied tightly together to prevent escape. Again, when penning birds, I notice a steward in a hurry with the bird carried by the legs head downwards, and swinging to the walk of the handler.

Let us see what happens when incorrect handling is practised. First of all I have always made the rule never to handle hens or pullets by the legs—great is the harm done by this all too common practice. If one refers to Plate 12 to see the placings of the internal organs, one can understand the damage through squeezing these organs together by closing the legs. The ideal layer stands wide between the legs, solely because she cannot help herself. She has a wide back and a full abdomen, and such compel her to stand wide—these items go together. Directly you squeeze the legs together you also cramp up the ovary and oviduct and the heart. A hen or pullet that is a weed or very small in frame will be very narrow between the legs, across the back, and in abdomen, and would not come to much harm by having her internals pinched up by the closing up of the legs. But I strongly advise one and all to leave the legs alone.

If I am holding a bird I keep well to the front and away from the rear, where the important organs are, avoiding the abdomen at all costs. By holding the head down and the tail up, the bird does not struggle like a bird held with the tail lower than the head. When first taken in hand, the open right hand should be on one side of the bird and the open left hand on the other; the thumb will be on the shoulder part of the wing, and the flat fingers will be along the side of the bird and partly on the closed wing, the bird being tilted as advised, with head a little down. The bird can then be passed along towards the loose left arm, while the left hand goes underneath the bird, with the fingers of the left hand between the legs high up to the breast-bone. If the bird is narrow between the legs, then two fingers may suffice; but if wide, one may have to place the four fingers in support between the legs. The head and neck of the bird can rest in the socket of the loose left arm, so that the bird can see the ground behind through the aperture, and she will then be quiet. Such a method of carrying can be adopted when taking a bird to her basket at a Show for dispatch home, or when carrying a hen any distance—her shoulders and side can rest on the operator's left arm. Or she can be carried in the first position with left open hand on one side and the right hand on the other, gripping the shoulders and wings, and holding the head a little lower than the tail. Male birds can be carried in the same way, the palm of the left hand becoming, so to speak, a resting-plank between the thighs for the breast-bone.

Supposing I am taking a hen out of a show-pen—my first point is to approach the bird and pen as if I were going to stroke the bird, and not catch it for market to break its neck. There is to be no sudden diving for legs, which will put the "wind" up the bird. I note the temperament of the exhibit at the same time, because that may decide the mode of approach. As I open the pen-door and hesitate a little, I notice the good-tempered bird comes towards the open door, and her sweet temper and docility I see written large in her eye. I pass my open right hand over the bird, and gently place it half heart-shaped over the shoulder furthest from me, and gently drag the bird towards the open door. She is then in a position for removal with the two open hands held full heart-shaped, one on each side of the body, taking the closed wings as already mentioned.

The next bird I handle from the pen may be a medium-tempered bird—one of those which can easily be tamed. She can be handled in the same way, but I must have more time for her. She may refuse coaxing, and the hand must be placed well above her—a long arm helping when handling from pens, also height; but anyone lacking reach must use a platform or chair to stand on. Naturally one should stand at the most convenient height, and Show officials should pen their utility exhibits with judgment, and not place heavy-breed cocks three tiers up. The medium-tempered bird can be stroked before she is taken out, and should not be withdrawn quickly to make things worse.

Finally we have the bad-tempered bird, which is no friend of mine; and I hope you will also detest such birds, because they so often lack utility qualities. Such a bird ruffles his or her neck-hackle like a broody hen, and will show fright in the eye. When one attempts to catch it, up goes the neck-hackle, and the bird begins to fly all over the place even when you look at it. I suggest that the best way to catch such a bird is as gently as possible by the nearest wing, obtaining a good hold upon it near the junction, and trying to hold the whole of the closed wing. Then take it in the two hands as advised. Catching birds by the wing in this way will not do harm, but one must not carry them in that way, making the wing support the body; rather use the method to get first hold upon a bird when other methods are not successful. I am not against catching a hen gently by the tail, so long as the whole tail is taken in the hand, and it is just the first method to capture the bird prior to proper handling. But that does not excuse a man when he seizes two tail-feathers only, and holds on until they are withdrawn by the hen's struggling; or when he holds but a few feathers of the wing. If a false hold is made, let the bird go and try again. The wing and tail methods are useful when catching birds in a house as they pass by, and particularly docile birds, which will not struggle when so held. Often one wishes to catch a bird and she just keeps out of the way; but by stretching the arm out one can get her tail or wing, and it is then that the methods mentioned can be practised. But when one can get the open right hand across the bird's back and round the closed wing furthest from you, then that is the ideal plan, as such a bird can be gently moved into position for a complete handle for lifting up.

Another dangerous phase of handling by the legs, and placing pressure upon the internal organs, is that many birds will struggle and force out the cloaca or oviduct. This may be followed by other hens pecking the exposed organ, and ending in death. If a peck does not follow, exposure may ulcerate the parts and cause a death. I hope every poultry-keeper who practises handling will remember my advice that the birds are not cast-iron articles. The execution of handling must be very gently and carefully carried out, and I am sure many birds are injured at Utility Shows by rough handling. I am quite sure, too, that very many deaths result at Laying Tests and at home from rough handling, catching and trapping. Deaths, too, by people who are not practising my handling system, but just catching birds in the ordinary way of farm routine. I am constantly amazed when I see how some owners catch their birds, and how they " do " them rough. No wonder, with such delicate organs inside as the ovary and oviduct, that deaths happen from slipped yolks, cysts, shelled eggs within, etc.

When handling, beware of the bird which struggles violently when being handled. She will not be your best layer, because the birds of this noble brand are always docile and sweet-tempered, and even like to be handled and fondled. I have never found it otherwise, and I handle tens of thousands of birds every year. More birds go through my hands every season than any other's. As you take her in hand you will see how readily the cloaca is expelled, just as it is sometimes when a bird is killed for the table. Especially will this be so with a fat hen whose abdomen is blown out to the full with excessive internal fat. Do not apply rush tactics, but let her settle in the hand first. Again, when a bird arrives at a Show while you are judging, let her settle down before you judge her, and if a bird escapes from her pen allow her time to settle. Allow her time, too, if she is laying at judging-time.

For general catching work on the farm I strongly advise a catching-box, for both adults and growing youngsters.. If used for the latter they will be quite accustomed to the boxes when they are grown up. Many use a catching-net very much like a string fishing-net and on the roomy side. I have no good word for this type of catcher. If you can picture trying it in a flock of Anconas, Black Leghorns, Bresse or most light breeds, you will see why I do not support it. You have to chase a bird like you would a butterfly, and bring down the net at the right moment. If inexperienced, you may give the bird a hit with the wooden handle of the catcher, but certainly you will upset all the other birds in the pen.

Catching-boxes can be of any suitable design so long as they are effective. I will give you a rough idea of the one I favour. In the first place I prefer a wooden bottom, to prevent a bird's legs from being held up between wire or battens. The two sides and one end will be of wire-netting of 1-inch mesh, secured to a stout framework. The other end, [illegible] has to be placed against the house and the pop-hole, will best be in the nature of a slide-up or slide-along wire-netting door. The top will consist of two wire-netting slide-along doors, the one above the other, so that birds can be removed at either end of the box. I like a lengthy box, say 5 feet long by 3 feet high and 3 feet to 4 feet wide. Smaller and lighter patterns can be made for use when one is handling small pens of birds, the top consisting of wire-netting with a wooden slide-in door in the centre, working along grooves. I prefer stoutly-built boxes, as they will last, and one can readily transport them by means of a wheelbarrow from one house to another. Naturally, gates to all runs must be wide; but they always should be on the poultry farm, to allow a free passage of barrows, etc. For preference make the box in sections for storing away easily, and regularly creosote each out; observe utter cleanliness as you will with all utensils used for fowls if you are wise.

Supposing we wish to handle a pen of one hundred hens. The birds will best be handled in the afternoon, when most will have laid. Do not shut them in on a hot day from early morning, and cause sweating, chills and colds, and, maybe, roup! Get them into their house by food as far as possible, letting the breakfast be a little less, so that the birds are ready for more grain in the early afternoon. This is far preferable to the usual method of giving the manager instructions to keep all birds in for the day while one goes the rounds. Often they get sadly neglected that way, and more so in warm weather, when birds should be out of the houses as much as possible. Keeping them in on days when it is cool may not be so harmful, but always use discretion and common sense. Even when they are shut in for a time that does not excuse the manager for neglecting to " water " them.

The box will be placed against the pop-hole on the outside of the house and the sliding wire front withdrawn. Next, the pop-hole steps or ladder (if in use) will be dropped down to rest nicely in the catching-box. The latter will show a nice fit against the poultry-house, to prevent any birds from escaping at the sides; and to help matters the ground will be level. One operator will next enter the house and very gently drive through the pop-hole and into the catching-box a batch of birds, say ten or fifteen. To save time and labour the operator can fit up a slide on the inside of the house, to cover the pop-hole when the birds have been driven through, and to prevent any from returning. He can then withdraw from the house to handle those birds in the catching-box. Having handled them and rung them, each can be dropped in the outside run, and more birds driven through the pop-hole into the box for handling. It is always useful to have a help with you to take down any notes you may call out concerning any bird. You may wish to have a note that hen No. 25 has a deformed back, or that No. 80 has an egg in the abdomen. They can be rung with red rings "for danger," and can easily be fetched from the house at night, *i.e.*, from the perches. I always have a few baskets tucked away behind the house, and labelled properly, so that if I find any birds with colds or ailing they can be kept in hand and sent to the hen-hospital. If I suspect any of not laying, I place them in a special basket to be taken to the trap-nesting house to be tested. Each man will have different methods, but I just give suggestions.

The general method of catching birds from a flock, as far as my observation goes, seems to be going into the house and closing the door. A dive is then made for any bird that happens to be nearest, and there is a general disturbance of hens and litter. If one of the last birds to be caught happens to be artful, and to keep well under the drop-board, then a handful of litter is used to fetch her out, without any thought for damaging her sight, etc., or putting her in the hen-hospital for a month or two, with the subsequent loss of eggs. Again, chasing sees the operator often on top of the bird, and that means trouble. Birds, too, escape into the run and miss handling. Adopt the catching-box on every farm; progress will then be made.

Many who trap-nest often prove bad handlers, as constant visits to farms will show anyone who is observant. On many farms I have seen trap-fronts nailed on to small nest-boxes, compelling the bird to squeeze in; when removed she would need to be pulled out by her legs. Because one is sold narrow trap-fronts, that does not mean that the width of each nest should be the same as the front. Why nail up the fronts when apart from releasing the birds there is the all-important part of cleaning out the nests. I impress upon every reader the importance of having roomy nest-boxes. The average box has apparently to do for all classes of hens, whether Bantams or Brahmas. Surely a Light Sussex calls for a roomier nest-box than a Leghorn or Ancona. Also any breed likely to put on fat and to get fatty hearts, like the heavy table breeds, need roomy and well-ventilated nesting-places if losses in hot weather from heart and brain trouble are to be avoided.

One should have hinged fronts to trap-nests, and should encourage the birds to walk out of the nests on to a wide platform, which acts as the "fly-up" to the nests. It is in trap-nesting where the women excel, seeing that they have every patience and handle the birds with care. By careful

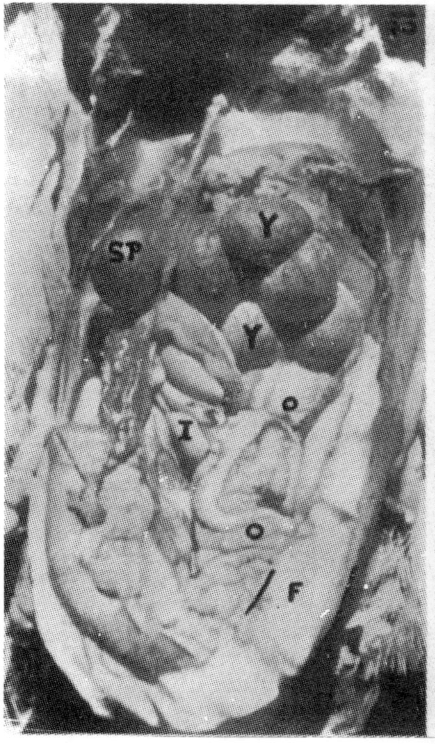
SP
Y
Y
O
I
O
F

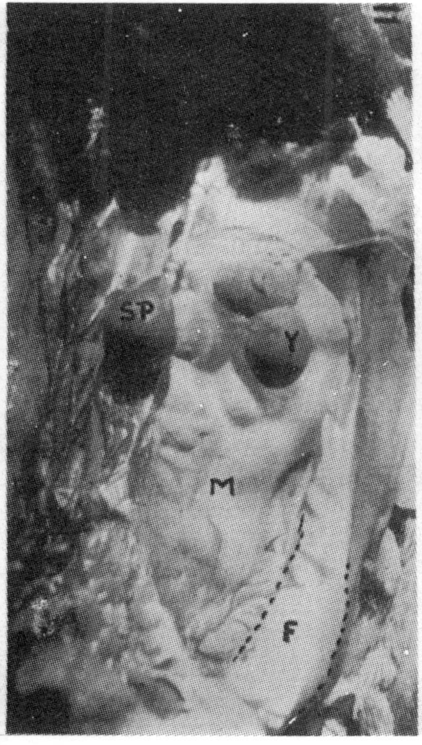
SP
Y
M
F

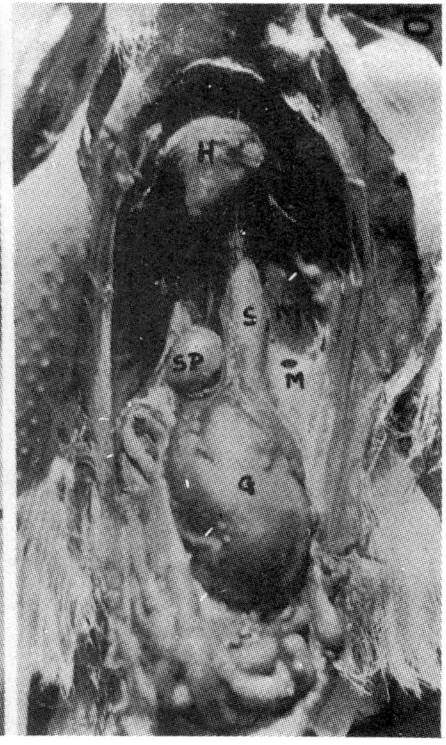
H
S
SP
M
G

treatment one can teach the hens to come out on to one's hand, and can take the ring-number without handling the bird. Arrange the trap-front, therefore, to enable it to swing clear when opened, or to lift well out of the way. When hinged at the side to open outwards, the operator has a better chance of preventing any hen from escaping; and that should be avoided as you do not want to have to chase a bird to get her ring-number. That is another reason why I never send a flighty bird to a Laying Test. You can picture her escaping from the nest, and the lad chasing her up and down the run to get her ring-number, and ending by throwing or kicking litter at her when she is under the drop-board. When caught, she may find her little body flattened out under the lad's knee, and the next morning she escapes she may get a good kick to stop her flight as she passes, an occurrence I witnessed at a Laying Test recently; and when the kick had stopped her, she was picked up and thrown over the six feet netting of her run, as if the culprit was releasing pigeons—and she did not land on her feet. I can name my witnesses if this is challenged.* Anyone who loses his temper with the fowls cannot hope to get the best results; in fact his heart cannot be with them. Hens like the company of the owner, and he should at all times enjoy theirs. Even when the broody is found sitting on a nest of broken eggs the owner should not be cruel to the hen. Remember from the start that the hen has very little brain, and is not a "patch" on the duck for cleverness. A rough handle or a kick, and you can soon put a laying hen out of action. I can show you a certain spot in the middle of the back where a slight knock from a judging-stick will cripple the bird. Remember, then, to handle birds gently and with every care when taking them from the traps. Take them with the right hand on one side and the left hand on the other, with closed wings included, which will show one the need for roomy nests to allow the operator every chance. See that the legs do not catch in any aperture when you take a bird from show-pen or trap-nest, as it is easy to break or damage the legs and thighs or toes. Use rings that are plainly numbered, and, if metal, let them be embossed. The aluminium rings are very light, but there can always be a method of ringing in addition by colours, so that the bird's number can be read without handling the bird. For instance, white could stand for 1, blue for 2, and green for 3, etc., and I suggest this method for small flocks.

There are other methods adopted of handling birds. For instance, some hold them between their legs; but such a plan means the application of pressure upon the bird's frame, and

* *An enquiry has followed my allegation, and the man was dismissed.*—AUTHOR.

I am strongly against that. Others place them in a sack and tie their legs, but that means unnecessary pressure. To carry them by their feet with head hanging downwards is the worst form, because the weight of the body must draw the legs together and cramp the internal organs. To watch some breeders catching young stock is too bad for words. There is a general round-up of the youngsters into one corner, and as many as can be held in the fingers of each hand are transported to the place destined for them. If fourteen legs can be grasped, leaving the other fourteen loose, that is O.K. I suggest that the catching-box is the only ideal way for the youngsters, because a batch can be quickly caught and transferred to fresh quarters, and another lot fetched. Or they can first be caught in the box and then transferred in baskets.

Often one desires to catch a bird in the pen, noticing something wrong with it. It is here that the use of coloured rings will be useful in addition to the numbered rings. The value of the latter lies in the fact that each bird on the farm has a numbered ring for life. By this number she is known, her full pedigree recorded, and her offspring located. Hence the essential part of using a numbered ring for each bird. But when in the pen one cannot ascertain the number of a hen without handling the bird. We certainly desire a method which will enable us to know a bird quickly in case she is off her food at feeding-time, or is ailing. I advise the use, therefore, of coloured rings as well, whether for ducks or hens. Supposing I have small pens, I can easily arrange a system of ringing by different colours, and each pen can have the same system. Should a bird fly over into the next pen wearing a blue ring, we shall need to handle at night, when the birds are on the perches, only the two blue-rung hens to put them into the right pens again. Their ring-numbers will guide us as to their exact pens. If any bird is noted to be ill or needs to be marked there should be a note made against it on a slip of paper, which should be taken into the food-store. This note should then be copied on to a sheet, so that the persons responsible can visit the perches at night and catch those hens down on the list.

Doing the catching in this way is best, although often a hen has to be caught at the moment it is noticed. I suggest, therefore, that in one corner of each run a wattled hurdle be erected, with a wire-netting door that can easily be brought into action, to make a small enclosure into which the bird can be gently driven and caught. One can readily drive a bird into a corner of the run.

I would like to warn readers of the importance of careful handling at night, when the birds are on the perches. At night the yolks will be travelling down the oviduct, and late in the evening can be felt in the shelling department. Rough handling might therefore do more harm than good. It is always a problem to say how long the egg takes to be made, passed down the oviduct, and laid. If a hen fails to lay to-day I generally find an egg already in the shelling department early in the evening, this egg being laid the next morning. Naturally, in the morning the hens will be carrying their eggs, but these will be shelled, and not so much harm could be expected as in early evening with the soft unshelled egg to contend with. In the early afternoon, with the eggs laid, one might handle birds as a practice, that being the ideal time. Handling at night should be confined to catching birds which have been marked down for that purpose. Whenever a hen is being taken from the trap-nest after laying her egg, the operator should handle her for laying condition, etc.

When hens and pullets are not in lay they can be handled safer than when producing eggs, but one should handle with care at all times. The fact that a broody hen is out of lay should not warrant rough handling, and I have seen broodies bundled out of the nest-boxes very unceremoniously for daring to try a broody-rest instead of laying.

Ducks also require special handling, and one should at all times avoid catching hold of the legs, which are the weakest part of the duck. Catch the duck by the neck always. Remember, too, that at night-time ducks are very scared if they see a light, and one should not visit the duck-house by night with a lantern. Ducks can be brought up to become very tame, especially if at four or five months old they are taken in hand. Training should be started early, and if they are to be trapped when grown up they should be got used to trap-nests when four or five months old by being driven gently therein for the night, even if several occupy the same nesting-section. The lad who whistles all day long and runs through the flock at fifty miles an hour is not the person to have charge of nervous ducks. The latter will need to be under the care of a quiet person, and preferably the same employee. Ducks do well when looked after by women, showing that management of flocks of laying and especially trapped ducks counts for very much. I also contend that ducks in a quiet field do better by way of egg-results than in an open field where many pass and where there are constant disturbances.

When catching or handling a hen, one should always aim at the least disturbance of the flock. Supposing a hen has gone over into the wrong pen and she is noticed. Do not start chasing all the birds until she is caught and the rest are left panting. Make a note of the fact that blue-rung hen has to be caught in pen 1 and taken to pen 2. Maybe such a note can be fixed to the trap-nest in that house, and then, when she is found in the nest after laying, she can be taken back to her proper quarters. Or one can wait until evening, when the birds are perching.

Another very important part of handling is to avoid annoyance to the male bird in the pen during the breeding season. A good male soons gets alarmed when his hens are handled, and, if constantly annoyed, such a bird may become spiteful and be dangerous to young folks about the place who may enter the run. When handling hens from the traps, drop down the shutter over the pop-hole, so that the male bird cannot enter the house and attack the operator. I have often said that a good stock cock is cheap at fifty pounds, whereas a bad male is expensive at fivepence! How often do we see the vigorous male spoiled because of a little forethought. As far as possible when handling hens keep the male bird out of the way. Drop the shutter of the pop-hole, therefore, and leave the male in the run when handling his hens inside the house. The cockerel or cock which shows no interest when his hens are handled is not fit to be in a breeding pen; he should have gone to table long since.

CHAPTER VIII.

THE "POWELL-OWEN" SCORE-CARD.

HAVING now studied the make-up of the hen from within, the reader will be able to follow the better my score-card. I have endeavoured to give a system of scoring to 200 points by taking the external build of the utility bird and allotting certain points for the different items. It is a very simple card, although my critics declare so often that it is complicated. Naturally, you are not going to learn the card and its application in five minutes, and it behoves all who are interested to get right down to the items and to master the execution on the card. Instead of being complicated you will find it possible to read from a bird's card every good and bad point, without having the bird in the hand.

We start off with the *capacity* side, and as will be seen I allot 70 points or marks. You will notice, too, that I go all over the bird for my capacity, and yet one so often sees a sarcastic reference to capacity as being useless when the writer means abdominal capacity alone. As stated, the late Mr. Walter Hogan fashioned abdominal capacity originally by taking that as his capacity aim in conjunction with the thickness or fineness of the pelvis bones. As a result, he drew up an egg-type schedule showing numbers per annum which these two items stood for. He measured the thickness or fineness of the pelvis bones by taking the bone between the thumb and finger and including all fat and gristle found on the bone. Supposing a White Leghorn pullet had a four-finger abdomen with each pelvis bone one-sixteenth of an inch thick, she would be recorded as a 220-egg-per-annum type. Every sixteenth of an inch added in thickness would mean so many eggs off, and every finger less in abdomen would mean a lower egg-type.

I have never been able to ascertain exactly what the late Mr. Walter Hogan intended to teach. For some time I thought that he wished to give breeders a rough guide to follow, but he often graded to these types by the two items mentioned, and no can deny that while he often went astray in totals actually laid by the birds and his version of what they would lay in eggs per annum, he made some very accurate forecasts. But forecasting on these two lines of handling could be very dangerous in the hands of any but the most expert,

THE "POWELL-OWEN" UTILITY SCORE-CARD.

(Originated by W. Powell-Owen).

Name of Show .... *Breed* .... *Sex*..............

Date of Show .... *Class No.* .... *Pen No.* ......

Capacity.	Fingers.	Points Possible.	Points Awarded.	Capability.		Points Possible.	Points Awarded.
End of breast bone to pelvic	1	4		Pelvic bone	v. thick	3	
	2	8			thick	6	
	3	12			thin	9	
	4	16			v. thin	12	
	5	20			If straight add	3	
Between pelvic bones	2	2½		Flesh	coarse	5	
	3	5			medium	10	
					fine	15	
Between pelvic and tail bone	1	5		Vent	size	5	
	2	10			fineness	5	
Width of back	medium	3		Head points	medium	3	
	good	6			good	6	
	v. good	10			v. good	10	
Length of back	medium	2½		Bone and horn	coarse	3	
	good	5			medium	6	
					v. good	10	
Width between legs	medium	3		Feather	excess	3	
	good	6			medium	6	
	v. good	10			ideal	10	
Length, depth and width of abdomen	medium	3		Exhibitor			
	good	6					
	v. good	10					
	Total...	70			Total...	70	

Breed Characters, etc.	Points Possible.	Points Awarded.	Grand Total.	Points Possible.	Points Awarded.
Show condition	10		Capacity	70	
Health	10		Capability......................	70	
Size (ideal utility)	10		Breed characters, etc.......	60	
Breed characters	30				
Total...	60		Total...	200	

Prize Awarded .... *Judge's Signature* ....

and no one could possibly be accurate to totals laid seeing that so many factors enter poultry routine which would upset matters. For instance, we have but to shift to new quarters a pullet when in lay and the chances are that she will stop production and have a partial winter moult, which will mean so many eggs off her type number. Again, a pullet that is too fine in the bone would lay well in a backyard run where she would be protected from the elements, whereas free range on the farm would soon find her weak spots. A pullet, too, would be more likely on free range to lay an increased number of eggs in Devon than in Northumberland. Then we have to consider the different feeding methods of two poultry-keepers. Even mashes vary despite using the same ingredients and in just the same weights. Mr. H. A. Hussey, the popular secretary of the Tottenham Utility Poultry Society, often makes me up two mashes which I demonstrate with at my lectures, the one with the best ingredients and the other with inferior foods. I can assure you that the difference would never be credited, although the same foods are used in each and in the same proportions. The poultry-keeper who knows how to grade out the foods is in a happy position indeed, and if this is your known weak spot please take my advice and strengthen your knowledge. Maybe your middlings differ in quality from time to time, and if you have a bad sample then the usual mash will be less likely to get the results than when you are using the best sample. Inferior middlings would dictate to the clever feeder that a little more "body" in the form of Sussex ground oats, or the like, should be added to the mash. But is every poultry-keeper a clever feeder and fully observant to such details?

Many tell you that the trap-nest is the best guide to types in an endeavour to combat my handling system, but the trap-nest merely says what number of eggs a pullet has laid. But what a pullet lays in her first season is no guide to her second-year record! And the second is no guide to the third year's total! If I keep a pullet in a small pen by herself she will lay more eggs per annum than if she is running with a small pen of six birds—and if she is in a flock of sixty I expect less from her, and so on; as the size of the flock increases the tendency is for each pullet to produce fewer eggs per annum. Consequently, egg-type can soon alter according to the size of the flock, and the trap-nest record of a pullet kept on her own might be 280, as against, say, 250 in a small pen of ten birds. That is because temperament plays its part in production. When the single-pen section was first tried at a Test in this country, so sure was I of this temperament part that I openly stated that the single-pen pullets would be the first to lay the 100 eggs, also the 200, and that there would be a greater percentage of birds at the time with 100 and 200 eggs to their credit respectively in the single-pen section than in any other. I also stated that it would be in the single-pen section that the layer of the most eggs would be found at the end of the Laying Test. These surmises, at first criticised, happened exactly as I stated, and you can trace these points at the next Test where there is a single-pen section to see how far environment alters egg-types.

We can see how egg-types alter when birds on a farm are compared with those in a backyard run. The best experiment I can quote is of a dozen crossbred pullets taken from a Lincolnshire farm and housed in a London backyard run. They averaged over 200 eggs for the year, whereas had they kept to their country home and been fed and tended by the farmer the average would have been less than 120 eggs for the same period. Even trap-nest records have to be read with care and allowances made. In short, throughout poultry-keeping we need broader minds than we have had in the past, and we very much lack the observation that should be our strong point.

To-day, far too many see only the *abdominal capacity* part in a utility bird. Some have been guided that way by the two Hogan items, while others have only got as far as the capacity side of my card and then given up the rest, starting, of course, as "doubting Thomases." If you cannot get down to details with my system then you had better give up handling. *Some will never learn handling, just as they would fail at anything else, even if it were playing marbles.* Everywhere one goes one hears abdominal capacity shouted as if it were the only point of a layer, and it is amusing to hear one poultry-keeper refer to a pullet as a four-finger or a five-finger bird as if such stood for 250 eggs per annum even on half rations! At Reeth I found on my visit that capacity had been the ruling method of selection, and I had to put them right on my card. The proper execution of a card or system is, after all the main thing that matters. When judging a class at Tottenham Show I did not give a prize to a pelvis-measurement instrument, which surprised one member, because he thought it would just be the thing desired. The operator had to place the spanner arrangement over each pelvis bone and give a few turns, when the exact width of the pelvis bone could be read on the attached scale of measurements in sixteenths of an inch. It was very clever, but what happened with a flighty bird when she escaped with the instrument firmly fixed to her pelvis bone? What happened with operators who were short-sighted? What would be the result if a beginner squeezed the two bones together to measure the thickness of the two in order to divide by two, as would be the rule when one bone happened to be thicker than the other? Seeing that so many think they have learned all there is to be known in handling when they have placed the open hand against the abdomen, I have devoted ample space to this matter.

I will give a few sample egg-charts as drawn up by the late Mr. Hogan, not to be followed, but for the interest of readers. I have no wish to run the late Mr. Walter Hogan

CHART 3 (Female). Three-finger Abdomen.

Pelvic Bone Thickness.	Egg-Type.
1/16"	180
1/8"	166
3/16"	152
1/4"	138
5/16"	124
3/8"	110
7/16"	96

CHART 4 (Female). Four-finger Abdomen.

Pelvic Bone Thickness.	Egg-Type.
1/16"	220
1/8"	205
3/16"	190
1/4"	175
5/16"	160
3/8"	145
7/16"	130

CHART 5 (Female). Five-finger Abdomen.

Pelvic Bone Thickness.	Egg-Type.
1/16"	250
1/8"	235
3/16"	220
1/4"	205
5/16"	190
3/8"	175
7/16"	160

CHART 6 (Female). Six-finger Abdomen.

Pelvic Bone Thickness.	Egg-Type.
1/16"	280
1/8"	265
3/16"	250
1/4"	235
5/16"	220
3/8"	205
7/16"	190

CHART "C" (Male). Two-finger Abdomen.

Pelvic Bone Thickness.	Egg-Type.
1/16"	180
1/8"	166
3/16"	152
1/4"	138
5/16"	124
3/8"	110
7/16"	96

CHART "D" (Male). Two-and-a-half-finger Abdomen.

Pelvic Bone Thickness.	Egg-Type.
1/16"	200
1/8"	185
3/16"	171
1/4"	156
5/16"	142
3/8"	127
7/16"	113

CHART "E" (Male). Three-finger Abdomen.

Pelvic Bone Thickness.	Egg-Type.
1/16"	235
1/8"	220
3/16"	205
1/4"	190
5/16"	175
3/8"	160
7/16"	145

CHART "F" (Male). Three-and-a-half-finger Abdomen.

Pelvic Bone Thickness.	Egg-Type.
1/16"	257
1/8"	242
3/16"	227
1/4"	212
5/16"	197
3/8"	182
7/16"	167

CHART "G" (Male). Four-finger Abdomen.

Pelvic Bone Thickness.	Egg-Type.
1/16"	280
1/8"	265
3/16"	250
1/4"	235
5/16"	220
3/8"	205
7/16"	190

down, far from it, because his name will always be with us as one of the first to work for graded flocks. I had the pleasure of reading the first copy of his system to reach this country. Had I said "yes," it would have been published in this country when a publisher was offered the English rights. My advice was that the time for handling was not ripe, because people would have to be gradually accustomed to handling hens at abdomen, etc.; that the utility side of the industry had not then made sufficient strides to warrant a large sale to pay the costs; that having tested it I could not work out the tables with accuracy with types constantly varying in the birds according to management, feeding, environment, and so on. The system was not, therefore, published. All due credit, however, to the system and its originator for setting me working in other directions bent on mastering this absorbing subject.

Capacity stands for the size of egg, stamina, breeding and hatching power, rearability, and maintenance of heavy-laying strains. It stands also for sound constitutions to withstand illnesses and epidemics. It means also capacity to contain numbers of eggs. I cannot impress too strongly upon all who wish to learn hand-grading to my score-card system the importance of balancing a bird, and understanding clearly what the various items mean. A little knowledge is often dangerous, and I want particularly to see those using my system who will be broad-minded in all things, and be capable of giving sound decisions. Often I see a breeder take a bird in hand and measure up the abdomen as four fingers, and declare the bird to be a poor layer, as if nothing else counted. Again, I have heard a man criticise one of my winning birds at a Show with the remark:—"She is a good pullet, but rather coarse in the WATTLE." It is to help one and all to balance a bird that I planned my score-card, and you will not go far wrong if you read a bird throughout in the order on the card, and in the proportion of marks mentioned on the card.

Capability comes next, and I allot an equal number of marks, viz.: 70, to that. Incidentally, I have an item for size, where I award 10 marks as an ideal, and deduct for a bird that is either too big or too little. We will, for utility purposes, keep to the two parts of the card, viz.: capacity and capability. Capability means NUMBER OF EGGS LAID, and I leave it to you to know how important that section of the card is. At one time all poultry-keepers who tried to master handling went solely by capacity, and mainly abdominal capacity at that. My card has put most right, but there are some now who have gone to the other extreme, and think only of capability. Hold the balance of every bird when she is being handled, and do not go by any given item, or you will go wrong, and say there is nothing in my handling system. Not a few get into the habit of saying that every bird with high tail carriage is a good layer; whereas the high or squirrel tail may mean that the back of the bird is deformed. Others go solely by the head, and say that all birds without a good "laying" head are low-grade producers; whereas they may have used a male with a bad head, and, being an external, this fault has been carried along to the progeny. External faults of any kind will soon be passed on from father to progeny, as you can check up with your own strain. Not a few will handle a bird, and say they like large combs because such an item stands for freedom from broodiness; whereas the excessive comb on the male will have caused that. If you get into the habit of confining your attention to any one point you will soon be trying to convince yourself that feathers on the shank stand for heavy laying; no doubt because they keep the bird's feet warm in the winter-time! Hence the warning to go all over the bird, to take your time when giving a decision, and to let the latter be a well-considered one.

CHAPTER IX.

WHAT CAPACITY REALLY IS.

THE first measurement in ascertaining the capacity of a bird is the distance in fingers between the end of the breast-bone and the pelvis bones. Take the bird in hand by having the open right hand on one side and the left on the other (see Plate 49), including the wing by placing the thumb over the folded wing at the shoulder. The little finger can go underneath the bird under the thigh. Now pass the bird on to the left arm (see Plate 51) so that the breast and head rest at elbow, allowing the bird to look through the aperture between arm and body, which will keep her quiet. Let the open left hand go under the breast-bone and between the legs, with the thumb remaining at the side of the bird, the latter being in a tilted position with head down and tail up.

With the bird held this way the right hand will be free for taking the measurements. For the purpose of description we will assume that the bird being handled is in full lay. If, say, a White Wyandotte pullet, we should be able to get the five fingers in the space between the end of breast-bone and the pelvis bones, the latter being found on either side of the vent. The hand will be held in an upright position, so to speak (see Plate 50). How little is known about my handling system may be gathered when I mention that I have seen persons trying to get the five fingers between the two pelvis bones, thinking that the measurement was taken there.

On my card I allot 4 points per finger, so that a bird measuring four fingers will take 16 marks, and one going four-and-a-half fingers will be scored 18, out of 20. Throughout the card give a bird any division of points you think fair, whether it be 6, 7, 8, or 10 for head, etc. Many critics try and pull me up for my finger-measurements, declaring that all fingers are not standardised for size. True, but when challenged over this point I ask for a show of hands, and most of those present are found with hands that are the same for width, but may vary in length. My own finger-measurements are as follows:—1st finger, 1 in.; 1st and 2nd fingers, $1\frac{3}{4}$ ins.; three fingers, $2\frac{3}{4}$ ins.; four fingers, $3\frac{1}{2}$ ins.; and five fingers, $4\frac{3}{4}$ ins. Such are taken near the palm, across the third joints, just as the birds are measured. The average man's hand can be taken as a standardised one, but ladies' will vary. Still, you will remember my rule that you must not lay stress on any one item, and the measurement side is only a part of my card. If you have small fingers, just score them up at the same rate as on the card, and allow a bird 24 if she measures six fingers at abdomen; because when you have scored all birds they will stand in the same order as if I had measured them up. My only reason for including fingers is to let readers see the right proportion of marks to allow for the abdominal capacity, and to save hand-graders when beginning from looking for a twelve-finger ideal.

It is very essential to measure up the abdomen in the proper way, because it is inside where the eggs are made. If we handle a hen in moult, we find the abdomen empty; it is so, too, in a hen that is broody or sitting on eggs, and in pullets that are growing and not coming into lay. But directly egg-making is active, and the ovarian organs are in action, the abdomen is full. Such is caused by the extra digestion taking place within, due to the egg-making. Naturally we need in our layers, when they are producing, all the abdominal capacity we can get. I would here remind readers that when they place the fingers against the abdomen, between end of breast-bone and pelvis, and call the bird a four or five fingered pullet, they have only taken the depth. But we have to remember that length and width must be considered along with the depth before we obtain an accurate idea of the total abdominal capacity. Consequently I have a double check on my card by duplicating abdominal capacity and allowing 10 points for the length, depth and width of abdomen.

The length of abdomen along the side of bird is taken from the thigh-socket, where the leg enters the body, to the end of the abdomen in *flesh* and not feather. Remember that we are dealing with trunk and frame by handling, and not appearances. Depth taken at the rear runs from socket of tail, or parson's nose, to bottom of abdomen (in a straight line with the breast-bone). By placing the palm of the right hand against the abdomen, with the second finger under the parson's nose or tail-socket, we can see how deep the abdomen is to the base, or end of breast-bone. As stated, the bird will be held in the left hand, with the fingers between the legs, and the protruding middle finger of the left hand (see Plate 51) can be the " pointer " on to the open palm of the right hand, to show exactly what the depth is.

Width of abdomen, again taken at rear, is gauged by placing the thumb of the right hand on one side of the abdomen, and the second finger on the other side, as if one were grasping the abdomen. Take no notice of *feather*, but measure the abdomen in actual *flesh.* Abdominal capacity as here dealt with, apart from standing for capacity for egg-making, is a safety valve against internal fat in excess. For instance, if you allow your hens to become fat without being aware of it, those birds with the greater abdomens will take longer to go wrong and obtain the excess of internal fat, thus giving the poultry-keeper a longer period in which to ascertain that his management is faulty, and to put things right again. One should always be looking out for overfat condition, and due warning will be given by the eggs laid by the birds. At first, when the hens are getting fat, they will lay extra well, but with the climax come abnormally large or small eggs, followed by badly-shaped eggs, thin-shelled ones, and, finally, soft eggs and egg-eating.

Apart from large abdomens standing for splendid capacity for digesting food *ad lib.* for egg-making, they prevent the cramping of the internals in a short time. Hens that become fat are readily affected, because the gizzard is pushed up by the internal fat on to the shelling department and oviduct, and if the abdomen is small such mishaps will follow more quickly than where the abdomen is capacious.

My next measurement in capacity is between the pelvis bones (see Plate 52), where, in my ideal pullet (in full lay), I am able to place the three fingers between the two bones in an upright position, *i.e.*, with the palm of the right hand against the abdomen and the finger-tips upwards near the tail. I have in mind two things here:—(1) the width of the pelvic arch and (2) the width of the egg laid by the bird. One has to remember that the vent is a little above the pelvis bones, and yet the egg has to pass between the two bones. If it happens to be a large egg as regards width, then I contend that it can have easy expulsion, although it does not follow that all hens giving three fingers between the two bones when in lay will produce a large egg. But I do find that hens with but two-finger space between the two pelvis bones so often lay narrow eggs, especially if the bones are thick and hard-set; if they are fine and pliable, then the egg can push itself out even if wide, unless the bones are coated with fat. One can see how fat on pelvis narrows the channel, and results in smaller eggs. One very important and practical advantage to be derived from a three-finger space between the two pelvis bones is the fact that one obtains an idea of the width of the pelvic arch, from which trunk the pelvis bones project. I find that if we have a wide pelvic arch we have a hen with a good wide back, wide abdomen, and a nice frame; also a bird likely to lay good-sized eggs and become a good breeder in her second year. On the other hand, a pullet *in full lay* with a narrow arch is likely to be small in frame, and also to tend towards small-egg laying, while she may be a splendid layer without being a breeding hen—a finished article. I always check up pullets with narrow pelvic arches for long thin beaks, spindle shanks, cut-away front, and light weights indicative of being too fine in the bone.

Passing on, we place the first finger of the right hand below the tail, so that it rests along the base of the parson's nose, in the socket, so to speak, *i.e.*, beneath the tail. An endeavour is made to get the second finger under the first before the pelvis bones are reached. I have in mind not only depth of abdomen, but also depth for the egg in the shelling department. With this measurement I rely, with other checks, for size of egg, the pullet showing, when in lay, but one finger or so from tail-bone socket to pelvis, so often leaning towards small eggs.

Width of back will follow, and here we must obtain as great a width as is possible across the broadest part near the wings (see Plate 55). In the first place, great width stands for stamina, and in a male bird vigour and fertility. It also means a good-sized frame on the bird—no mean asset in a breeding hen. If we study the Plates we notice that the heart and ovary have to be placed in the back, where the width is. Once, when lecturing on "The Hen from Within," I cut up an Ancona that had only a 1½-inch width of back, despite being in full lay. In the first place, I have to remove the feathers from the back, which must greatly reduce the width; then I have to find room for the heart and the stomach, also the yolks in the ovary, including the five large ones that are ripening. I want all the space for these organs that I can have, for many common-sense reasons. In the first place, the yolks can be small when developed if the width for the large ripening yolks is on the short side, and small eggs can follow. Width of back is also a measurement for size of egg. Next we need room for the heart, and wish to prevent the stomach from being pushed up on to the ovary, especially if, as in overfat birds, room has to be found for a lot of internal fat.

Width between the legs is a cross-checking measurement, because the wide back and the full abdomen follow, with few exceptions, providing the shanks are straight and medium in length. When looking at a pullet in lay as she is in the pen or on the ground, you can take it for granted that she has a wide back if she stands wide between the legs. Width between legs must be taken with the bird standing on a firm table or box (see Plate 56) by placing the open hand up near the breast-bone and between the thighs. Personally I aim in my ideal at five fingers between the legs, measured with the open palm of the right hand uppermost against the breast-bone. I have often seen poultry-keepers (and critics) take a bird in the hand and stretch her legs wide, at the same time placing the open hand between the legs. Try this, and you will find that nearly every bird shows a ten-finger measurement. The bird MUST be standing naturally on a firm base when measured for this item! Width, then, means room for heart action, and for the ovary and internal organs mentioned. It is useless to have a wide back to obtain room for the organs if their action is curtailed by absence of room between the legs, but such happens, should the back be wide, in birds that carry an excess of bone, and are so tall that they stand knock-kneed. Such birds must be cut for excessive bone, length of leg being equal to bulk of shank.

Measurements are difficult to give by rule of thumb, because one must balance the bird. For instance, in heavy breeds, I might suggest that the abdomen from socket of parson's nose to base of abdomen, in a straight line with the breast-bone, should be about $5\frac{1}{2}$ ins. to 6 ins., with a width of abdomen of around $4\frac{1}{2}$ ins. to 5 ins. (*i.e.*, span or side to side), and a length from thigh-socket to end of abdomen in flesh of 4 ins. to 5 ins. It is not difficult to handle a bird for length of abdomen, because the left hand will be under the bird, and the first finger can be pushed through against the socket of the thigh to protrude on the outside. The right hand can be pressed against the abdomen at the end, to push in the feathers and to show the length from the "pointer" first finger of the left hand to the palm of the right hand. By swinging the "pointer" from the thigh back again between the legs, one can tell the depth at the rear from the tail-socket down to the spot where the "pointer" finger of the left hand touches the palm of the right hand.

Length of back finishes the capacity part of my card, but it is important from the view-point of the laying organs. The oviduct and the shelling department need all the room they can get when in full working order. You cannot have room for a shelled egg in the shelling department of the oviduct, and for the latter itself, unless the bird has a nice long back. The oviduct measures 2 feet or so, and you can see from the Plates where the egg is when it is being shelled. The egg points from tail to ovary, and while a short back may be sound so long as normal-sized eggs are laid, the time will come when a double-yolked egg passes down only to touch the ripening yolks, and to put the laying organs out of gear, as illustrated in my Plates. A long back will help matters, and prove a safeguard should larger eggs than usual find their way into the oviduct's shelling section. Personally I aim in my ideal laying

13

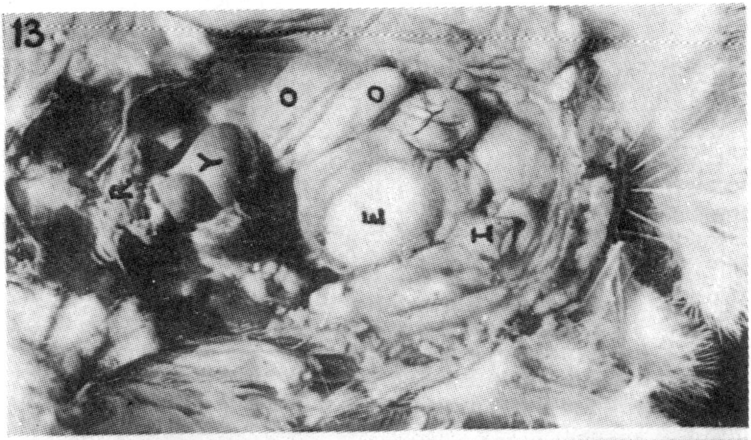

14

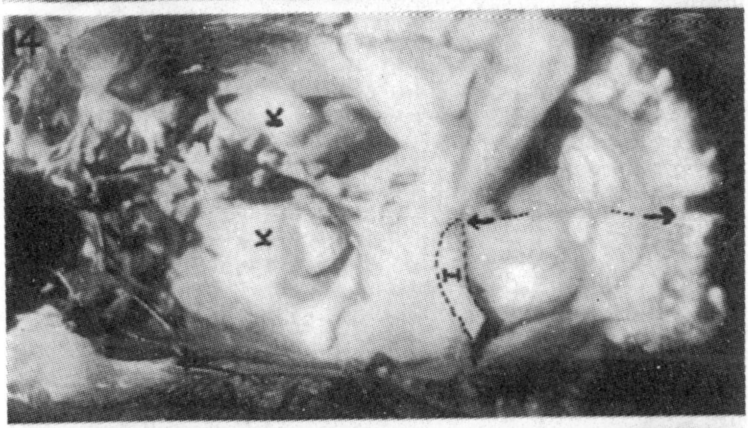

15

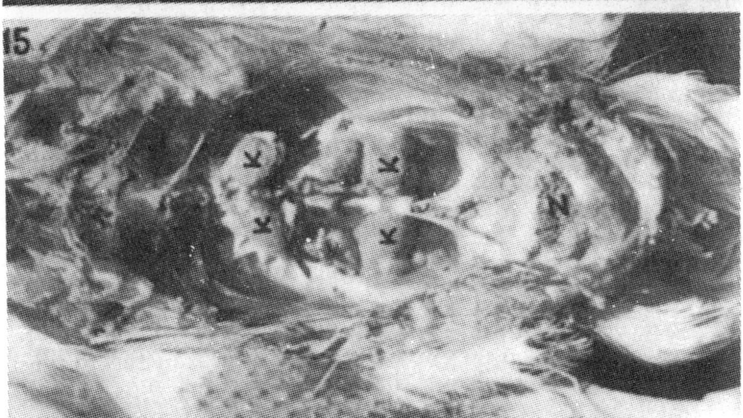

pullet at a back measuring 5 ins. from the widest part to the parson's nose (see Plate 54, B to E). This measurement is best taken (as is the width of back) with the bird held in the left hand, which releases the right hand for the measuring. If the left foot is placed on a low box, the bird can be supported underneath with the left hand while the breast (see Plate 55) rests on the operator's thigh—head down and tail up. With the thumb of the right hand one locates the end of the back or frame by placing the thumb at the base of the parson's nose and lifting the tail with the first finger. One needs to get at the actual frame, and not to allow for the length of tail. Often a pullet looks long in the back because the tail goes along further in a straight line, but we have to lift the tail of all birds to arrive at the length of back in framework, which is what interests me.

When scoring up a bird for width between legs, the handler can allow 2 marks for each finger distance, a five-finger bird taking the maximum 10, a four-finger pullet 8, and so on. Mark up to any simple system you like, and after you have gone out to master my card by scoring very many birds, you will begin to handle pullets and males, and to say with a rough handle whether or not they stand high on the card, without actually scoring them in detail. My card has the great advantage of being a "teaching" one, in that it will keep the handler to the many items one by one, and in their proper ratio. It will prevent the beginner from going off on side issues or individual items, which may see him misdirected as to the characters or indications of a layer. I have practised my card for so long that I am able to take a bird in the hand and to have its grading in a quick time; and you can do the same if you will get down to the card, and go through the practising stage of handling by scoring in detail plenty of birds of both sexes. Many say I have a happy knack in handling, or an extra sense for it, but I can assure readers that it is due to my having used my card so often, and learned therefrom how to *balance a bird correctly*. Therein lies the secret of successful hand-grading.

CHAPTER X.

WHAT CAPABILITY REALLY MEANS.

CAPABILITY is the one part of my score-card which must be very carefully studied, for the very reason that capability means the ABILITY TO LAY THE EGGS. How important that is I leave readers to answer, although, in passing, I draw special attention to it, seeing that so many critics only get as far as CAPACITY, no doubt because they find that the easier part of my card, or are unable quickly enough to arrive at the capability decisions under the various headings.

We will start with head points. Our ideal will first be a beak that is short but deep, to show nice medium bone. The eye will be high up in the skull, and very prominent, so that when viewed from both front and back of bird it will stand out from the face. The face will be smooth, and the skull narrow at the back, while the neck will be slender or swan-like, and drawn on rather medium lines. The pupil will be large and distinctive, while the iris will in most breeds be bright red, or yellow, or brown (red preferred), and the facial expression will be kindly. There will be a sparkle about the eye that tells of business and production, docility, kindliness, and without a care.

I have described a " quality " face, and I want to see the bird that is able to keep that splendid texture of the face for years. A two-year-old hen or cock with a fresh young face that I have described has quality in abundance, which cannot be said of the two-year-old which has a wrinkled face, etc. I will put it quite simply by saying that the hen with a pullet face, or the adult male with a cockerel's facial features and expression, should be our aims, as against the non-quality cockerel with an old cock's face, or a pullet that looks like an old hen. One has, however, to bear in mind that you have but to get your stock overfat and they will have wrinkled faces; and that is why one has to be so very careful in balancing a bird and proving items by cross-checking; a reason, too, why I do not tell beginners to go solely by head points.

The comb should be of nice medium texture, and in size proportionate to the bird's body, while the wattles should be of fine texture as against beef.

In Plate 44 we have an ideal head for a layer. The snapshot is of Mr. H. A. Hussey's Ancona pullet that has won several times under me, including the members' cup at Tottenham Show. She has proved a splendid layer, and been a good advertisement for my handling system and judging. Note the bold eye, which is placed high in the skull, and the short deep beak, with wattles close together and comb nicely carried. Plate 43 is of a Light Sussex pullet, which has won often under me for Mr. H. W. Honey, Pedigree Poultry Farm Alton, Hants. She laid 238 eggs in her pullet year, and one is able from my photograph to see the keen, bold eye and kindly facial expression of the bird. Wattles are close together, and the bird is of really nice Sussex and laying type with long back and depth at rear or abdomen. The tight feathering on back and thighs will be seen, together with the prominent breast, which, as stated, stands for stamina and good digestive power, which are followed by capability to produce eggs in plenty. She has that width of back and space between legs which indicate a good breeding hen. Neatness of type and medium bone in both beak and shank are also illustrated.

Contrast the ideal head with that of the broody or drone! The eye is sunken and carried low in the face; the skull at the back of the head is thick, preventing the eye from being seen from behind the bird. The sunken eye, too, cannot be seen standing out prominently from the front. The face is wrinkled, and the eyebrow heavy and sullen. The comb and wattles may be coarse, and the wattles wide apart. The eye lacks sparkle and life, and shows to the observant a lazy appearance. It behoves every reader to study the types of broodies which enter the broody-coops several times per annum, and an idea will readily be obtained of the drones. A hen may go broody often, for a short period each time, but such should be labelled with a small "b." I wish readers to study those hens with the big "B" against them for long broody-rests and frequent ones.

The head-points of the male should be studied in the same way. Plate 45 is of a White Wyandotte cock which has won under me, and stood second in a strong class at Westminster to the noted "Rodwell" cock, for Mr. Leslie Williams, of Easton Poultry Farm, Liss, Hants (late of Grantham). This male has bred some splendid layers, including the pullet which won first and cup, best in Show, at Yeovil, 1921, and first and cup, Tottenham, 1921. One will notice the excellent open face he has, with bold eye, smooth face, neat comb and wattles (all in proportion to size of body), and, above all, the capital breed points. He is a Wyandotte all over, with depth of body and cobbiness, a "bushy" tail and correct headgear.

Being snow white, and handling well, he has bred some first-class utility Wyandottes that boast of *beauty and utility*—the progressive utilitarian's ideal (or should be).

Feather may be taken next, and there should never be an excess. The feathering on the back right up to the parson's nose should be tight, so that when it is brushed the wrong way it springs back quickly into position again. The feathering on the thighs should also be tight, so that it keeps flat and level with the side of the bird, instead of "fluffing" out and looking baggy from the rear of the bird. In Plates 43 and 45 we have good examples of tight feathering.

An excess of feathering is very undesirable, because it sucks the abdomen, and prevents the latter from concentrating on egg-making. One can test this by handling a tight-feathered pullet, and noting the very full abdomen, as compared with the "fluffy" pullet, which has a greatly reduced abdomen, even when in full lay. The more feather on thighs, the narrower will be the abdomen and the smaller the intestines within, solely because the intestines have to feed the feather instead of making the eggs. Again, the feathery bird takes longer to moult, and also comes into production later than her tight-feathered sister; and both mean a loss of so many eggs. When mating, the excess of feathering should always be noted and kept out of a strain. Supposing I am looking at a feathery bird: she will look very full at abdomen, but handle empty. Many insist on feather to make up certain types, but the practical way is to reduce the feather, and let the abdomen itself make the full or "curvy" type. One has a good idea of feather in excess when handling such a bird for the first time: while the bird *looks* very heavy, she is quite light to *handle*.

Passing on to bone and horn, we should aim at medium shanks, both for length and for size. It is quite a simple matter to tell if a bird has an excess of shank, seeing that she will stand knock-kneed in the pen. To have the right length of leg for utility purposes the bird should be able to stand very wide at all times. Coming to the size of shank, we should aim always at soft sinews at the back of the shanks: the scales, too, should be small, numerous and tight, and the spur should be medium. The beak will so often help us when checking the bone. If it is very long and deep, like a vulture's, it will be too coarse, and with it one should find too coarse shanks and scales. On the other hand, if the beak is very long and thin, there will be insufficient bone, and one can look for a spindle shank; insufficient bone is quite as bad as an excess. The medium should be our aim, and it will be found in such birds where the beak is short but deep, the lower part of the beak also having nice depth; medium shanks will follow. Remember that when a pullet is ready for lay her shanks are full of fat, so to speak, which egg-laying will suck; and one needs to bear that in mind, and to picture the shanks as what they will be after heavy laying, as well as what they are at the time of handling.

The vent should be large and of fine texture in both sexes. I consider that the size of the vent of a bird in lay conveys an idea of the size of the egg for circumference, especially when taken into consideration with my other points. Taking the pullet in hand with the left hand underneath in support of the body, as already mentioned, place the second finger of the right hand below the vent, and with the first finger smooth open the vent until the latter is flat, when one can get an accurate idea of the size, circumference, and texture. This should always be judged with vent flat and fully stretched. The lips of the vent, if of good texture, will be soft and moist, and smooth, as against the dry, hard, and wrinkled vent of the coarser bird. For size take the outer ring of the vent.

The pelvis bones and abdominal flesh sections of my capability side of score-card call for special mention. In the first place, one has to bear in mind that these are the very points which change according to the seasons, and the high or low capability of the individual bird. Every poultry-keeper, when commencing to seek knowledge anent the pelvis bones, goes for the bird with a very fine bone, aided by the late Mr. Hogan's egg-type standard. Nothing could put the poultry-keeper wrong sooner than that. If you are after a sixteenth of an inch pelvis bone, you can always get it on a pullet ready to lay that is too fine in the bone, and a weed. Please understand that when the pullets are ready to start laying (or hens after the moult) the pelvis bones should be nicely coated with fat, so that each bone measures as wide as the depth of your little finger approximately. The abdomen will at the same time get a coating of fat, and the breast-bone will be well fleshed. Such a pullet is ready to fight colds and egg-laying and like strains. As eggs come they will play upon the fleshy parts mentioned, until the bird, after producing a hundred eggs, will handle pliable and fine at pelvis bones, and the abdomen will also be pliable and workable. Aim at a wide arch with fine, straight pelvis bones, but do bear in mind the conditioning of the birds according to the seasons. I will deal with this subject more fully in a later chapter.

The idea of my capability side of card is to have in every layer at least 50 marks out of the 70 total. I consider that 50 is the danger zone, where one begins to lose egg-production.

Supposing I am handling a pullet in full lay and at the height of her condition, and she fails to get my 50 minimum for capability, then I am not interested in her for laying purposes. I like the pullet which is well over 50 when ready to lay, and when she is handling at her worst (seeing that she will carry flesh at abdomen and pelvis), and then, as she lays eggs in plenty, she will improve in the abdomen and pelvis, and go right up to the 60's. The beauty of my card is that you can read a pullet once she is score-carded in detail. If she has a good head, ideal bone and feather, she will score high in capability, and then we have left only the pelvis bones and the abdominal flesh. It will be egg-laying and feeding that will keep these two items going along nicely. You will see from this how one spoils good layers by fattening them—by getting the pelvis bones coated with an excess of fat, and the abdomen the same. That is where laying capability is lost. Now the coarse bird, which scores low in bone, feather and head points, will have nothing to play with under the pelvis and abdominal flesh sections, and will grade out. Her excessive weight will prevent her from keeping over my 50 minimum for capability, and yet every bird, to lay well and up to sample, must keep well above my 50 ex 70 marks throughout the laying season, and not for one odd week or month, when a spurt of summer laying helps her to accomplish that. That is where the balancing of a bird comes in. The owner takes a bird in hand and finds she is handling well at the time, and thinks she must be a capital layer. No; he may have forgotten my rule No. I. —that weight comes first. If the bird is too heavy and coarse, she is handling false for the time being.

With these explanations readers will have had a little further insight into handling, and may be tempted to say that they did not know there was so much to learn about the subject. That is the right spirit, because you cannot in ten minutes learn the system, and be capable of handling a bird and grading it in or out.

For Shows I have 10 for show condition, 10 for health, and 30 for breed points, with another 10 for size; the latter after all being my chairman's casting vote for capability or texture. If a bird is on the border-line for capability, and is good for utility size, I may grade her in; if too heavy, she grades out.

CHAPTER XI.

RAW MATERIAL AND FINISHED ARTICLES.

TO understand my hand-grading system, the reader should bear in mind that there is such a thing as sound, common-sense breeding to be attached to utility poultry. Tracing the breeding of layers back to the start, one can see that heavy egg-laying was considered to be a matter of like breeding like. If a hen laid a lot of eggs she earned her place in the breeding pens solely because of that. The public were educated that way, and as a result, have called for stock that are phenomenal producers; 300-egg strains are spoken about by the poultry public as if they were common as dirt. The breeder has just to get a pullet to lay 290 eggs and his strain has become a 290-egg strain. There has been this chasing after phenomenal individual producers without a thought of the breeding hen. To offer a poultry-keeper a bird that has laid 200 2¼-oz. eggs is considered by him to be an insult; to offer him a bird that has laid 280 1¾-oz. eggs is thought by him to be correct. The trouble with utility poultry is that there has been a complete absence of sound breeding, and for too long has the tendency been to breed from finished articles and to judge the quality of a bird by the *number* of eggs only that she laid. As a result stamina becomes affected, and in endeavouring to call for better poultry I am working solely for the Utility Poultry Industry, which provides me with my bread and butter, just as it does those breeders who criticise my progressive actions.

Let us all work together for better utility poultry; in the past the cry has been for MORE instead of BETTER birds. There is not a flock in the country that cannot be severely graded, and, getting around the farms as I do, I ought to be in a position to know. It is not customary for me to mention names of farms I visit, but I will just mention a few happenings that have come my way of late, not by way of criticism, but to prove my point that we ought all to revise old systems of breeding and to get going again with sound, common-sense methods. The beauty of poultry is that you so quickly get results from severe grading and taking every care to breed from ideal hens and males. Two or three seasons of grading will accomplish wonderful improvements, as I have proved repeatedly. I have started with common White Leghorn hens of no known strain or breeding, and mated them to a tip-top cockerel out of a high-laying dam, and in two seasons bred pullets laying 280's to 307. And I have maintained high laying for the flocks by further matings. I have mated White Wyandotte pullets laying but 130 2½-oz. eggs in their first year, on the big side, but nice in hand, to a tip-top cockerel from a high-laying dam, and bred in one season daughters which in every case laid as many eggs, so to speak, as their dams, but in SIX WINTER MONTHS instead of the twelve months taken by their mothers. The pedigree cockerel that handles well, and out of a high-record dam, will carry plenty of laying merits on to his progeny, even when mated to the medium hens.

As I write I have in mind the Laying Tests that have just finished. At one some hundred birds died of roup, etc., in a few months; at another there have been several outbreaks of roup and losses; at a third the same has happened. How far is this associated with the old system of just mating up those hens which have been heavy layers, without regard to their handling or breeding qualities? The other day I was given an instance of thirty hens being bought from a big breeder with records of over 200 for their pullet year. Fifteen had to be sent back, and of the new birds replaced ten were dispatched back and finally two. Inside two weeks roup broke out, and the whole place became a hospital; three of the hens never laid an egg the second year, and at the end of twelve months only three live to tell the tale. Everywhere I go I hear such accounts, and I consider I am justified in calling for better breeding. A common cold can lead to roup epidemics, and we need stock vigorous enough to resist even colds, correct management being allowed for.

I will deal in brief with those farms where I have been to of late without giving names, but every incident can be proved:—

1.—Lecturing in N. Wales I sorted out my own birds to demonstrate with. Found six or seven Rhode Island Red cockerels and two or three pullets with badly-deformed backs and cripples. Ascertained they were from eggs bought at a high price from a big breeder. The latter's cockerel heading this pen was deformed, and not being handled for defects it went unnoticed.

2.—Graded out some White Leghorns, and had to turn down every one of about 70 cockerels for deformities, which were discovered in the original cock supplied for over £5 by a breeder. Like produces like in deformities certainly!

3.—Every son from one pen of Rhode Island Reds in Somerset was graded out for wry tails, roached backs, and deformed backs. Faulty back found in the male at head of that pen. Thirty cockerels or so thus became fit for table only, despite a high price paid for the pedigree male.

4.—Selecting, in Gloucestershire, for a Test, I found one pullet with severe diphtheritic roup in the flock of pullets.

5.—In Bedfordshire found a drake at head of fifty ducks; at one time thought to be a duck, and after ascertaining it was a drake was left in to eat food for no return.

6.—In Hants found three drakes in a flock of twelve, all of which were thought to be DUCKS *in lay*.

7.—In Essex, when grading, found a hen which could not walk for abdominal dropsy, the abdomen being full of fluid and the trouble not noticed.

8.—Had a like case to No. 7 in Bucks.

9.—Many flocks down with colds, which prevented my picking the birds for the Tests.

10.—Two birds found with shelled eggs in abdomen, and out of lay when killed and opened.

11.—In Leicestershire several hens with diphtheritic roup, and one that had not laid for six months through scrofulous liver.

12.—At one farm I visited I was told the interesting story that the young ducks and the old ones had got mixed, and one of the old ones had probably gone off to the Test as a "youngster."

I could quote dozens more, all met with this season.

At the Shows birds are exhibited under me with eggs in abdomen, also cysts, suffering from dropsy, deformed, blind, and with oviduct full of "hard-boiled" yolk-matter. I mention all this just to point the way to the importance of hand-grading, so that we can work on better lines than those that have ruled in the past or are in general use to-day.

The first thing to admit is that we have many types of birds. In the first place we have the beef pattern, which does not interest us except from the table point of view. We can grade these out for being too coarse and too heavy. Then we have the raw-material bird, which should be the favourite, although the majority of breeders and poultry public look down on such. I refer to the hen laying 180 to 220 LARGE eggs in her "pullet" twelve months. Next we have the layer, which is the finished article, laying like wild-fire without regard to size of egg, and just giving numbers. Finally we have the type which runs too fine throughout through super quality, which to-day is bred from, and which should be severely left alone, like the beef type.

In this chapter I wish to make an appeal for the "brood" type of hen, which should be the most valuable bird on the farm. She has just that frame which stands for large eggs and fewer of them, ensuring quality of egg and contents before quantity. Laying from 180 to 220 eggs per annum, she is so often despised because poultry-breeders think the number of eggs too low. It has been thinking in numbers and not in *quality* that has got us into so much trouble with our utility stocks. Laying fewer eggs, the "brood" hen has time to make richer yolks, thicker albumen and shells, and the resultant chicks are all the better for it. I regard this medium hen as my raw material for the breeding pens, and she will not let the owner down, whether as a breeder or builder-up of the finest laying strains. At first, when you start, you can reduce the standard to 160 eggs upwards, and each year as you improve the standard can be increased. Naturally, by selective breeding you will get the factors established.

Breeding layers is quite a simple thing, because you can get them in a season or so, but what really counts is maintaining the laying strain, and to see that accomplished I contend that the breeder has ever to be building, making, and maintaining. It is the medium hen of nice frame and the layer of large eggs that is going to help in the building-up, because she is the raw material. With utility poultry we must remember that management, feeding and environment are responsible for a given number of eggs laid by each bird. These factors are so often rejected because they do not represent scientific sections of breeding. But it is easy enough to have good quality birds and to ruin them as layers, just as it is to have lower-grade stock, and by good management to make them lay better. I think I would back the lower-grade birds, well fed, to beat the better-class ones that are badly managed. Admitting, then, that management does come into the argument, we have a personal side of poultry-keeping which under no circumstances can be reduced to breeding equivalents. I can give one man six pullets, and another a like number of sisters, and yet each at the end of the year will show me a different yield. I wish to point out the supreme value of the personal side, which should be strong in every poultry-keeper. It has been the desire to strengthen the personal touch that has always tempted me to teach poultry-keepers, especially the small producers, my handling system. I have endeavoured to get them to see the interesting side of poultry, and to know the hen from within before attempting to wrestle with the problems of feeding, etc.

Taking for granted good housing, feeding, environment, and management, we have just two types in a way, namely, the bird that is a beef type, and the other of the laying pattern. We utility men are after the latter. How best can we breed and maintain it?

In every kind of production there is the raw material and the finished article, and I suggest that we have in the medium hens the raw material, and in the " fliers " the finished article. Up to date we have known just the finished articles and bred from them for all we could, and in that direction must lie small eggs, lack of stamina, heavy losses in chick mortality and poor hatching results, together wth epidemics. Where one is able to get £5 5s. per cockerel for stock, and ten guineas a dozen for hatching eggs, one can stand such losses; but are not such breeders in the minority? I have watched beginners start high in the Laying Tests with plenty of large eggs, and after doing well for two seasons or so they have gone back to an excess of small eggs and heavy mortality. For this I blame the breeding from FINISHED ARTICLES. The winning pen in the Test is bred from, and being finished articles one is breeding from hens that may be quite unsuited to breeding. Some Laying Tests after all are won by the number of eggs laid by the pen without regard so much to size, and one can be pardoned for sending in such pullets that will put up the numbers, because to send other types would not help a win. I repeatedly select pullets for Tests which I know will never make good breeding dams, but one is compelled to study each Test and to send the right type to have a chance to win. It should be well known before this that small-eggers will always put up the highest total of eggs, and where the Test is scored on values you need not worry so much about the size of the egg. If Tests are on aggregate, I should just send pullets to lay around 1¾ to 2 oz. eggs; and even where egg-size is taken into account a pen of five pullets has 100 second-grade eggs allowed them, although ONE pullet in the pen can have the lot to herself if the others produce first-grade eggs.

Let us come down to actual selections to see where some layers are found. I suppose I handle as many birds in a week as most do in a year and many do in a lifetime. For weeks on end I often grade at the rate of several thousands per week. A breeder whom I may be visiting for the first time constantly reminds me of pedigree, telling me, when handling a pullet, exactly how she is bred and what her dam laid, etc. I ignore this now, and grade according to the individual bird I have in hand at the time. I have tried all methods of selection by pedigree to win Tests, sending sisters out of the same hen, those with the best pedigree, and so on, early-hatched and late-hatched, but to-day I go solely by the individual bird and how she handles. For the purposes of data, however, I always look up the pedigrees of the birds I have selected AFTER I have completed the selections. My data goes to prove that tip-top layers come from the general pens, which contain the medium hens. The general pens are often listed in the breeders' catalogue as "Containing well-developed hens of nice laying quality, specially selected and mated to a cockerel from a 250-egg dam." They are so described because, while they have been trap-nested, they have just the medium yearly records of from 160 to 220 all nice-sized eggs, and their records are not mentioned in the catalogue, because the public would fight shy of buying eggs or chicks.

One year I handled about 1,000 pullets bred from very many pens, and I was selecting about fifty pullets for the several Laying Tests. After making my selections and comparing pedigrees I found what I had anticipated, that only four pullets were from the best individually bred pens, while the rest were from GENERAL PENS, where the high-bred male had carried the quality through to the progeny for numbers of eggs, while the "brood" dams had carried on the frame to their daughters, and with it large eggs. The majority of these pullets I selected did their "over 200 eggs" for the year at the Tests, and the winter average for six winter months was about 100 eggs.

I should never hesitate to buy eggs and chicks from the general pens if I wanted layers, and females for foundation purposes. I am after sound, good-framed brood dams that handle well.

But when it comes to the cockerel, then I want the best, and I do not mind what I pay for him so long as he is the "goods" and handles well. I prefer him to be from a hen laying 240 eggs or over for the twelve months, and bred in line to such high numbers for several seasons, so that high records will be in the back blood. I should always buy sittings or chicks from the breeder's best pens, even if they were twenty guineas a dozen for eggs, *i.e.*, for my cockerels. Where several can join hands the cockerels can be reared, and the best divided up when maturity is reached. A bad quality cockerel is dear at fourpence, whereas a good male is cheap at £40, and should be taken great care of for many years.

In breeding and maintaining strains of layers the first rule to follow is that size comes from the dam side. Therefore we must use in our breeding pens good-framed hens; and these are the mediums which produce 160 eggs upwards in twelve months, and every egg a large one. They are generally a bit coarse here and there, and go broody once or twice, and have a reasonable rest in the autumn moult. These are the reasons why they are medium layers and produce large eggs. It is a common-sense contention of mine that the largest bird lays the least, while the smallest pullet lays the most eggs, and, as a sequence, breeding from the prolific small-framed layers tends to get Bantams and small eggs. Prolific layers, to my mind, are the finished articles. If you are commercial poultry-farmers, or interested solely in egg-production, I do not mind what you keep so long as they lay marketable eggs. You want numbers. But I refer to stock used for breeding, and here the raw-material hen is the bird to be proud to possess, instead of to despise, as is the general custom to-day.

We can refer back to my score-card to follow the types of the various birds. If a hen scores 67 for capacity and 35 for capability, the card tells us that she is a coarse cart-horse, likely to be a breeder and layer of large eggs (capacity) but a poor layer. We grade her out. The next bird scores 40 for capacity and 60 for capability, which tells us that she will lay like smoke, and yet never be a good breeding dam. Our aim should be for the pullet which in full lay after a good winter output will be high for both capacity and capability—say 60 and 60 as an ideal. Keep up the capacity side while still retaining the texture or capability. Such a pullet would be my ideal, and I think we should aim at breeder-layers for the breeding pens, such birds only being included which will lay well, and large eggs, and which have the good breeding frame. Such breeder-layers would have to be bred for, because they do not exist on a given farm in plenty. This because they have never been aimed for. You cannot breed breeder-layers from Bantams.

To study more closely texture you will, after a deal of experience in handling, begin to see that capacity and capability are direct enemies. When you lose capacity you get better capability, and when you lose capability you increase the capacity section. With present stocks we have to take as our raw-material or breeding hens those pullets which have plenty of capacity and which score fifty or more for capability when they are on the point of lay, and which will keep above the fifty minimum all the year through, so to speak. Such will make the good breeding hens in their second season, and will lay from 160 to 220 eggs in their first year. Armed with this information we commence to handle our cockerels, and find that the male, when matured, which scores 60 capacity lacks capability, and scores below the fifty for capability. Such will be the beef type, and is not desirable to breed layers; and his dam will be found to be just a medium producer. As we begin to lose in capacity we gain in capability, and the cockerel's dams increase in the number of eggs they laid. When we get up to dams laying 260 eggs, we find we have super texture and low capacity. You will see from this that we are playing with fire, seeing that the more eggs a bird lays and the greater the capability and the lower the capacity—a tendency to "bantamisation."

Capacity stands for stamina, size of egg, freedom from mortality, breeding power, and maintenance of strain; and the problem that has always interested me has been how to make the best use of it. Many breeders try to get back size by using heavy or beef-type males, but you cannot make progress that way. We must have size on the hen side, which means that we must use hens that are high up in capacity and yet have nice capability and handle well. We will call them breeding hens. The care comes in when mating them, because if mated to a beef or table type of male one will just breed "breeders" and some table stock. We must therefore bring in males with texture or high capability, even losing a little size or capacity, knowing that we shall have the capacity on the hen side. We do not want to use Bantam cockerels just to get high capability, but to get the males reasonably large without losing the texture, which is put on to the bird by his high-laying dam.

It is extraordinary how quickly the trained hand senses the kind of dam that bred the cockerel one is handling. Texture of feather appears to present difficulties to many, but with experience of comparisons it becomes easy to detect like chalk from cheese. Imagine on the right of you there is a sheep, and, closing your eyes, you place your right hand into the wool, you would have a coarse, woolly material. Such is the equal to a male (or hen) with coarse feathering. On the left of you will be a silver-grey rabbit in finest fettle, and your left hand will wander down into the fur and will be passed over the animal. Here you will have a waxy feel, which is like the ideal silky feathering we are after in our poultry. A table cockerel will have baggy, woolly feathering—the sheep—as will the coarse hen; the cockerel from a high-record dam will have tight, silky feathering—the rabbit—as will the quality pullet. Then we have the medium brand, which may be inclined to be silky but in excess, not tight but baggy.

Run your open left hand over a pullet and down her thigh-feathering, and you will note the tight, waxy handle of the "rabbit" kind. She will respond as high to other sections of my capability side, and lay like smoke—just the bird to send to a Laying Test to exceed 220 easily, providing she handles plump and is not superfine. But note how low she scores for capacity or breeding power. This is the finished article already referred to, and she performs but one duty, to my mind, namely, cockerel-breeding. Mate her up to a male from a high dam that laid over 220 eggs, and breed cockerels of high capability, expecting low capacity, which you have strong on the hen side of your every mating.

My present methods, then, are to bring in frame on the hen's side and texture on the male's. In addition I have a pen mated up to breed cockerels, where each has high capability. As the best male is used for each pen, a few cockerel-breeding hens can, if desired, be placed in the same pen as the medium laying hens. At the same time do not forget to have a few large hens laying the big eggs which, mated up to a cockerel from a medium hen, will throw you a few breeders to carry on with. Such a male will be higher in capacity, and yet handle well for capability. Weight or size is a question of bone, and while we do not want to waste time over cart-horses, we always want to have some big-framed birds on the place, to hold the balance and to help keep up size in the strain. That is what I term maintaining. Laying is a matter of bone, because if we have too much we lose number of eggs and get larger ones and stronger chicks; if we have too little we lose size of egg and bird and get more eggs but lose in hatchability and rearability. By a judicious use of bone in the breeding pens we can hold the balance, avoiding in the strain either the extremes of too much or too little bone.

Whenever I find hens or males very high in capability, and yet with nice frames, I give them preference, and think of such hens as breeder-layers. That really should be the ideal to go for gradually. I often get a pullet with a breeding hen's frame and yet the texture (rabbit) of a small bird and I earmark such to be trapped for cockerel-breeding. In like manner I am always after the cockerel with good size, and yet high handling for capability. To get texture is easy when you breed from the layers of the most eggs, because you lose capacity, and the small birds have the higher texture. The art lies in getting the texture with the frame, and such should not be missed as an ideal. I have outlined a good way of breeding which is sound, namely, size on the hen side and texture on the male, but good capability in the hens as well as the high capacity.

CHAPTER XII.

THE EGG TO INCUBATE.

NOWADAYS poultry-keeping is very artificial, and poultry-keepers as a rule seem to think solely of getting as many eggs as they can from each bird while she is alive. They have but one thought for a hen, namely, EGGS. All along it is quantity before *quality*, whether with eggs, chicks or adults. Is there no such thing as quality? The majority seem to have an idea that the egg is especially laid to adorn the breakfast-table instead of to produce a chick that will be vigorous and grow up to be a healthy adult, fit to lay and breed from.

When it comes to incubation, I want every reader to think only of PROGENY. The shell has its part to play in the making of the chick, for which reason every egg incubated should have a thick shell. The yolk feeds the chick, and is really the latter, seeing that every chick has to absorb its yolk if it is to flourish. But there are all shades of yolks, from rich and full-bodied to pale and anæmic ones. Again, the albumen has its part to play, and the white or albumen can be sound and thick or thin and watery, the latter brand not being desirable if strong chicks are the objective.

There should then be severe grading of the eggs, without a thought as to the dam which laid them. It is this knowledge of what a hen has laid that has done so much mischief, because the owner is at once biassed, and it is human nature to be so. He will not throw out unsuitable eggs because they are not up to sample if he is acquainted with the fact that the dam laid 100 eggs in 100 days, or that the eggs are from hen No. 16, which is known all over the farm as the first pullet to begin to lay. In due course trap-nesting and pedigree work will be left to the specialist breeders, who will be the suppliers of males of high breeding. On the average egg-farm hand-grading will be the main method of selection. At least I think so, and I am not often wrong when seeing ahead. I said years ago that the Rhode Island Red would be very popular one day, and singled out the intensive house for laying flocks. I consider that the latter alone, through its artificial nature, has made egg-laying jump from 200 upwards per bird per annum. The reason I advocate my hand-grading methods for

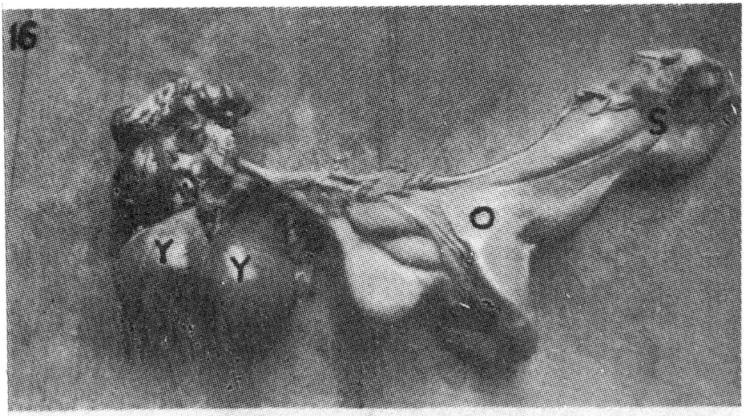
16
Y
Y
O
S

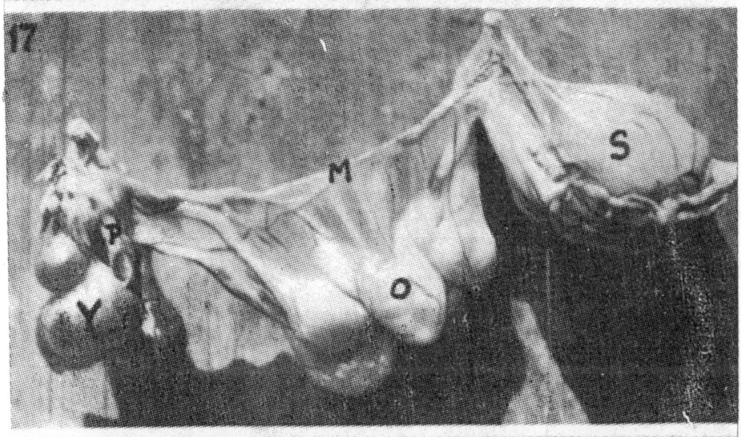
17
M
S
P
Y
O

18
C
O

general work is because by it you can be without bias, and you work and breed for a type and strong, healthy stock without being pegged down by NUMBER of eggs, and the number of each bird. You have a good handle of each bird, and grade any out without fear or favour if it does not fit in with your aims and objective.

Let us take the lesson that hand-grading offers, and we are able better to get flock-averages. To-day everybody asks how many eggs a bird has laid, and if you reply, "220 large ones," the enquirer turns up his nose as if the bird were worth fourpence! Reply, "280," and forget to add, "little ones," and he says what a good bird she is, without even having a look at her. This chasing after the sprinter or the high-record individual hen should be stopped, and attention focused on what actually counts to all who keep hens for utility purposes and for the production of eggs for table, namely, flock-average. The latter decides the profit every time, and you cannot found high flock-averages if you work solely with the high individuals. Flock-averages have to be built up and improved, and one must always be maintaining. You will find every time that your best friend for such building is the medium hen which you have perhaps so often despised.

First of all there is a type in hatching eggs, just as there is a type in everything. A hatching-egg type can best be followed after opening a few eggs which have gone the full time, but have failed to hatch off. We find therein dead chicks almost fully developed. In the larger end we have the head twisted in order to find room for it in so small a space. Then in the smaller end we have the legs which have been turned up to get them in. Naturally common sense tells us that the hatching-egg type is one where there is ample capacity within. We should aim at an egg that is very wide from side to side, with good circumference around each end. Just examine a few eggs and see how the width from side to side holds the key to type. Hand me an egg that is narrow and it will lack circumference at top and bottom. How often have you marvelled at a chick hatching from a small egg and remarked: "How did such a large chick get out of so small an egg?" Solely because that little egg had the type, and while it was not very long it was short and dumpy—wide across and good in circumference at top and bottom.

How are you told to select eggs for hatching? Weight? But I can give you plenty of eggs which weigh two ounces that are narrow and long, and which I would not set. Keep to weight by all means, but keep type in mind. Have a careful look at eggs as gathered from the nests, and see how severely they can be graded, and yet we set incubator after incubator without troubling to grade out any eggs. There is one word that has been sadly omitted from the poultry-keeper's dictionary, namely, *quality*. Everybody tells you not to put a hundred eggs into a hundred-size machine, but forgets to remind you to set only *quality* eggs. You are told to mate ten hens to a cockerel without a thought as to *quality*. Why not be told to mate ten GOOD hens to the BEST cockerel you can get? Again, you are told to give hens two ounces of meal without a word being mentioned about *quality*. I could hand you a sample of meal of which not a third would be feeding value, and another sample where 100 per cent. would be nourishing. Is there no difference to be drawn? I appeal now for a little more thought for QUALITY and a little less for QUANTITY, considering that the latter has done enough harm to utility poultry-keeping.

Having got our type fixed we will concentrate on the shell. This should be thick, and you can readily tell the thickness or otherwise of the shell by gently flicking it with the nail of the first finger as one would flick away a fly after killing the poor creature. If it sounds tinny, then it may be suspected as being thin; if the note is deep-sounding, then the shell may be regarded as thick. Often when applying this test the finger-nail will be put right through the shell, showing how thin it really is, so if you always practise my method you will ruin many eggs that would otherwise be set, and they are undesirable.

A thin-shelled egg generally gets broken under the hen when she turns the eggs, and the weakly one gets crushed against the strong ones. This means extra labour in cleaning the nest, and maybe the hen will give trouble in other directions. In an incubator, when the chick attempts to break through, the shell crumbles and one loses chick and egg, with the labour, time and cost. We must, with poultry, keep an eye on waste and prevent it as much as we can—a thing that does not receive sufficient attention. A man will trap haphazardly 700 pullets—all he has—instead of attending in detail to 100 picked birds. Another man will have dirty nests, and just for the neglect of clean nesting-material he will laboriously sponge and wipe every dirty egg before marketing. How labour and time are wasted on most farms is obvious.

We will also have an egg with a smooth shell and without any lumps thereon. Each egg set will be of nice shape and not abnormally large or small. It will not be misshapened in any way, whether flat-sided, crinkly-shelled, or ridged. I am always very careful over the texture of the shell, because this is another important side of my grading system. When hens are fat they show the loss of condition by the texture of the eggs they lay. Supposing we are incubating in the spring, we may rest assured that the fat hens are those which have not been laying well during the winter, and their eggs will be noted for the bad texture of their shells. Those hens which have been laying well, and which have got into their stride through the winter months, will, by the hatching season, be laying good-shelled eggs. It is thus a further check to handling.

Next we turn to colour, and here I had better say very little. One can no longer say that such and such a breed lays a brown egg, because most birds of that breed will show you a very pale egg, and not a few will produce white eggs. But one should always set the good-coloured eggs, especially for breeding cockerels. On pedigree farms the sample egg of each hen might be blown of its contents, and the shell filed away in a cabinet for use in pedigree breeding. Mr. Leslie Williams does this, so that he has the history of every bird on the farm. Each bird is carded and her weight of egg recorded, and every minute detail is known of her.

One should always keep an eye out for eggs that are poorly coloured, because they denote blood disorder if the hen is accustomed to lay a good-coloured egg, and such whitish products need not be set.

Coming to the contents of the egg, we probe again into the depths of quality. If you are a pedigree breeder you will have to admit that the eggs of many of the high-record hens do not hatch, and the chicks are not reared should a few come out. It happens on those farms I am interested in, so why not on yours? With me, many of the high-record hens are lost, but on some farms they never die! Some high-laying hens lay abominable eggs which never yield a chick, and yet this does not open our eyes to the fact that we are playing with fire, or remind us that there are indeed two types: the layer and the breeder—production and reproduction are never separated, and yet they are to-day absolutely distinct. The better the former and the more inferior the latter. Here and there we get a hen that lays on for ever, and every egg she produces gives a strong chick; and, while she is the exception, trap-nesting has the real advantage of locating her; but only where one records the hatchability and rearability of every egg, and how many do that? You can count them on your fingers. That is the work of the pedigree breeder, and in compensation for his extra labour we should all pay an extra price for stock of this class, because it is the graded article so long as reliability is there. I would like to see stock-breeders concentrate on such essentials as size of egg, hatchability, longevity of laying, etc., instead of going after the "fliers." The public should pay the price for graded stock, naturally, and they would soon be educated to do so. But they have been so blindly led as to the *alleged* value of the sprinter and the high flier, that they have only from experience just awoke to a different ideal. You can soon touch a man's brain, and get him thinking, when he has one chick left to tell the tale of what happened to the other 499! Odds on the odd one will be a cockerel, which took some killing! Utility poultry-keeping is not picking up dead chicks by the buckets every morning, as many find it to be.

The high flier, then, does not wait long enough to get the ideal contents put into her seven-a-week eggs, and we must look elsewhere for quality. In the first place, the feeding of the breeding stock must not be neglected, as I will show later.

The moral is that we should keep up the capacity side of utility stock, leaving high capability or texture to answer for the number of eggs. In short, let every hen bred from have a breeding frame and not be an undersized Bantam. Keep in mind size of egg, hatchability, rearability, and a clean bill of health. Build up your strains from raw-material hens mated to the best cockerels, and when you are losing the frame and getting to Bantams, remember that you are backing the producer and not the reproducer. One cannot keep strictly to number of eggs a breeder-layer should produce, because it all depends on the individual's breeding frame (capacity) and her back breeding. The sounder and longer you breed by selection the higher, within reason, for egg-totals you can go. The higher you go and the more careful you should be in grading for stamina, size of egg, hatchability, and size of frame in the hens bred from. A hen that handles wonderfully for texture or capability, but is a shade small, will often breed excellent cockerels. This is a very important subject even if controversial, but if my views cause some to think, they will do good. To-day one pays a lower price for the dams of competition winners than for the latter, whereas the former surely should be the better value as reproducers of more competition winners. One pays less, too, for their sittings and chicks. Number of eggs laid (production) seems to hold the key to quality in reproduction. Is this the best policy to follow?

CHAPTER XIII.

CARE OF THE BREEDERS.

TO obtain the "quality" egg for hatching, the personal element comes in; and first of all you must make up your mind that you agree with me that production and reproduction must be graded out into separate groups. Secondly, you must agree that there is such a thing as a hatching egg, as proved by the contents. How, then, can the hatchable egg be produced? Solely by common-sense methods and the above broad-minded admissions.

In the first place, we must bear in mind that feeding plays a most important part in obtaining the hatchable egg. The first rule to make is this, that quality eggs have to be fed for; cheap feeding or the use of inferior foods will not give us what we want. Secondly, all breeding stock must have plenty of raw greenery and some maize to get that rich full-bodied yolk. Then comes drinking-water, which must be there always and not in snatches. There must be mineral salts in the foods, and, fortunately, feeding problems are now receiving greater attention than in the past. Personally, I think we are weak in feeding, because it seems nobody's duty to thrash the problems out; but I hope for better things in due course. Salubrene has of late come into vogue as a vegetable to provide mineral salts, and to have a sustaining and invigorating effect on hens that are asked to produce and reproduce. All who use it speak well of it to me for poultry of all ages.

Next we have to think of the quality egg before getting the quantity, and plan our management accordingly. We have always made the great mistake of treating our breeding hens purely as layers, and not as breeders. If we take trap-nesting, we trap the birds in their first or pullet year, and go on forcing some of the best during their second season; AND BREED FROM THEM during the same period. We continue to force for eggs, just to see if the hen which laid 250 in her first year will complete 500 for two seasons and 1,000 eggs in four! Take the question, too, of early-hatched pullets, some of which happen to be grand for frames and size. Such birds would make ideal breeders in their first season, but they must not be used because they have no winter records. Having started to lay early, say in August, they later on enjoy a partial moult (usually the neck and breast) and rest, not coming into lay again till December. They then come up nice and vigorous for breeding purposes, even in their first season; but just because they missed the October and November months when enjoying a rest they MUST NOT be bred from. What an undesirable rule, and the public follow blindly, and would not buy eggs from such birds on that score alone, and some would say " No " if the eggs were offered to them gratis.

Take again the question of the Laying Tests, where they now for some idiotic reason start in November. By November all the best pullets are in lay (and they ought to be by October), and one has to send late-hatched stock or pullets that are backward. When such birds come home they surely cannot be the best breeding material. And no wonder they have a tendency so often to lay small eggs. I prefer to breed from a hen because she is a quality bird first, and secondly because she is actually a breeding type and one likely to give the strong progeny I want. I see in every hen not the bird I have in hand, but the future offspring from her when properly mated. We need just common-sense methods and a ton of observation. Oddly enough, the fancier has plenty of observation to offer, and those who have been brought up in the school of exhibition poultry seem to come out on top with utility poultry. I can mention many names, but just instance three, namely: Miss Nellie H. Bell, Mr. R. Rodwell, and Mr. L. H. Wace. I suppose the fancier masters observation because he is always looking at externals. His duty is to handle stock to see if the best cockerels are throwing a few feathers on the legs, which will send them post haste to table. Then when he thinks he has a Dairy winner he watches it daily in case the comb should flop over at the last moment. Anyway, observation is what is wanted for utility, together with a knowledge of the hen from within.

To get the quality egg, then, I treat breeding hens as breeders. I do not care what you do with your laying flocks, because I hope you will get all the eggs you can from them by any sound means. But I appeal to your common sense to treat the reproducer on totally different lines. Do not put the lights on to make some of them lay two eggs daily; do not force them for eggs—numbers again!—and do not keep them in large intensive houses managed for egg-production like the layers. We have reached that stage of greediness when we try to blot out broodiness entirely. I am no lover of the bird that goes broody every month, but I do like to see a pullet have a rest and come up fresh as a second-year hen for breeding. I suggest that we use our discretion, and let a hen rest whenever we think it will do her good and build her up, so long as we are sure that the said hen is a quality one and a breeder.

Many of the best hens do not go broody till late in the year, when they have put up a big total. I think we should then let them have a batch of eggs, the chicks going to the table; it will provide just that rest which is needed to build up the young hen for reproduction and the laying of the hatchable egg. I know my methods are against the usual ones recommended, but that is why I am placing all my cards on the table, to make each reader have a quiet think, and to let common sense reign instead of automatic methods.

It is well known that the early moulter is the worst layer, and we can grade her out; but while the late moulter is generally the layer of the most eggs (I will not say "best" layer as quantity is not the only key to quality in a bird), she lays on so long and moults so late that her eggs are not fertile till late in the following spring, and she can be a source of trouble. She moults in an unnatural season, when everything is against her, and she cannot be expected to build herself up again quickly and be fit and strong for early breeding. Yet breeders try and force her to be a breeder, and cannot see why she fails. Hatchability should be the only test for the dam whose sons we need. Many a hen will give six chicks from every twelve eggs put down, while but two will be reared to maturity. We do not want that undesirable type. Another hen will yield ten strong chicks from twelve eggs, and eight will be brought to maturity. From such a dam comes the brand of stock cockerel we need every time, even if we have to pay a goodly sum for him.

I may not want to bother with pullets that moult in June and July, but do not think so much of those which moult in, say, November and December, and are out of form for breeding —generally the flier. Let us hit a happy medium, and try to avoid both extremes. I think that the best breeding hen is the one which gets over the moult and comes into lay around December, thereby coming in for early mating and fertility. The question arises as to whether or not we should put all hens into a moult in good time to prepare them for breeding in their second year. I have tried this plan with excellent results, although I agree that with some hens it is difficult to stop them laying, and they lay away merrily while dropping their feathers at the same time. The hen which so often gives trouble with hatching results is the bird which commences to lay in October and carries on round the twelve months without wishing to start the moult. Laying well into the next autumn, the bad weather catches her when she wishes to moult in November and December, and she often hangs in the moult and does not prove very fertile early in her second year. Maybe it would be to the advantage of high-record hens of this brand to make them moult by a change in housing and environment, with a reduced ration. Or, if one wanted a quick return, then the hen could be sent to an autumn Utility Show, where she might be started into the moult very readily if loose in her feathers.

One has to bear in mind the reasons why one bird lays more eggs than another in twelve months. The large, coarse bird will take a very long time to mature, will have several broody spells, and, while going into an early moult, will take months and months to get over it and into lay again. The high flier will begin to lay early, will go through the year without a broody period, and will moult late; and many will lay on while moulting, just sapping their systems and weakening their constitutions for breeding the next year, *i.e.*, for reproduction. The problem is very important, and the owner should decide very carefully where production ends and reproduction begins.

Another vital item concerns the management of the hens during their breeding year. Since the intensive type of laying-shed came into vogue it pushed aside the old-time pattern of housing as then employed for the breeding pens. In the old days, before modern housing became popular, each breeding pen of birds occupied a small roosting-house, to which was attached a covered scratching-shed. The birds were kept on range, so to speak, and many did not even trouble about scratching-sheds. Of late years it has become the modern plan to keep hens intensively, even if they are to be bred from, whereas I suggest that extensive methods are very desirable. Intensive housing without a doubt makes for high egg-production, and must be adopted where the owner is after egg-production—quantity. But all hens earmarked for breeding from should, in my view, have free range, even though they should be provided with sound and comfortable quarters. Intensive methods mean too many eggs, whereas extensive management keeps the birds laying quality eggs instead of quantity. I suggest, therefore, that while we should see that the breeding hens are not subjected to every set-back from bad weather conditions we should let them have as much range as possible, and just keep them laying. The extensive or free-range plans I recommend will slow down egg-laying a bit, and make for better eggs as regards contents from the view-point of hatching and rearing.

I like a type of house as illustrated in Plate 47, which design is made by Messrs. Blakslev and Blyth, of Pedigree Poultry Farm, West Moors, Dorset, and used on their farm. It is a "double" pattern house, 12 ft. long by 10 ft. wide, and is divided into two sections across the centre by a wooden and wire-netting partition containing a door. There are glass shutters to the front, with outside nest-boxes below, and glass panels underneath to light up the floor and litter. At the end of each section there is a pop-hole or exit with a door at each end of the building, and perches and drop-boards at the back of the house.

The advantage of such a double breeding-house lies in the fact that prior to mating up the pens I can run twenty hens in the house as one flock, with the door in the centre partition left open. As I shall have double grass runs or enclosures, I can open the pop-hole shutter to one of the two runs while the other enclosure is resting. When I mate up I can close the division door and allow ten hens in each half headed by a male, and then we have two mated pens. Each pen of birds will use its outer grass enclosure. After the breeding season I can withdraw the males and let the twenty hens run as a flock, thereby cutting down expenses and labour in feeding and attention, and the runs can be rested in turn.

Other advantages are that I can run in the house twenty hens for moulting out when they do require care and attention and shelter from the winds, etc. I can also run twenty pullets for trap-nesting as a flock. Breeding hens, or even pullets being trapped, I would allow out in all weathers except the worst, the birds going in for a scratch when they felt so disposed. In this double house, too, I can run up to thirty hens, which allows me to withdraw any inferior ones at the last moment if I so wish. I can even grade out all except the very best ten, which I can mate up, leaving the others unmated in the other half-house. When in a tight corner I can devote a house to hens in one half and pullets in the other, or even have a pen of heavies in one side and lights in the other section. It is an accommodation house which will be found very useful, as will all houses that can be divided as advised. If I should be pushed for housing accommodation for a time, I can clear one half of the house and use it for maturing pullets, as, for instance, those picked for Laying Tests. As I have pop-holes at each end of front and back I can, if desired, let the birds out on free range at the back and rest the runs.

The question of the best system of management for the pullets being trap-nested for their first year of laying calls for attention. I contend that the kind of care devoted to them, then, decides their value as second-year breeders. If they are forced for egg-laying, just to put up high records on paper, then they can be easily ruined as breeders. I have known birds come back from certain laying competitions which were utterly worthless for breeding purposes, because they had been so forced for eggs to get the records from them. I suggest that it would be better to go without a few eggs in order to let the pullets have some free range, etc., so long as they are not stopped in their laying by the very bitter elements. The same applies to rearing, in that we must not coddle the youngsters, and must have both eyes on strong and healthy stock. In a nutshell do what you will within reason with the layers, but do rear and manage the breeders on sound common-sense lines, thinking always that they are the reproducers.

One matter of importance to me is the choice of an ideal site for the breeding pens. I contend that they do much better when they have a quiet spot, such as an orchard. In Plate 48 we see some of Mr. H. W. Honey's White Leghorn breeders in his orchard at Alton, Hants. Some of the hens in my photograph are four and five years old, and laid very well in their fourth year. They have bred some of the best stock in this country that have done well at Utility Shows and in Laying Tests. I always believe in keeping hardy valuable hens for breeding so long as they throw fertiles and robust chicks. I never believe in parting with quality brood hens. A four-year-old hen may prove a wonderful stock-getter, especially in light breeds, if mated to the right cockerel; and if one goes by ages only, then the usual plan is to discard hens when finishing their second year of lay. By all means grade out birds exactly when they cease to be useful, even before they lay if they are weeds. But do not sacrifice quality for age, whether in males or females, but take every care of adult stock and keep on the best only for several seasons. Let quality decide.

Adult stock of three or four years of age cannot, however, be expected to be commercial propositions, *i.e.*, you must not expect to sell hundreds of hatching eggs from the pen. Rely on your early pullets and second-year hens to vigorous cockerels for the commercial sale of sittings and chicks, and just get a few nice chicks from the older stock. Adult birds that are kept on must, of course, retain their vigour, and should always be mated to vigorous cockerels or young males, so that you have age on one side only. Help them over the moult as they get older, and let them be mated up only when fit.

The advantage of a sheltered spot for the breeders cannot be overlooked, especially for early work. You cannot expect to have the best hatching results early in the year if the birds are kept in a bleak and exposed field where there is not a particle of shelter. The benefits of an orchard for breeders come in here, because the birds have protection from wind (which by the way fowls of any age do not appreciate) and shelter from the hot sun in the summer. The sun will exhaust hens, and shade is very important. It is a good idea to let the breeding-pen enclosures run into a belt of trees or a hedge, the hens enjoying the shelter provided. One can help matters, too, by having the wire divisions to the runs boarded up for a distance from the bottom to keep the ground winds away; or thatched hurdles can be used, or zinc sheets, or even asbestos. Until the runs are properly boarded round as described, a barrier of wattled hurdles can be erected to advantage right across the run near the house, behind which the hens can shelter. There is nothing like ground winds to cause infertility or bad hatching and rearing results.

If small roosting-houses are used for breeders, one can attach to each house a small, dry, covered run, which can be littered down; or one can erect a nice scratching section at the side of the house. This can soon be erected, and will need just a roof, back, and end, and can be open in the front except for a board at the bottom two feet or so up. If necessary, there can be just a roof on supports, with a two-foot board at back, front, and end to keep out the wind and ground-draughts. A section of the front can be left unboarded, or better still, a pop-hole made for the fowls to gain entrance to the shed or shelter. The latter must be rain-proof (the roof can project well over the front and back), and the structure need not be high, and can be littered down, the grain being scattered therein. The fowls can use the shed when they like, and will not be confined to it.

Breeding stock, too, should be fed on different lines to the layers. They should not have so much fish-meal as the latter, and should not be forced. If the breeders are fed on more grain than mash to keep them in hard condition, that will be an advantage from the view-point of vigorous chicks. And all breeding stock should be kept free of excessive internal fat, seeing that overfat hens give so much trouble from infertiles and dead-in-shell, also weakly chicks. Scratching exercise is imperative.

When the hens are in moult they should be taken care of to see that they moult out properly. When dropping their feathers they need protection from the severe weather, as wind and bitter elements will make them "hang" in the moult. They can be specially fed when moulting, having more fat in the diet by way of melted suet, cod liver oil, or boiled linseed. On some farms all the hens can be collected and placed in a flock in a large house just for moulting out, as they can be better looked after when in special flocks. The best methods of management will depend of course upon each poultry-breeder's circumstances and requirements.

The care of the male birds calls for attention here, because so many seem to neglect the males, no matter how valuable they may be. The old plan used to be to place the cockerels in fresh quarters directly the breeding pens were broken up, but I am not sure if that is the soundest plan. My view is that the pens should be mated up very early, say in November or December, so that there can be from six to eight weeks of mating before the first eggs are set. I am against quick matings because males so often predominate in the offspring, and if we only wear the male down a bit by a longer mating there will be more pullets. In the ordinary way when pens are mated up in, say, January, and eggs are set within a short time, cockerels are to the fore at first and pullets predominate later. We have been inclined to think that pullets by Nature's laws predominate later in the season, instead of giving as the likely cause a longer mating up which wears down the male birds somewhat. I am all in favour of early mating and allowing ample time before eggs are set.

If the season is to be a long one, then it will be easy to give the male a rest while another is employed. The question now arises as to whether or not the males should be taken away and housed separately after the breeding pens have been broken up. I rather incline to the plan that it is best to let the males run with the hens till well into the summer, and just house them separately while they are moulting. It cannot be sound to let the males run with the hens all through the year, but I do contend that it is natural to let the male function and be mated for the greater part of the year. Many males are bad breeders in their second year because they are kept for so long on their own, and their organs become dormant, so to speak. On the other hand, they should not be mated throughout the whole year, to include the moulting season.

Take the case of matured young males which are hatched in January and February, and I find that far too many fail as stock-getters because they are not mated up till they are a good age, say bordering on the full year. I consider that directly the male is matured and full of fire he should be mated up to a few hens, not for good, but just to function the organs. Since in my research work I have adopted this plan with the early males I have very little trouble. Another point to bear in mind is that males should not be mated till they are fully grown and furnished, and hens mated to the male should be in lay or near production, so that they will receive his attentions and not upset the pen. High-strung "squeaking" hens should never be mated as they only upset the male and the pen. For the best hatching results the family or pen should be happy and be quartered in a quiet spot.

When deciding the number of hens to mate to a male, please avoid strict rules, and try to suit the individual bird. I have often found bad fertility from a vigorous White Leghorn male mated to a pen of ten hens, but directly I allowed the one male to look after two pens of hens he has proved very fertile. We always want the vigorous male, so that we must have both eyes on the number of hens placed with him.

The question of condition also comes into play with the male. During the breeding season he must be kept very fit, and that means he must always go to roost with a full crop or have a good daily feed. I find it a good plan soon after mating to visit the roost at night and to examine the crops of all males. Any with empty crops night after night should be especially marked for special feeding, the others one need not worry about. I always like a male that feeds well, because so many are the other way. The good breeder is usually fond of his hens and plays the gentleman to such a degree that he goes without his food, calling the hens and often letting them go in first to roost while he follows in the rear. That may be gentlemanly, but it does not get the fertile eggs and the vigorous chicks. Every male, therefore, that is a poor feeder must be taken out and fed by himself every morning if the best results are to follow. Shut him out in the run by himself for twenty minutes or so and regularly give him a few scraps of meat.

I am a great believer in dusting the males regularly with Izal disinfectant powder when they are resting, in order to keep the lice down. The same can be done during the breeding season, because lice worry the birds and loss of condition follows. Dust each male regularly with the Izal powder, rubbing it well into the abdominal fluff. Hens can be treated in the same way. It is well to remember that lice attack all birds that are overfat or overthin, *i.e.*, out of condition.

Therefore, when pullets are ready for laying, and hens are through the moult and ready to commence production again, I see that they have plenty of Izal powder rubbed into the abdominal parts. Being fat, and built up for laying, the lice will come, and if the birds are dusted the lice will be kept down, until by reason of the hens laying well and getting into pliable condition at abdomen the lice will cease to attack, as they do not worry about birds that are in a really nice healthy condition. Hens that have lice at abdomen get irritated, and may readily drop immatured yolks or go wrong in their laying organs through that irritation. And lice, to my mind, bring many other pests and troubles.

Often when handling male birds that are resting I find them very inflamed at abdomen through being attacked by a real host of lice (the louse is a grandfather in twenty-four hours) without the owner ever suspecting it, showing that handling is very important as holding the key to condition of the stock. During the moult, too, I often find the males struggling to get their new feathers without being helped. Often the feathers will become quill-bound, and are prevented from coming through. Such birds need more fat in their diet, such as melted suet, cod liver oil, or boiled linseed. And each feather may need to be loosened with a needle, care being taken to prevent bleeding. Young maturing cockerels also have difficulty sometimes in getting their tail feathers through, and need help in the same way. Often I meet with a flock of young cockerels, none of which has a tail, and the owner has not troubled to handle a few and ascertain the cause. Particularly is this met with in warm and dry seasons, and a little fat in the diet will put things right.

When males are isolated for resting we need a good type of house or cockerel-box. Plate 46 illustrates a good pattern of cockerel-box, the design for which I gave to Messrs. Blaksley and Blyth, of West Moors, Dorset, and which they make. It has many advantages, in that it can be employed in innumerable ways. I am always against one-purpose plant on the poultry farm because I consider such to be a waste and an investment in " dead " stock, whereas it is the live stock on the farm which brings in the returns. Supposing I am working with indoor heated hovers, I can place one of them in this cockerel-box until the chicks are old enough to leave it. The hover is then removed and used elsewhere, and the chicks remain in the cockerel-box. Later on the young cockerels can be graded out into one box and the pullets into another. Later still, perches can be placed inside, together with dropboard, and the centre division can be slipped into place and we can run adult males therein while they are resting; or half a dozen cockerels in each half for fattening for table; or half a dozen pullets intended for sale or Laying Test; or one can keep therein a few birds that are intended for Shows. At any time, too, a hen and her brood of chicks can be placed in each half, or one can wash a few males that are for sale and keep them clean till dispatch with the floor well bedded down with clean straw or litter. Such a cockerel-box, then, fits in with farm routine, and can be used in a hundred different ways for which I so planned it.

When males are in separate quarters, remember that they should not be cooped entirely. Let them have a run out in turns, and be sure and have their quarters bedded down with litter so that all grain can be buried therein to promote healthy scratching exercise, which is as necessary for males as for layers to ensure condition.

CHAPTER XIV.

THE CHICK TO REAR.

THE poultry-keeper can only expect the best results from rearing if he starts with the hatchable egg. If the breeding stock have failed to produce the hatchable egg containing the strong germ, then I contend that the robust chick is out of the question. Many are unfair to themselves by not taking enough pains with the breeders and the eggs incubated. Hens may be bred from which are too short on the leg, thereby dragging their abdomens in the long grass and having intestinal disorders which will affect fertility. An excess of abdominal fluff in either sex may give infertiles, and one should clip the feathers during the breeding season. Again, hens with dropped abdomens, or hens that are high strung or frightened, will prove infertile. Take the case of the male, too, where the mash hopper or trough is so narrow that he does not like to get his head in it, and, in consequence, goes without his meals. Again, take the case of large open vessels where the male gets his comb right into the drinking-water, to be knocked out of condition when he gets his comb attacked by the wind.

Backyard poultry-keepers endeavour to get fertility from hens kept and bred intensively, with just fair results, in many cases because one can only get the best results where the breeders are on grass; results would be better if the male, each year, was bought in and had been reared on grass. Backyard hens, too, may be run short of raw greenery, and that will hinder fertility. One other point concerns males with excessive combs, which play on the brain, are subjected to the bitter elements, and cause loss of condition in the male. All these things have to be considered when it comes to the rearable chick. Many a male bird should have his comb removed if only for his health, but, better still, do not breed males with excessive headgear, but cut it out of the strain by selective breeding. Apply the same methods to excessive abdominal feathering—prevention is better than a cure. Hens with combs hanging over and obstructing the eye are made nervous, and so often prove infertile; these should be removed from the mated pen, as should all hens which are ill, on and off, and which are not in laying condition throughout the breeding season. Let each mated pen be a happy one!

Once you make up your mind that the hatchable egg is imperative, you will get much nearer to the rearable chick. Make a rule to incubate only perfectly fresh eggs, and ideal ones for shape, shell, etc. Next, see that the pens are mated up in good time and that the breeders are in good condition for breeding, also that the male is vigorous to the last letter and the right brand of bird to head a breeding pen. The waste in eggs is appalling on many farms, because sufficient attention is not paid to getting the fertile and hatchable egg before incubation is due to commence. Eggs need to be collected very frequently in cold weather, and the nest-boxes should be liberally lined with material. When stored, the eggs should be kept in a warm place and be protected from the elements. The eggs dispatched for hatching early in the season should also be packed extra well in wooden boxes and be well covered with warm material and ample packing. If the cold gets a free passage to the germs, then I do not expect to get the rearable chicks, and again I drop the hint that eggs are laid to produce young, and not for the table. Their contents are delicate and need every care and attention on sound lines. Have a proper storing-place, and see that the eggs are well protected from the cold until they are incubated. I have often found eggs stored in a cold outhouse in an ordinary orange-box type of cabinet. Eggs can be better cared for in some instances if the cabinet is wrapped round with an overcoat of sacking or blanket, which idea I generally apply to incubators early in the season if they show a "jumpy" temperature, or fail to reach the desired degree of heat and hold it.

With regard to incubation, you will be able to obtain instructions, *re* this and other problems, from my "Poultry-keeping on Money-making Lines" and "Poultry-keeping on Small Lines," so that here I will deal only with special points. When the eggs are in the incubator your management will decide the rearability of the chicks that hatch out. If you allow the machine to go from one extreme to the other, or run it too high throughout the hatch, then the chicks will be weaklings, and you cannot rear that brand even on the best of food and management. When you get the machine going steady, let it cool down before you place the eggs in, or the germs will be affected. You must not subject eggs to extremes, and placing cold eggs into a very warm chamber is damaging. Do nothing in a hurry, and do not risk valuable eggs until you are sure that the hen is really broody or the incubator is steady, and that both mean business. Make sure, too, that the eggs are really fertile. It is wrong to go by fertility on

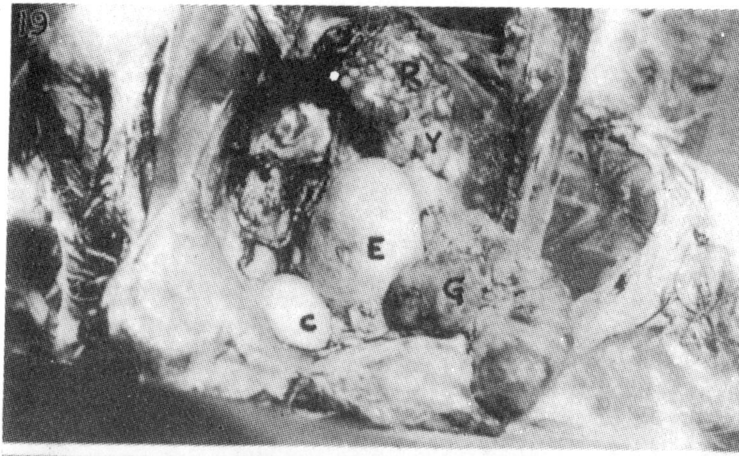
19
R
Y
E
G
C

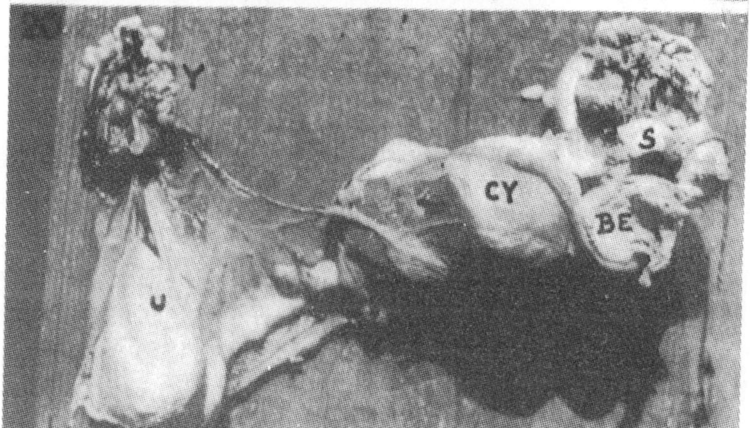
Y
S
CY
BE
U

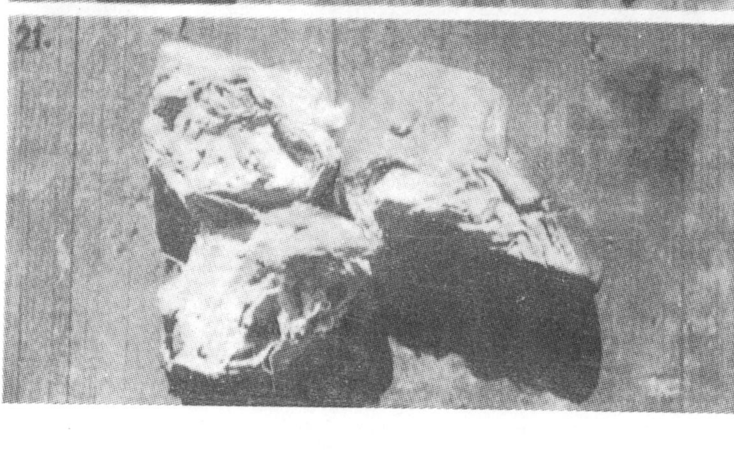
21.

the seventh day, seeing that there are strong and weak germs, and other stages have to be passed successfully before live and rearable chicks appear. I refer to dead-in-shell and addled eggs. Many just test on the seventh day, and finding developed germs (they never distinguish between strong and weak germs) in the majority, start selling eggs for hatching, only to find later that the clients complain of dead-in-shell, *ad lib.*, and few chicks. Others place a few eggs under a trial broody hen, and getting chicks off, start the machines and sell eggs for hatching. Broody hens will hatch off eggs when those from the same pens are hatching badly in the incubators, and reliance should not be placed on broodies in that way. It is here where there is so much waste of valuable eggs. If I sell a host of hatching eggs and the clients return all clears for replacements, I not only displease my customers, but get put back in my hatching, as I have to send other eggs. Do nothing in a hurry, and mate up in good time, going all out for the fertile egg. Get trial hatches out very early and rely upon the number of chicks out and doing well rather than by the number of fertiles on the seventh day. This is the only reliable fertility test of a breeding pen.

By mating up early one can grade out any male which is giving trouble from infertility or weakly germs, and can ascertain the reasons. One can also tell which pens are doing well, and can go ahead with their eggs for sale and hatching. Waste is thus avoided and delay prevented. If one has to replace a lot of clear eggs or infertiles, it may end in late broods for the owner, because he has to allow other eggs to go in order not to disappoint customers, and goes without himself. The poultry-keeper must think of his own stock; and get the eggs fertile early so that he can get plenty down when they are not in greatest demand. So many make the mistake of selling all the eggs they can and then come along with late chicks which they have to keep for themselves and which never do well for them. This is a very weak policy in poultry-farming.

I here make a plea for the early chick; the late ones do not pay. Get them out early every time, and stop hatching at a given time so that those already out and growing can receive every attention. In other words, keep to a system and always be too previous rather than behind one's schedules. Every year is much the same on a poultry farm, so make out a schedule and do not depart from it. Have definite dates to mate up, to get the first incubator down, and to stop incubation. Make up your mind what number of chicks you will rear to replenish your stocks and to develop the farm—and under no circumstances depart from your plans. It is the lack of planning that sees so many with a lot of ungraded flocks by late summer, and maximum results cannot well be expected from such an assorted lot. The funny part is that a poultry-keeper will decide the next year to put things right, and then goes along the same path again with equally disastrous results.

I particularly wish every reader to practise my rule of *breed well and kill well;* along that road lies success. Do not help weakly chicks out of the shells, and make a rule to close down hatching, say, twenty-four hours after hatching has started in earnest. The chicks that just get out late are usually but a source of trouble and a waste of valuable time, food, labour, and costs. You must learn that it is impossible to turn a weakling into a robust chick, no matter how good the feeding, etc. Often a whole batch from some mishap will come out late, but I do not refer to such. In a normal hatch the chicks we want are those which are out to time and without any help from the owner.

When testing a pen for fertility, the first hatch may find rather too many clears or infertiles, which will be accompanied by too many dead-in-shell and addled eggs. As fertility improves, the number of chicks hatched out will increase, while the dead-in-shell, addled, and clear eggs will decrease. Therefore, I like to see in my hatches but a few clears, and then with good management of the machine I can expect a good hatch of strong chicks. If with clears few there is an excess of dead-in-shell and addled eggs, with too few live chicks out, I am always inclined to blame the owner for his mismanagement of the machine. Pass each pen in for fertility, then, when the chicks out are high in number and the clears low.

Many poultry-breeders make the mistake of persevering with a mating, hoping every week for the results to alter for the good. My rule is to put my finger on the real cause, and if I find that the male in a given pen is to blame, I quickly try another. Delay only makes one behind in the season's work, which must not happen under any circumstances. Some males never will be good breeders or producers of fertile eggs except later in the season, so that one should make sure by using those males that do not give trouble. Vigorous cockerels generally succeed early in the season, but not all adult males. The necessity for keeping a few extra males in reserve cannot be too strongly urged upon my readers. It is asking for trouble to keep but four cockerels when one has but four mated pens, because anything can happen, and the results are affected. Many think that they can buy in a male at a moment's notice, but that is dangerous, as you may get a male that has already failed the seller, if he is unscrupulous. Again, breeders do not keep hundreds of cockerels eating their heads off throughout the season, and every reader should order males early and take delivery in time. Also, one should see that several are held in reserve. Do not argue that they cost so much to feed, for no service rendered, because it is the insurance policy that I am after. Take care of your males, too, and see that they do not knock each other out of condition between seasons. Where they have separate quarters they will keep fit, but prevent them from pecking out each other's eyes or injuring each other seriously, as it is too risky to lose a valuable stock-getting adult male. I should always place in separate quarters my best males, and the rest can go in a flock. I can always place a male bird in a flock of young growing cockerels and he will keep them all in order and act the policeman, putting an end to fights. If I run the males together other than the best, I can tie each up. Fasten a length of cord to one shank, and leaving a little between the legs, tie the other end to the other shank. Each bird can then walk, but not run or use its spurs to knock the other out of condition. After a time, the males will agree, and they can have the cords cut, except on any which continue to fight. Place the birds together in a spacious run so that they can escape if fighting starts; do not place them in a house where they are handicapped for room.

Young cockerels and pullets when placed in new houses should be put down inside the house, because they will then find their way out through the pop-hole and readily go in again. If placed in the run, one will have to drive them in at night. This will save labour, but I should always put hens in the run, also cocks that are likely to fight, for the reason stated. Young cockerels do not fight so long as they are reared together and not shifted as they are maturing. Do your grading early, and leave several in a flock of the same pedigree so that should a man want three brothers at a later age they will be friendly when dispatched. The cockerel-box will be useful to take six brothers in case any three of the same pedigree are wanted by one man. Often a commercial egg-farmer wants four or six males all from one good hen, and when they arrive they will be kept together, so that it is a real advantage to have them friendly when dispatched. Remember this always when grading cockerels and planning their quarters. Later on some cockerels get full of fight and bully others, and they should be graded out and housed by themselves, or dropped in with a few hens, or an adult cock should be put in among the youngsters to keep them in order and to stop squabbling.

When the breeding season approaches, the males will be full of fight, and will often "scrap" between the wires of adjacent pens to such an extent that both get knocked out of condition. Such must be prevented at all costs, because just when eggs are needed for incubation the males will be out of tone and robust chicks will be out of the question. Early mating will see them get fit again should such fighting between wires take place. I prefer, however, to have the divisions to runs boarded up for a given distance to prevent fighting, while at the same time shelter and protection will be provided which will help fertility and the condition of the breeders. Wattled hurdles or zinc sheets can be used if desired. Another way would be to have double runs to each house, so that the birds had a grass run resting in between.

For the best results I am always in favour of small breeding pens headed by one male, care being taken to see that he has just that number of hens he can individually manage, based on his vigour. To mate up a hundred hens to ten males is rather of the "hit-or-miss" kind, seeing that the proper males to use are the vigorous ones that are always fighting. If one used weaklings there would be peace because one—the "boss"—would have control of the hundred hens, while the others spent their time on the perches afraid to come down. I consider the ideal, therefore, is to have a properly laid-out breeding section with small mated pens headed each by a single male. If one pen fails, the others can be persevered with. Watch a large mated flock, and apart from fighting, you will see that directly a male treads a hen other males immediately knock him off. If large-flock mating must be employed, then never mate up less than three males, so that when two are squabbling the other can attend to the hens. Again, have the house some distance from the outer wire-netting so that there is ground behind the shed; also have small bushes planted near the house so that each male can be in charge of a lot of hens and is unobserved by other males when attending to the hens.

Another point of importance concerns the spurs on adult males, which grow to a great length and become dangerous to owner, the bird itself, and to the hens. The latter can easily be injured in the back. Just as I should dub the male, so should I remove the spurs of the adult male when they became too long and dangerous. To remove them is quite a simple matter. Wrap the shank up with a damp cloth, just leaving the spur clear for treatment. Next get a steaming-hot baked potato and push it right on to the spur and down on to the damp rag, which is to protect the shank. Leave it on for a short time and then you can twist the spur off from the roots. Some file the spur through, or cut it and then file off the edges to make them smooth, but the other method is very simple and efficient. When dubbed, or with spurs removed, the birds are not exhibited at Shows.

When the chicks come out we will be looking for weaklings, and not thinking so much as in the past of the number actually out. Many sleep, so to speak, in the incubator house to see if any more come out! Let us have *quality before quantity!* In the first place we will get some chicks that are faulty as regards breed points, so do not argue that breed points are nothing to do with laying merits. *Breed better poultry*, and let that be your aim always; if you are a stock-breeder my notes concern you all the more. By all means let us have eggs in plenty, but that does not excuse us from neglecting purity of breed points. We steal the breed name, and should keep the stock up to true characters or call our birds white or red or buff layers, and omit any reference to variety. Many chicks may have feathered shanks or single combs instead of the true rosecomb. Mark these when hatched and place them apart for rearing up for table or cheap sale. Some poultry-keepers want just layers, and do not mind where they fail in breed characters; such persons will buy mismarked birds at cheap prices. And if commercial egg-farmers want mongrels, or first crosses, or impure stock, we cannot interfere with them because that is their business. But those who sell stock have a duty to the industry to perform, namely, to breed for purity as well as egg-laying. If they sell a cockerel it should breed true progeny, which I am afraid has not been the case up to now, as a general rule. Still, many now are seeing my view—*utility with beauty*—and the public are calling for that stock, so that those breeders who wish to keep up with the demand and the times, and not to be left behind, will take note and breed to some standard—and charge extra for it. All who want graded articles should pay for same, naturally. But for breeders to make their own standards and not to heed what the public have to say, and require, is contrary to common-sense rules. I have always been honest in my ideas of utility first and beauty next, and that bird the public want. No one can find fault with such an ideal, because we eliminate the mongrel-looking layer and the monstrosity which has just breed points and no utility. Happy mediums are the objective. And they are possible without losing anything by way of egg-laying, so long as when mating one has in mind beauty all the time. Because a hen lays 280 eggs and has a wry tail I can never see why she should be mated to breed a lot of ugly and deformed progeny—it is like mating up a rumpless hen because she lays well. Breed by all means from the good layers, but give preference to those which possess and will breed true characters.

Always incubate White Wyandotte and White Leghorn eggs separately, where possible, especially if the chicks are for sale as day-olds. Then any single-combed chicks from the Wyandotte eggs can be marked and known and sent to table if cockerels, and sold cheaply as layers if pullets. They will not then be sent to customers as Leghorns to cause disappointment, and to give the breeder a bad name. At the Shows I constantly have single-combed Wyandottes exhibited under me, the owners thinking that they are true Wyandottes. I remember once having four such pullets entered in one class when I had to pass them all; here the owner bought a sitting of eggs as pure Wyandottes and the resultant chicks had come with single combs, and he did not know even what comb a Wyandotte should have. At one Show I also had several single-combed Wyandottes under me, and the owner had even gone so far as to show a trio of single-combed Wyandottes in the breeding trio class. Some try to persuade me that they are White Plymouth Rocks, Rhode Island Whites, etc., and enter them in A.O.V. classes—and not a few breeders openly mate them up to pure Wyandotte males, ignoring the single combs.

Here one should know exactly what a rearable chick looks like at a day old, because the same type holds good throughout the bird's life. In the first place, the chick walks on the pads of the feet with well-developed front and abdomen. Throughout it is well balanced. The weakling walks on its toes, with front flat and abdomen cut away, giving it a stilted or unbalanced walk. Its shoulders may be raised to meet the head, while the weakling will constantly be crying or peeping as if it is cold and miserable. Observation will show us which chicks to kill, so that they do not spread trouble to others in the brood.

CHAPTER XV.

CROSS-CHECKING ON THE "POWELL-OWEN" SYSTEM.

IN order to understand my score-card one should go all out to have plenty of experience in handling with trap-nested stock with known records—high, good, medium, and low. By adopting this plan the operator will obtain that observation which makes for good hand-grading. After all, it is the *execution of my system* that counts. Many err in commencing to grade right away without knowing what is their objective, and without endeavouring to balance a bird. It is *the balancing of a bird* that makes for success in grading, because one must have an idea of what the bird has been and may be in the future, as well as what it is at the time of handling. Is she handling at her best or will she improve? Is the hen growing on to make a heavy bird in excess of our ideal? Experience alone can teach the operator the answer to such problems. Readers, therefore, must bear in mind that *hand-grading cannot be learned in two minutes*, and that the way to master my system properly is to practise on hens and pullets with known laying records, also with males individually bred from dams whose records are known. It took me years to complete my score-card, and it was based entirely upon wide experience in handling birds with known records. At first a practice card, I brought it out publicly when I submitted it by request to a committee of the N.U.P.S., who had to examine all utility score-cards for judging utility birds. My card was accepted of those sent in, and since then I have done everything possible to explain its workings to the poultry public. It was also accepted by the Poultry Club. My critics have attacked the card at every possible moment and in every kind of way. All I wish to add is that the card remains exactly the same as when I completed it, and has not been altered. My critics thought that I had merely put pen to paper and hurriedly drafted the card, and thought they would be able to pull it to pieces. Sufficient to say that all other cards have been altered, and many of my own original points have been added to the others. Still, they are welcome to them. I mention this so that all who are interested in hand-grading will just carry on in a straight line and reap the benefits, not turning away because they have doubt in their mind as put there by the critics and those with an axe to grind. *All I am interested in is teaching the good of hand-grading to all who care to listen and to study it.*

What attracts me at first when handling is weight, because I am not interested in a bird which is too heavy or too light. I want a good medium bird every time. In the first place I am after a compact bird which, when taken in my hand as in Plate 49, does not struggle but is easily held. My method of taking a bird in hand is here seen, and if the bird is compact and well balanced she will hold her head up and tuck her legs up under her for preference. That is the best key to weight, because it means a well-balanced bird. Now if I take a bird in hand, as illustrated, and down goes her head, I know that she is angular and unbalanced with her excessive weight, or length of shank (excessive bone). Every time she is moved her head will go down. Practise this, and you will then have your first lesson in what I call compactness.

If the bird is excessively heavy I am not interested in her, nor am I if she is too small and light. My card is specially prepared so that one can check the points of every bird. For instance, if she is a good hen she will continue to score up, but if a low-grade bird she will be marked back throughout the card. We have to bear in mind that *bone holds the key in the utility bird;* if in excess, then the bird will be too heavy, and if too fine it will be the other extreme. We must keep in mind our objective of the medium bird in each sex, and medium bone is desired. Weight in excess is too much bone, although it is not generally believed to be so. I often ask those in the audience to guess my weight but they are always very wide because they do not know bone weighs. The average person thinks that the stout person weighs the most and the lean one the least, whereas bone alone holds the key. I wish all graders to bear in mind that one can check up the bone in a bird by its beak and shanks. If a beak is very long and deep like a vulture's beak, one can check up the shanks to see if they, too, are too coarse and the scales the same. One can go further and check up the breast-bone and the pelvis bones, which may be like iron if in excess.

Just as we grade out the coarse-boned type, so do we treat the bird too fine in bone. Such a bird will have a long, thin beak, pins and needle shanks, and will walk on its toes, with front flat and abdomen small and cut away. It is the lack of symmetry that unbalances the bird. Seeking an ideal, we always go for a breeding shank which is the medium. If I am handling a White Wyandotte pullet I aim at a shank that is true to the Wyandotte or a heavy breed and one which by observation tells me that when the pullet has laid well she will make a good second-year breeding hen. If I find on a Wyandotte or heavy breed a shank fine enough for a Leghorn or Ancona, then I regard the bird as too fine. But if you handle many birds you will find that quite a number of pullets when ready to lay have very thin shanks. The fine shanked (to excess) pullet will lack weight and plumpness, will not make a breeding hen, and is likely to stop laying after a batch of eggs in order to build up again for the next batch. She is too fine and lacks stamina, and even if she lays well she is a finished article. If she is flat in breast and small in abdomen (cut away) she may have gone too fine for laying, but if she is good fronted she may be on the border-line, although still undesirable from the view-point of building, breeding, or maintenance. Such a pullet often lays a great number of eggs when she is kept well fed and housed as on intensive and forcing lines, and may breed good males so long as they are not Bantams for size. As a valuable unit of the farm aiming at flock-average and stamina, with hatchability, rearability, and maintenance of health and heavy laying, she is not a business proposition as a breeding hen. Pullets bred on the fine side, too, will do better in a backyard run than on an exposed farm, and where they are owned by a man who knows how to feed for eggs, as he puts in by way of food what is lacking in the direction of stamina and breeding.

A pullet that is too fine will on my score-card be backmarked under "bone" as too fine, under "head" for too fine a beak, and under "size" for being too small. What is more, she will score low for capacity. The compact pullet will have a well-developed front and abdomen with medium bone throughout. In Plate 49 the proper method of handling a bird is illustrated, the small finger going underneath the thigh, with the bird's head down and tail up. When the hand-grader has become experienced he will make mental note of each bird when handled like this, especially as he will know exactly what type of bird he is handling for. *I aim at having a heart-shaped bird*, with the two thumbs meeting over the back and the finger-tips of each hand meeting under the bird at the breast-bone, the full abdomen and roomy frame bulging out the fingers to make the heart shape. I am not keen on the triangle-shaped bird, where the fingers are not bulged out by the bird's construction, but where the thumbs meet on the back and the finger-tips at the breast-bone to form an inverted triangle with flat, cut-away sides. The Barnevelder pullet being handled belongs to Mrs. J. M. Walker, of The Ferry, Chatteris, Cambs.

When my score-card was being attacked, a kindly correspondent sent me some valuable data of the system of score-carding employed by him at his insurance business-place. When you answer questions on the approved forms you are supplying information for the score-carders. At one time the doctors used to be in control, and if they gave it as their opinion that you would live thirty years, your insurance policy was planned accordingly. After the doctors had made their report, the score-carders used to add a report, but under a cloud of criticism their word ranked second to the doctors until a noted case came forward. In this instance the man was insured for a vast sum, which business the firm refused to turn down because of the good advertisement that would accrue. The doctors on examining the man said he would live thirty years more, and then the score-carders gave him up to two years. He died within six months, and before all the insurance had been placed elsewhere. As a result, the score-carders won the day and are now supreme. Naturally, I was pleased to have this data and went into matters very fully. The card allows deductions and additions like my card, but in number of years and not in points. For instance, if no relative has died from consumption you may have ten years thrown in on your card, whereas one case may mean the deduction of two years. But what interested me more than anything else was that weight started off just as it does in my handling system. Weight was taken into account, together with height and age, and the heavy person had very many years deducted, meaning a higher premium for him or her to pay. *After all, for a long life of activeness a compact weight is the ideal, and you must insist upon the same with layers.*

Several years ago when I judged the first utility section put on at the Dairy Show, I passed a good handling pullet because she broke my first rule of excessive weight. The owner was rather disappointed because she handled so well, and was a bird that would have deceived those graders who only see capacity and think they know all about selection. That pullet was bought on my recommendation by a breeder, and she laid under 140 eggs, but mated up to a utility cockerel her daughters laid as many eggs as the dam but in six winter months. I remember that the top weight was over 140 eggs for the winter, and the others were 130 and over. She was too heavy as a layer but just useful raw material to breed layers when mated to the right textured high-bred male. She laid 2½-oz. eggs, and her daughters not only had true Wyandotte type, but laid eggs over 2 oz. There is a use for the heavy hens so long as they handle nicely for head, texture, etc., and are not coarse, provided they are mated correctly. If one mates a large male to them then one breeds just table birds. Utility poultry-keepers should always remember that *there is such a thing as breeding*, and that one must think of what a hen and male will breed when mated together, and not what each is when taken separately. You cannot breed two-ounce-egg layers from Bantams, nor can you breed heavy layers from " cart-horse " types on both sides of the mating. But seeing that heavy laying tends to reduce the bone you can use one or two very large-framed hens to supply the strain with bone, so long as they handle well for capability. But do not use " table " males if you want to breed layers, because every time one should have *size on the hen's side and texture on the male's*, as stated previously.

On my card, if a bird is too heavy, it will lose points for coarseness of head, feather and bone, and excessive size, being marked " too coarse," just as the opposite type is marked down as " too fine." You can always read a bird from the card when once taken, and see which hens will be the best mates to such and such a male, bearing in mind his score-card. The latter can be read in detail as regards every good and bad point.

Supposing we now study types of pullets. In Plate 60 we have a pullet that has capital tight feathering, a splendid eye for both boldness and good temper, together with a face that is smooth and fine in texture. By many she would be called coarse in the bone, but on the other hand she has a nice medium beak (short and deep) and medium shanks, such as one should have on a pullet that will make a good breeding hen. She has a splendid front and good depth at rear, with full abdomen. We want such points in our breeding hens, and this pullet is the type to lay well, and large eggs, and breed the right stock when mated to a textured cockerel; she may be termed my 180 to 220 egg bird. She has the desired breeding frame, and in addition handles well for capability; she is compact and active, and good tempered or docile.

In Plate 58 we have an Australorp pullet owned by the Australorps Farms, Ltd., Street Court, Kingsland, Herefordshire. She too has tight, silky feathering, a bold, bright and docile eye, smooth face, and wonderful handling for capability. But she is a little fine in the shank and beak, and in weight, and would be just the bird to put up a very high record and win a Laying Test, and yet might not thicken out into a grand dame as the desired breeding hen. She has, however, a well-developed front and abdomen, and consequently has the necessary stamina and is far from being a Bantam. I should use her as a cockerel-breeder to get that grand texture we want in her sons, and the latter would not be Bantams. In other words, you can lose a little capacity in the cockerel-breeding hens and on their sons because one gets the extra texture for it, but one keeps the capacity in the breeding hens with as much good handling or texture as one can get. The dam of this pullet laid 298 eggs, and she herself laid over 220 eggs in twelve months after importation.

After weight comes temperament, and the observant will notice how friendly and docile are those pullets which put up the good laying records. In other words, those who are not able to get the measure of capability can at first be guided by temperament. *I like the pullet which talks to you* all the time, and which is easily handled. She gets in your way when you enter the poultry house, and comes to see what you are doing when you are hammering a nail home inside the house. Many in their ignorance and lacking observation never learn the lessons of this friendliness, and kick aside such birds which get in their way. *The friendly, good-tempered pullet is the layer and the payer*, and when she enters the trap-nest she will remain happy and contented within the nesting-box until released, and will not struggle and get hurt. She will receive the attention of the male, and will get her crop filled at every meal-time. When grading a flock and picking some pullets for the Tests, the pullet I am seen with in Plate 57 had a lot to say for herself, and I just placed her on my arm and had her "snapped." Such is the pullet to send to a Laying Test or to trap-nest for future use in the breeding pen.

When grading I am never interested in any pullet which is bad tempered, highly strung, nervous, or which struggles when being handled. Never hold on when a pullet in hand is struggling violently, or the intestines or oviduct may be expelled and the bird may be lost. If she will not allow herself to be handled then grade her out as of undesirable type for breeding, trap-nesting, or Laying Tests. The same applies to the male, so select the one which has a lot of talking to oblige you with, calls the hens at feeding time, shows fight, and yet is friendly when in hand. It is high capability which is denoted by the best docility and friendliness in both sexes.

I often have to pick birds out and do grading after dark just aided by a storm-lamp, and I go a lot then by temperament followed by my thorough handling. If you will place the lamp just outside the poultry house and enter the roost where the birds are perching you will have a few lessons on docility and grading, but do not take the lamp inside or all the birds may commence to move about. The flighty, nervous, or highly-strung birds will be noticed to drop from the perch and to run along the drop-board at the back, or to take a spring from the board on to the floor. Before you get so far you may find a few birds of the real flighty kind sleeping in the nest-box, where they hope to find peace and be free from bullying by others. Now pass along the perch, gently patting each bird on the back, when the most docile will not make a protest, whereas the frightened ones will squeal and arch their neck-hackles when handled, and fright will be seen in their eye. They may even peck the handler in bad-tempered style, believing to get one home first before they are killed. I am sure that these frightened squeakers think they are going to be slaughtered. They continue to squeal and will not be pacified. When feeding, another hen has but to look at them and away they go, and in consequence their crops are invariably empty. When in the nests laying they soon fly off if another hen just looks at them.

Cockerels of this frightened type are of the same kidney, ruffling their hackles like a cockatoo and squealing, with fright in the eye. They spend most of their time on the perches out of the way if several males run with the flock.

Allowances have to be made for such breeds as Anconas and Black Leghorns, which often are wild, but plenty will be found among them to be very docile. With these breeds, however, study observation. Drop those which struggle violently and refuse to be handled, and concentrate on those which settle down after first protesting. I have graded pullets out of a wood and found plenty to be friendly, although they had not been handled before. When writing of docility I am often told that trap-nesting and handling are alone responsible, but those who speak in this way have done little handling. It is experience in handling very many birds each week which teaches you right from wrong. Even before they have laid, some pullets show nice dispositions, while others are wild and bad tempered. It is the trained eye which decides correctly the temperament of the pullet or bird being handled, and there is a type of fighting pullet which I would not send to a Laying Test because she is quarrelsome and ready to fight any bird. You can detect her by her " bossy " talk and actions. You may have seen such a pullet and just commented upon her as showing traces of game blood. She can bully her sisters that are with her in the same pen at the laying competition, but might in a flock get a good thrashing herself and lose condition over it. To my mind, pullets that have the laying capability strong within them inherit naturally the good and docile temper, whereas those low in capability are born with nasty tempers. The nervous or frightened ones lay either 28 or 280 eggs just according to their environment and the number of birds running in the flock. It is temperament which makes one pullet when kept on her own as in a single-pen competition exceed the average of six pullets; six will beat sixty, and so on. The larger the flock and the usual rule says that the lower shall be the expected average of production.

I have tested my contention, *re* docility, too many times and for too many years to refrain from pointing it out as part of the grading system I practise. For instance, I have at very many Shows lectured on temperament and docility, and have for my demonstrations asked a steward to bring me any of my winning birds. Now these have been picked out by my detailed card and not by temperament, and yet I have never made an error in that direction, all winners answering with good docility, whether in lay or not, trap-nested or not, trained or not. Many critics at Shows when this is pointed out to them state right away—and criticism is on the tips of their tongues—that training is responsible, but utility men do not train their birds like exhibition breeders, and as mentioned, the same is met with when I am grading on the farms, even if the pullets have just been brought in from a wood or from free range.

When I took the members of Bethnal Green Society over Major Tom Potter's farm, a White Wyandotte pullet followed us round from pen to pen, and even across a field. When I was on Mr. W. Collins' farm at Bulwell, a White Leghorn followed us round by flying over the wire-netting divisions to the many runs. When we stopped she remained handy, and finished up where we all did. We had been grading the pullets, and she carried a top quality ring. When on the Worcestershire Poultry Farm after dark I entered a house and said I would pick out a good laying hen by docility. I just patted each on the back, brought out a hen I liked best, and, striking a match we took her number and later compared her record. She had laid over 250 eggs in twelve months, and was the best producer in the flock. Last year I picked out in the dark a pen of Sicilian Buttercups for a Test, and I see for the tenth month one of them leads the section with nearly 200 eggs for the ten months. Remember the breed mentioned—which is not the White Wyandotte—and I went on docility as far as passing over quickly all which did not respond. Recently I was grading for Mr. Leslie Williams, and wanted several nice hens, but everyone I selected could not be spared. The manager suggested that I should nominate a hen which was not on the lists for retention, and one that had not been a good layer, so I picked out a bad-tempered hen and had it right first time. Remember that the breeder had trap-nested all the hens and had their records, and I was grading after production without knowledge of what each hen had done. I wanted, naturally, his best layers of large eggs and good typed birds, and was telling him what had taken him twelve months of trap-nesting, labour, and expense to find out. This grading, too, was after dark, aided by a storm-lamp.

There are two kinds of docility, the one which is inherited, and the other that is acquired, as in the pullet which is on the border-line. The latter type grades in, of course, such pullets settling down when handled and becoming good tempered when trap-nested for a time. The former do not want handling to bring out the docility merit, but the latter brand can be improved if pullets are handled when growing. In fact, the owner should be with his maturing pullets as often as possible, and should let them eat from his hand and from the bucket while he stands near. The owner who makes his birds docile and lets quietude and friendliness run through each poultry house will get the best returns. Even Black Leghorns and Anconas will in most cases respond to handling when in the growing stages if quality is there or high capability, and it is a point owners should not neglect in these and like breeds. But the frightened and bad-tempered pullet or male bird in any breed will not be tamed, and I discard the same from the breeding pens if I have the other sort. When I am grading I have ample opportunities to handle the parent stock as well as the progeny, and I have traced the handing down of the frightened temperament from dam or sire to sons and daughters. In fact, when grading I start with the breeding pen so that I can memorise its good points and failings when handling the progeny. I thus trace deformities, temperaments, and short breast-bones through to the progeny and I base my tracings on such personal experiences and not on theories.

It is interesting to watch the docile pullet in the moult; she just carries on laying and moults at the same time. The highly-strung bird on the other hand is one mass of quills, and looks like a porcupine, suddenly stopping laying and going into a full moult. As a pullet she will not bother about a neck moult, but will go through a full moult if the pullets are put out of tone and a false or partial moult sets in among the flock during the winter. I have noticed, however, that hens in the moult, and unproductive, do not in some breeds respond so readily to docility as when in full flush of lay or just coming into production, but it is at such times I handle for this merit point. Otherwise, one must use discretion, which, after all, is the art of handling. I have often been asked why temperament is not on my score-card, as I attach importance to it. My reply is that it is allowed for, and very much so, seeing that docility stands for capability, the latter being half of the utility side of my card.

Next I handle for any deformities, and will devote a chapter to this part of hand-grading. Finally, before I start score-carding a bird, I insist on a good type of breast-bone. I have always stood out for a reasonably long breast-bone, as against the excessively short one favoured by all, until I hit home my pattern, and even now supported by those who have had little experience in grading by hand. Often you see a letter written to the Poultry Press saying how absurd is my contention, but what experiences in handling does the writer put up to prove I am wrong? A pullet is so built that there are four muscles at abdomen, one running from one pelvis bone to the end of the breast-bone, a second from the other pelvis bone to breast-bone, a third from pelvis bone to ribs, and a fourth from the other pelvis bone to ribs. These four wide bands of muscle have one main duty to perform, namely, to expel the eggs and the droppings. In the ordinary way, before the hen laid 200 to 300 eggs per annum, these muscles would not be heavily worked, nor would they be in the expulsion of droppings. To-day, however, the duty of expulsion is very great, and the security for heavy laying depends upon the bird keeping the muscles active and the abdomen in control. This is possible only where the breast-bone holds up the abdomen. If the breast-bone is very short, the abdominal muscles have to hold up the abdomen as well as expel the eggs, and, if the internal organs become coated with heavy fat, the muscles are stretched to such an extent that they lose their elasticity and refrain from coming back into position. As a result, the bird can go wrong and begin to lay soft eggs, or to have yolks slipped before they are ripe into the abdomen, or to produce double-yolked eggs. She can be infertile also, and have abdominal dropsy while the egg-organs go wrong through the bird constantly straining to lay the " lump " which she seems to feel at abdomen. I have looked at the bird with the dropped abdomen from all view-points, but cannot find a good word for her. The dropped abdomen, too, can drag in the long, wet grass and set up intestinal disorders, while the bird can become constipated, retaining the droppings and dying from internal poisoning. To me, a bird with a sagging or dropped abdomen lays 28 or 280 eggs per annum, and I prefer to take it as an undesirable type and to grade it out as a troublesome type. Such birds usually refuse the attentions of the male, and, while it takes six months or so of laying for abdomen to drop, it needs but a month or so to go wrong after the first real moult. Consequently a hen can lay well the first year and

22

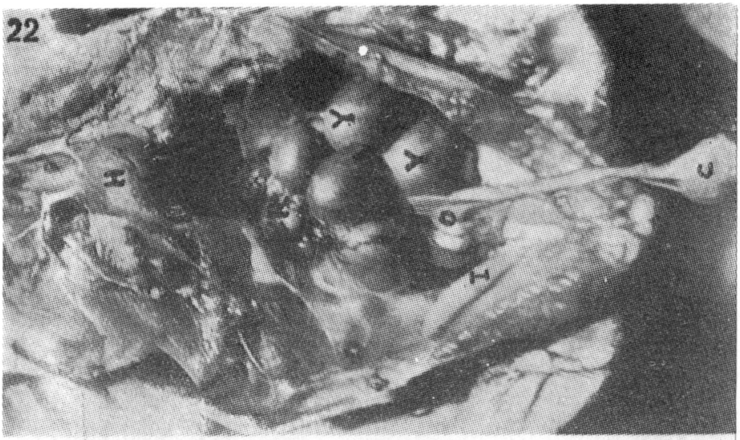

23

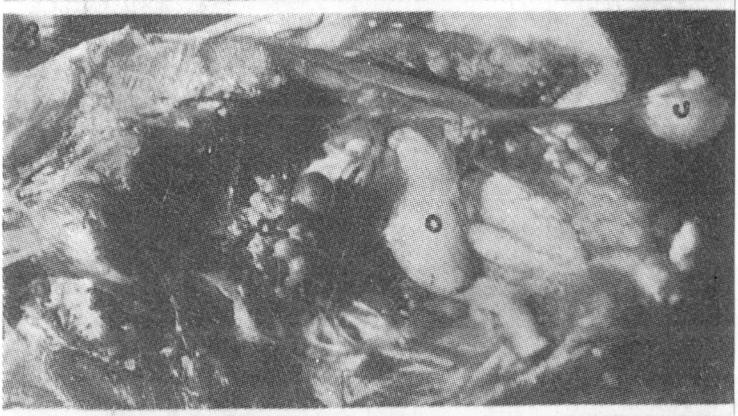

24

badly the second. The abdomen, too, may fill with fluid, causing abdominal dropsy. I have found pullets go wrong after laying but a few eggs through short breast-bones, abdominal dropsy setting in, shelled eggs being retained in the oviduct, or soft eggs being laid from the start through the hen straining the oviduct walls and muscles in attempting to "lay" the sagging and irritating abdomen. Very readily is the heart weakened, as noticed when such birds are handled and gasp.

Heavy layers are apt to have, sooner or later, crooked breast-bones due to prolonged production and the finer bone found in utility stock, and, if the breast-bone is on the short side, then the abdomen may drop and laying be interfered with. A slight curve downwards of the end of the breast-bone is sometimes sufficient to make the abdomen sag. A longer breast-bone is a security against any kinking or twisting of the breast-bone that may follow a period of heavy laying. Many judge a layer's quality by the number of eggs she lays in one stretch. You will hear a man refer to a hen as being his best layer because she has laid twenty-eight eggs in a like number of days; what counts is the longevity of production throughout the individual bird's lifetime. Even if a bird lays a large number of eggs in her first year, she loses her usefulness as a unit if she goes wrong in her second, and produces but a few eggs. Often spring or summer laying is taken as a key to quality, and a man will say that his hens have been splendid layers, producing four eggs weekly for the last three months—when such were summer months—when anything in feathers ought to lay. No special period should be singled out, but both eyes kept on longevity, together with full winter production, if the birds are hatched to time.

I remember grading some Columbian Wyandottes for Mr. L. H. Wace, the popular Secretary of the Columbian Wyandotte Club, and earmarked several birds as likely to do less than standard the second year, and they confirmed my contention by laying spasmodically through the short breast-bone failing. Such a bird often lays well one month and down comes the abdomen; when it produces only a few eggs the next and the abdomen recovers, the bird produces well the following month—until the abdomen finally gives way or breaks down and then laying is bad. Another case I can quote concerns a poultry-breeder who selected all his birds with very short breast-bones because he was told by a Lancashire breeder that such was correct. After winning at the Test he made a tour of breeders' yards and, by request, I went up to see him to get him over his troubles. Having sold all the decent breast-boned birds, he was left with all the short-breasted ones, and when I called

I found dozens of birds with diarrhœa through the long, wet grass. Another hint on grading he had obtained was that short-legged birds laid the best; the shorter the leg, the better the layer. Yet a third hint was that birds with white eyes always laid well and did not give trouble from broodiness. At the same time several pens were egg-eating, undoubtedly due to a few producing soft eggs as a result of their short breast-bones. At the Tests his birds were dying, but after putting him right he was back among the leaders in two seasons at the several competitions he entered.

In the ordinary way the heaviest bird is most readily affected by too short a breast-bone, a Light Sussex sooner than a Wyandotte, and the latter before a Leghorn. The only way one can help such birds is to keep the abdomen light and pliable, so that excessive and heavy internal fat is kept out of the abdomen. Then the muscles do not become so easily ruined. But one cannot treat any special bird in a flock, and so I grade out such types from the breeding pens in order not to have a lot of progeny with the same failing. If the owner cares to keep such birds together in a small house, and feeds them on grain only to keep them pliable at abdomen, he will get better results from them. But do not breed from them. The breeder I mentioned had bred the failing into his progeny. The greater the capacity and the weightier the abdominal contents, the easier is the bird affected.

I like a type of breast-bone then that supports the abdomen, leaving the abdominal muscles to do their proper work of expelling eggs and droppings. I insist, however, in having a straight breast-bone. When I place the right hand against the abdomen, between end of breast-bone and pelvis bones, as in Plates 50 and 51, I like to have it in an almost upright position. Place the palm of the right hand in position without going right into the abdomen, but rather as if a piece of wood had been placed on top of the pelvis bones and rested on the end point of breast-bone. Then turn the bird down in the position that it would walk and the hand should be on a slight slant not almost flat. When one places the fingers under the abdomen in such a position, or when the bird is on the perch, and draws the fingers along the breast-bone, the abdomen should not be felt sagging below the end of the breast-bone. Birds suffering from dropped abdomens can be noticed when perching by reason of the abdomen sagging well below the perch. On the drop-board under such hens will often be found soft eggs dropped at night from the perch.

In Plates 50 and 51 I am seen measuring up a hen at abdomen. She is a four-fingered bird, as shown by my four fingers which are between the pelvis bones (on either side of vent) and end of breast-bone (at small finger). The angle shown is the desired one. Note how in Plate 51 I am holding the bird with the left hand under the breast, and fingers between the legs while she rests in the socket of my arm. In Plate 50 a side view is given of the same handling at abdomen.

Held in the same way, I next place my three fingers between the pelvis bones in an upright position, as shown in Plate 52.

It will be seen from my method of handling for depth of abdomen, in an almost upright position, that I secure real depth. The shorter the breast-bone the less the actual depth, as you can prove with a few simple tests. What counts in a layer is the depth, together with width and length of the abdomen. Now I may get a hen with seven fingers through a very short breast-bone, but she has no depth, so you may blame the handling for most fingers between the pelvis and breast-bone as responsible for very short breast-bones being popular. The handler, in his ignorance, thinks he has greater depth at abdomen in a seven-finger bird over a five-finger one. Had he kept to my card, and not to a single point (most graders go wrong here), he would have seen that I allot twenty points to a five-finger bird, which is given as the maximum. Naturally, when I wrote that down, I had in mind the fact that a five-finger bird in an almost upright position was about the most I found over a wide experience in ideal utility types. As a rule, heavy-breed females go five fingers here, and light-breed females four, although often one gets a full five in a light breed.

When judging Moore Show the other day, Mr. H. Woulds mentioned the case of a Columbian Wyandotte cockerel with which he had done much winning, and even in utility (except under me). The bird had a very short breast-bone, and measured six fingers at abdomen in the space mentioned. The bird was never a good breeder or fertile, and suddenly died from abdominal dropsy. I should not trouble to trap-nest such excessively short birds, and should never breed from them or send them to Tests, nor would I send to a Test a bird just on the border-line. She might lay well for six months and then fail badly, and, remember, I have learned a lot about handling by examining minutely birds I have picked for Tests on their return, and others on their return that have been sent by breeders in an endeavour to find out why they laid well or badly. Often I have found a bird go wrong through a crooked breast-bone setting in and letting down the abdomen. A bird can have a perfect breast-bone for eighteen months and then it will become crooked.

Never use a male with a very short breast-bone. Take no notice of those who criticise me, endeavouring to make out that

I want a breast-bone that is the other extreme and which goes inches beyond the pelvis bones, or that I need a breast-bone which curves up and supports the abdomen, representing a two-finger hen. I merely want true depth of abdomen, with the full and weighty contents supported by the breast-bone, and not the abdominal muscles. Bear in mind, too, that you can easily make a good breast-bone short if you so fatten out the bird at abdomen that the skin and muscles get stretched to such an extent that, not coming back, the muscles lose their elasticity and the abdomen sags.

I am going into detail over the handling part of my card, because sound execution is the main thing. Few, if any, of my critics have seen me handling, except at a distance, or know my system correctly. The width of back is a measurement that is original in the way it is taken, and it is practical. I often hear a man say a bird has a wide back when I should call it the reverse, because there is only one correct way to measure a back. In Plate 54 I have photographed two birds to illustrate my point. I want you to picture the back, from wings to parson's nose, as a bottle with the neck part at the top (L), B to E being the base. One takes the width of back at the widest part which is marked B. The question now arises as to the width and also the length of the neck part of the bottle. This is very important, naturally, because the heart is within. In the first place I want great width at the neck of the bottle as at L, and I require less depth of neck than at N. The reason I need width of neck is to get room for lung and heart action. Plate 53 shows a very good snapshot of me taking this measurement, photographed by my friend Mr. A. Rice, of Wollaston, Wellingborough, Northants. In Plate 53 I have placed my two fingers in the neck part, with the second finger right up against the socket of the wing. The wide part of back starts where my first finger ends. I allow two fingers for this neck part as a maximum, but like even less; in other words, I prefer to have the wide part of back well up to the wing sockets. The man with the broad chest has great power, and it is the chest part of a bird I am measuring in this way. I aim at depth through the chest of bird to get stamina.

I can find you plenty of birds on your farm with very narrow and long necks and in some cases I could place four fingers in the narrow neck part. These are the birds which will be proved to have lung and heart troubles, roup, colds, bronchitis, catarrh, etc. You can check me up very readily with the next bird you find bronchial. I always stand to be tested on my methods, and for years have been handling birds for breeders and selecting their pullets for the Tests. Had there been nothing in my system you would have heard of it from them.

I have never hesitated when at a breeder's farm to take a bird in hand and to have myself tested. I remember at Mr. Oscar Brown's place, in a friendly way, I said I would have a try at placing his ducks in their order, and I placed them just according to the number of eggs they had laid at Bentley Test. I was picking after they had done the laying, and with my experience could have saved myself the trap-nesting or recording of the ducks for twelve months. The pedigree breeder must trap-nest, but look what hand-grading must mean to the commercial man or the thousands of small poultry-keepers up and down the country who work in the daytime and have their birds as a business side-line; miners, too, who have night shifts, and attend to their birds by day, or brothers who work together, one working by day and the other by night, so that the fowls can be looked after in turn. One has to get about the country to see the innumerable ways poultry-farming is carried on. When I was at Miss Barbara Raye's farm last year, I suggested which I considered her best ducks in their order, and was correct. When on my first visit to the Australorps Farms I went into every pen and picked out the best bird in each; only one slip when I had the second best in a pen. I do not do this to "swank" as all who know me are aware that is foreign to my nature, but I have done it each and every day to get converts to my hand-grading system, because I know its value to poultry-keepers in general, and the poultry industry in particular. On another pedigree farm we had a friendly argument on short breast-bones, and, although after dark, I suggested we should get a lamp and visit any house containing trap-nested hens. This we did, and I found one short-breasted hen which was apparently in lay, but I suggested that she had not laid for many months and the records proved I was right. Had I opened her I should have found coagulated yolks in the abdomen. Then I discovered another short-breasted hen with abdominal dropsy, and we could not find an egg recorded to her for months back. Trap-nesting was continuing, but the non-laying of such birds would probably be put down to moulting as it was in September, but had I run through that flock many would have gone to the table, and at a great saving in time, labour, and foodstuffs. Where I am handicapped is that I cannot well mention names of the breeders publicly for all to read, as in their ignorance many might think these breeders had no good quality stock. I am out only to convince one and all that there is no flock in the country that would not be the better for severe grading. If a critic should challenge me I am quite prepared to prove my statement on his own farm.

All birds bred from and sent to the Tests, and especially male birds selected for the breeding pens, should possess a wide neck to the "bottle" of the back and a short neck. I always insist upon this for males, even if I let in a hen that is failing a little there. But I do not send such narrow-necked pullets to Tests where roup may attack them. Remember, then, I allow you as maximum a two-finger space in the neck. When measuring the width of the back, I always put on the card the number of fingers for the neck against the score for width of back, then I know the bird's good or bad point. If a bird goes three or four fingers at neck, even if the back is wide, I deduct points, or marks, for width of back.

Length of back is also governed by the neck of bottle, because you must take as your length from the wide part of back to the base of the parson's nose. In Plate 54, length of back is good in the one bird, running from B to E, whereas it is short in the other. The ovary is in the widest part, and we need to aim at about five inches as a safety measurement for the length of back for ovary and oviduct.

Plates 53 and 55 show to a nicety a good plan of handling a bird when taking the measurements of the back. With the left foot placed on a low box the bird is rested on its breast on the operator's thigh, being held or supported with the left hand, two fingers being between the legs and two on the other side of the bird's thigh. The legs must not be squeezed. The well-balanced bird will be easy to handle, being compact, as already explained.

Width between legs is taken when the bird is standing on a firm base, by placing the open palm of the right hand between the thighs, near the breast-bone, as in Plate 56. I have seen many take this measurement with the bird in hand, when one could press the legs so far apart that one could get seven to ten fingers between.

When cross-checking on my card it will be found that the bird with too much feather on thighs and at abdomen loses also in the length, depth, and width of abdomen, because the feathers in excess there suck the abdominal contents and prevent the intestines from making the eggs. Often type in breeds tends to encourage birds that are very fluffy, or feathery, in order to get curves, and eggs are lost accordingly, whereas, if my plan were followed, the full abdomen would make the desired curves, since with the loss of the feathers the abdomen would swell out in egg-making. Excessive length of leg (bone) will also suck the abdomen.

If a bird stands wide, like an Indian Game, the handler can consider that it has just the right length of leg. If when standing normally the bird is knock-kneed, then there is an excess of bone which should be cut for. Supposing with the right length of shank the bird stands narrow, then she will be narrow across the back, providing she is in lay or ready to lay, which will mean that she is matured. The width of pelvic arch will be wide if the bird has a good abdomen, a wide back and good width between legs. If the span of the arch, as taken with the thumb of the right hand on one side of the parson's nose and the first finger on the other, is narrow, then one can expect the shorter measurements in width of back, between legs, and in abdomen.

When handling a bird, if she (or he) becomes easily fagged, the bird may be suffering from heart-trouble, which is a thing to note especially in a male bird and in stock sent to Tests or to clients.

CHAPTER XVI.

GRADING FOR LAYING TESTS AND TRAP-NESTING.

EVEN where one has been careful in selecting the best possible breeding stock and in setting only hatchable eggs, the problems of rearing come along in their sequence. It has always been a controversial point as to who is the breeder, the man who mates or the person who rears. The majority vote for the breeder who mates up the pens, but I give the rearer half the honours. It is in the rearing that we make or mar our birds, and the pullet or cockerel at maturity is what the rearer has made it during the growing stages.

I want the utility poultry-keeper to make a fast rule not to bother with chicks that are not of the rearable kind. All weaklings should be killed when hatched, including all which are unable to stand or walk, but sprawl out their legs when they move. Let chicks settle down for a few days after hatching, and then go through them and kill any which you do not like the look of. The average rearer saves all chicks because he is not sure how to kill them. I have known poultry-keepers drown chicks because they did not make a success of the method of dislocating the neck. The best plan is to take the chick with the legs in the right hand and the head and neck in the left, then strike the head against a brick wall or a hard post sharply. Let "breed well and kill well" be your motto! It cannot help strong chicks if they run about with weaklings, and remember that the latter are the first to catch any trouble or epidemic and to spread it. I would go so far as to say that the laying of the matured pullets depends to a great extent on the high or low mortality of the broods from which they spring. If you have heavy mortality in a brood of chicks, those that remain will produce a low flock-average; if you have little loss and trouble, then the pullets that are reared from that brood will put up a good flock-average, other things being equal. All the more reason then why you should try and have the least possible mortality and trouble with the chicks in each brood.

Many losses are due to faulty incubation, but I will assume you have taken steps to get the hatchable egg and the rearable chick. The next stumbling-block comes in the rearing and care when the chicks are very young. I would warn all readers of the risks run through intestinal chills. I have known a poultry-keeper take day-olds straight from the incubator into the garden to have them photographed because the hatch was a record one, and yet saw nothing against it. You cannot play about with young chicks. I think, as with eggs, one should use sound and solid boxes for dispatch early in the season, with lighter receptacles later. Also the packing should be warmer. But day-old chicks should never be packed tightly because that will certainly lead to chilling. I have supervised exports of thousands of chicks and always put down the success to light packing so that the chicks could move about. Many breeders send out chicks a week old without the buyers noticing the age, and I have seen chicks with wing feathers sent out as day-olds. If you should get older chicks send them back, because you may otherwise have heavy losses within a week of delivery, although on arrival you may find them strong and healthy by appearances. Chills soon lead to heavy mortality.

Always take your time in getting the foster-mothers ready and going steady; at the start of the season have lamps burning in them for a time to dry out the internals and the litter. Speedy preparation of incubators and rearers is dangerous; work to schedules, and never overcrowd young chicks in the foster or let them out into the cold wind when but a day old. Use discretion always and study the weather.

The construction of the chick is such that it will not help you in the rearing during the first fortnight; in fact, it will hinder you. Nine-tenths of the epidemics which chicks die from in large numbers can be translated into a single complaint, viz., chronic indigestion. Many people start feeding chicks as if they were cast-iron creatures, and it is a pity that the "chick from within" is not better known, for that governs the feeding. When a chick hatches out it takes into its body the yolk, which is its natural food. It is this latter which allows us with safety to send chicks when hatched thousands of miles by rail in a small wooden or cardboard box. The chick goes along nicely while the yolk is melting away. The yolk is attached to the intestines by vessels or strings which have the duty of absorbing the yolk or passing it into the digestive organs. Now the yolk must be given a chance to get away, and there must be no hitch in digestion if that is to take place. Should the yolk not get clear then it hardens with the heat of the body, and nothing will make it melt, and every chick with such a particle of unabsorbed yolk inside the abdomen will die, even if it lives till six months of age, when you find it dead one day in the run. Sooner or later complications set in and death follows.

The first step towards getting the yolk absorbed is to withhold food until the chicks are 36 to 48 hours old. The next safety course is to adopt a common-sense system of rearing. I prefer to start the chicks off on lightly-boiled egg and bread-crumbs, moistened with warm milk or water, and dried off with fine oatmeal or Sussex ground oats. I give nothing but this for two days and on the third and fourth days I give one feed of chick seeds as well as the above mash. From the fifth day I drop the egg and bring in middlings and give two feeds of chick seeds daily. On the eighth day I introduce bran and then give mash and grain alternately. I am a great believer in plenty of melted suet or fat in the mash for chicks during the early stages in particular, and from the second week introduce greenery of a succulent kind, giving milk to drink now and then. I find that if we take a little care during the first two weeks and get the chicks safely on to their legs we have little trouble. You can nominate epidemics with long names as the causes of heavy chick mortality, but wrong feeding and chronic indigestion are mainly to blame, in my view.

Be sure and have dry litter, and keep chicks out of long, wet grass until they are long enough in the leg to raise their bodies out of the grass. Have the latter near the rearers cut short and let the chicks out on range when the sun has dried up the dew. Again, do not expose chicks to wind, etc., until they have become feathered; let them have plenty of shelters. Keep all utensils clean and foods up to sample and fresh to prevent irritants from doing damage. The sooner, too, the sexes are separated the better, because pullets cannot get along and make most growth if the males have all the food and bustle them. Males are left with pullet chicks too long, mainly because rearers are not sure of the sexes. Master early sexing, and at a young age get the sexes apart. In heavy breeds the cockerels at an early age show good enough signs to guide one in separating them out. The head and neck of the male are built on more solid lines, but usually the cockerels in heavy breeds are tall, and the older they become the longer the legs seem to get. They feather with difficulty too. The pullets are more symmetrical, have neater and shorter legs and feather up more quickly. The shank appears to be short and straight, whereas in the male the thigh shows. Do not worry if you make a few mistakes because they can be put right later on. Grade into sexes at or near eight weeks of age and the pullets will then get a good start and not look back. Bear in mind that the utility birds are medium in bone and therefore are not out to help you to get size, seeing that bone makes the weight. Therefore all depends upon quick growth and every encouragement if the pullets are to have nice frames before they lay the first egg. When the first egg arrives the pullet has finished her growth, and, if she is small in frame, then the eggs from her will tend to be small and she will be ruined as a breeding hen.

Rearing methods I am not dealing with in detail, but I would ask every poultry-keeper to do plenty of grading among the cockerels, so that as many as possible can be sent to the table and be prepared for killing early to get the best prices, making room for the pullets. Often I have seen hundreds of cockerels kept on at a heavy loss, the owner thinking he could sell them all for stock purposes at top prices. If you have no name then you will not sell many cockerels, no matter what their breeding or pedigree may be, and only the few best should be kept on. Decide what number you wish to have for your own purposes, so that you can keep on a few more than required to make the last choice therefrom as late as possible. Grade out for table any with faulty points. A Light Sussex may have feathered legs, or a White Wyandotte may have white shanks, etc., so do plenty of grading out. No pullet must be overcrowded for a minute during the rearing stages if it is to make the best growth. Later on some of the cockerels' combs begin to flop over, and other birds will go up on leg, showing no signs of utility; such can go. Deformed toes may put others out of the running, as may bad crooked breast-bones and deformities. So long as you have in mind the supreme need of getting to table early, and at the right time, all throw-out males, you will be working on the right lines. Keep ever in your mind the ideal matured male.

Pullets can be hand-graded at any age if the handler has had the experience. Always go for the compact heart-shaped pullet with prominent breast, width of back, ample space between legs, wide pelvic arch, neat beak and shanks, and the youngster which handles plump for age and walks on its pads. You can start by grading into a lower flock all pullets with flat chests, cut-away abdomens and triangle-shaped bodies, and which handle light for age—" bad doers." Those that walk on their toes come under this category. You will not be struck with the pullets which have narrow backs, narrow bodies, knock-knees, and no space between legs, or those with narrow pelvic arches. Aim at a good type of breast-bone and also straight pelvis bones. I always go for the bird which, when young, handles nicely for abdomen, and at the right angle with pelvis bones, straight, fine, and wide apart. I place them up in the top division, whether in males or females. I always keep in mind a straight breast-bone as against a rounded one which cuts away the abdominal depth. Exceptions there will be, but in hand-grading we keep to rules, and like all good birds and detest all low-grade ones. We try to pick out GOOD birds every time; if we miss a BETTER we do not worry as we have a graded flock, and have picked good ones, which is far better than ungraded lots. Symmetry is the key-note to utility quality and at all ages. The chicken which handles plump, and yet fine, is heart-shaped, and walks about in a well-balanced style as if the farm belonged to him or her, is our objective.

The time comes along when the bird sets in bone, and we test this by placing the thumb of the right hand on the back of the pullet near the parson's nose, putting the first and second fingers of the right hand beneath the end of the breast-bone. Gentle pressure will move the breast-bone up and down if the bird is not set in bone, and the pelvis bones will be easily bent together, all bones being supple. Directly one finds the breast-bone rigid, and the pelvis bones stiff and not easily squeezed together, the setting of the bones has taken place. At that stage the vent is very small and the abdomen empty. I want you to study this side of my handling system very carefully, because it will help you immensely in hand-grading. What is more, they are my original methods based on the construction of the hen and the internal workings.

A pullet has to set in bone before she can develop her ovary; at that stage the ovary is dormant and in appearance like bloater's roe. When she begins to make her ovary, and the yolks begin to separate into small individual balls, the vent begins to increase in size and the abdomen to fill with the intestines which are egg-making. At first the vent (stretched to the full) may be as large as a shilling-piece when the abdomen will be half-full. One can tell to a nicety when a pullet will start to lay from this part of my handling, and I have won many Laying Tests on this knowledge, especially where eggs are scored at market values; the selection depends to a great degree on sending pullets which will be timed to lay just as the Test starts, as the eggs will be scored at high winter prices. Send pullets which miss a month to such competitions and you will never pick up the lost time and values. A pullet picked on my system last year laid 143 eggs in six winter months and then went wrong and did not lay again. I think for the tenth month, with no eggs for four months, the pullet had gone down only to sixth or seventh.

I want you also to bear in mind that the pelvis bones and the breast-bone begin to get fleshed with the setting of the bone. Therefore it is up to you to feed liberally from the setting of the bone till the laying of the first egg if you are to get the pullets fleshed as they should be. Any sparse feeding will see the ovary develop without the bird fleshing, and when the first egg is laid the pullet will not be able to stand heavy laying which drags on its system while a partial moult may take place. We are always told by some of those who do not believe in handling, that when a few pullets start to lay the flock should be kept on grain and short rations and be shifted to fresh quarters to stop them laying, and you write and state that you tried it and still they laid. It is for you to watch, by my handling system, the forwardness or backwardness of the pullets by picking one or two up now and then. If any shifting and short rations are to be tried they must be carried out before the ovary goes. In like manner, if a pullet starts to lay do not get excited but handle a few likely pullets until she is located and shift her. Do not cut the rations down for all the others just as you should be building them up for laying. Any set-back just then will see much damage done. You must make the best growth up to the time the ovary starts to develop and not after.

We should have a proper schedule to work to, and I will mention my own. I consider that pullets which start to lay before September 15th will, in the majority, go into a partial moult; those starting to lay after will go right through the winter with production. Now I grade into two lots, because I know how dangerous it is to let those pullets which are coming into lay in October run with any pullets that are in a partial moult. The habit spreads and I run too great a risk of losing baskets of eggs. At Tottenham we ran our first Test and let all birds sent compete. Many were not set in the bone and others were in lay. We had a bad time of it because directly one bird started with a cold the younger pullets caught it and it took three to four months to get the flock going. This happens on many farms where grading is not done. I want you all to tighten up your grading of maturing pullets just to see how much better results you can obtain by my methods.

The first thing to plan for them is a house where the early laying pullets can be placed when graded out of the main flock. If you have no such place, then they will be better put with the old hens which will moult as usual and all can moult together. When handling reveals pullets in lay up to mid-September, grade them out from the flock and then concentrate on those which will start production from mid-September, say, to mid-October. These should be placed in a special house for winter laying. All pullets found on handling to be months out of lay can be placed in a house and be run for November laying. That is what I mean by graded flocks. A little practice soon makes one efficient in handling for the likely date of laying the first egg. If the vent is small and the abdomen empty, one can say that the bird is two months out of lay; if vent is as large as a shilling and abdomen is half-full, then in lay in a month; while if the vent is larger and abdomen full, then eggs in a week.

When pullets are about to lay one can get a good idea as to their qualities and they car be score-carded. At first they will be two fingers between the pelvis bones, with abdomen nearly full and vent large, and it is at this stage that they should be dispatched to a Test to commence production in about two weeks. Supposing the arch is wide and the pelvis bones measure two fingers between with four fingers from end of breast-bone to pelvis bones in a heavy-breed pullet, and three to four in a light-breed, you can reckon getting a further finger between the pelvis bones and also between pelvis and end of breast-bone. Laying makes a finger's difference approximately where mentioned. If the arch is narrow and the back narrow the bird may only go two fingers between pelvis bones when in lay, but you can only learn by experience in handling and I have given you the key.

The next thing is to see that the pullets are fleshed up for laying the first egg, and here one must know what laying condition really is. The short period from the setting of the bone to the production of the first egg does not allow for any waste of time or colds that will reduce the pullets in condition and see the flesh " melted " only to be built up again at the expense of a delayed start in egg-laying. A pullet has been properly built up for laying when, just as the first egg is due, the breast-bone is well fleshed and rounded, the pelvis bones are coated with flesh and soft fat, and the abdomen feels somewhat on the fat side. Only handling can reveal this condition. I am very much against fat laying stock, but make exception with pullets just coming into lay or hens that have moulted and are ready to begin production again. The breast should appear rounded as in a table chicken that is ripe for market, while the pelvis bones should be as thick, say, as the depth of one's little finger. All who fatten poultry for the table should practise handling for ripe condition. I have always been impressed by the very true statement made by Messrs. White, Tomkins and Courage, Ltd., that poultry-keepers waste thousands of tons of valuable flesh through not sending their poultry to market fully ripe. If one handles the young cockerels at breast, those that are rounded off well are ready for disposal, whereas those which are lightly fleshed need a longer time before they are marketed. One should grade table stock to this method of handling, because by so doing you will get the best prices, whereas if ungraded produce is marketed the lowest price will be paid by the salesman as based on the lowest quality bird sent. In like manner, when you are grading out birds they should be marketed when they are well fleshed. It is useless to grade out hens when they are in the moult and have lost their flesh and internal fat. Grade out while they are still ready for killing and then there is no loss. Again, when you decide to grade out any stock for table, be sure to see that they are given a course of extra fattening so that the best weights and prices are made possible. A short time in the fattening coop will add extra weight, and that will mean maximum returns. In a nutshell, then, you do not decide when a bird is fit for table by its age or by appearances, but by the way it handles at the parts where the flesh is desired, not forgetting the thighs, and you grade out all low-quality birds when they are handling ripe for table and not in seasons when they are naturally thinned of flesh and internal fat. It is surprising how well a bird for table will respond to a short course of fattening along with a few other birds in a small house and run, and with a final short spell in a fattening coop fed on the best foods.

One should be very careful to build up the pullets for laying, and it is here where so many fail and lose winter egg baskets. Supposing all the pullets are ripe for laying, and well built up in the parts mentioned, and the whole flock through neglect go down with colds, the flesh disappears and the pullets have to build up again, which means a heavy loss in eggs equal perhaps to 30 or 50 eggs per bird. It may take several months, in fact, to build them up again, and flesh their frames. Be ever so careful when the pullets are maturing to keep away colds and partial moults. I know of nothing to beat colds for knocking flocks of pullets out of tone and delaying egg-production. This is where personal management comes in, and where the careful poultry-keeper scores over the haphazard owner. When pullets are approaching maturity they should be graded out, as stated, to lay from mid-September onwards, etc., and the pullets in this category should be placed in their winter quarters the first week in September, so that they will settle down to their new houses with the last of the good weather and before the change comes in, and while they are settling down they can be built up for the commencement of egg-production. They should be graded into separate degrees of forwardness and backwardness so that the best results can be had per flock from the same feeding and management. It stands to reason that you cannot get the best egg-returns from an ungraded flock containing chicks, hens and pullets fed in the same way and housed as one flock.

Keep up the feeding and watch the birds very closely. When the time comes round, the laying houses should be got ready and any necessary alterations made, so that before the pullets are placed in the houses everything is ready, with litter already down, so that the pullets will not be disturbed once they are settling down. Too many get the pullets into their quarters in a hurry and then frighten them by throwing in the litter, hammering in nails, or attending to alterations.

Most careful supervision is necessary when the pullets are being settled in their winter houses. At first the nesting-boxes should be boarded up for two reasons: (i.) to prevent the pullets roosting in them at night, and (ii.) to prevent too early laying or encouraging it. When the pullets are placed in the houses in daylight they should be gently dropped inside the sheds so that they have to find their way out into the run through the pop-holes, and then they will usually return, which will save much chasing after them at night and driving them in. See that every step is taken to keep colds away from the flock. Many prefer to move pullets after dark, using baskets as carriers, and placing the birds straight on the perches. If one is compelled to drive any birds into their houses for the night, use tact and quiet methods and secure the help of others, seeing all birds safely on perches. At night after dark, visit the house and place any birds on the perches that are not already thereon; do this every evening until they learn what is expected of them. See that no pullets spend the night on the drop-board, or in the corner under the droppings-board, or in the nests. Also see that no birds are perching on any stays of the house inside or out. If you neglect the pullets in this direction then they will sweat, and chills and colds will start, or partial moulting set in. All the care in hatching and rearing will then be wasted, while the winter egg harvest will be ruined and many losses of birds may follow. When a pullet dies, remember that the loss must be deducted from any of the returns to be recorded.

I prefer to have the perches about a foot or more above the drop-board, especially in large laying houses, because this discourages the pullets from sleeping on the boards. If the perches are near the latter, the pullets slip off them to get the warmth of the other pullets that are on the perches, whereas with a good space below they keep on the perches better and are not anxious to sleep in the cold. Sweating and chills will, in a week, see the pullets reduced to skin and bone, and the building-up process that must follow will mean waiting for eggs for months.

All houses, too, should be made draught-proof before the pullets go in, because it is the draughts that cause colds and roup. Common colds are very dangerous, yet I find them taken very lightly by poultry-keepers. Roup is the result in the majority of cases, and the man who allows his birds to get colds just as they are maturing and reaching the laying stage

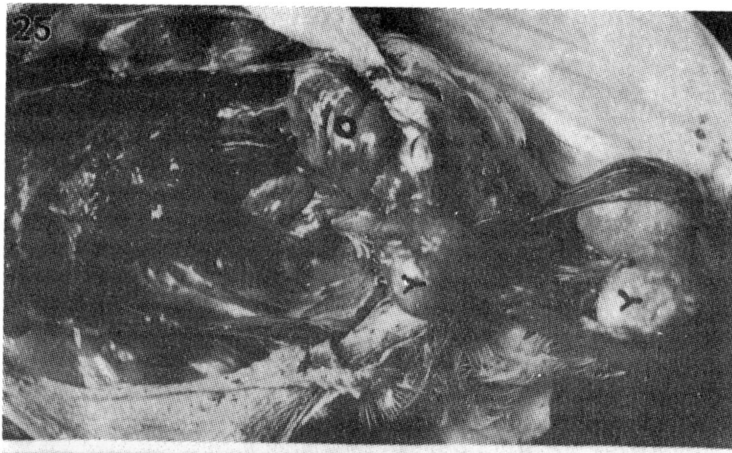
25

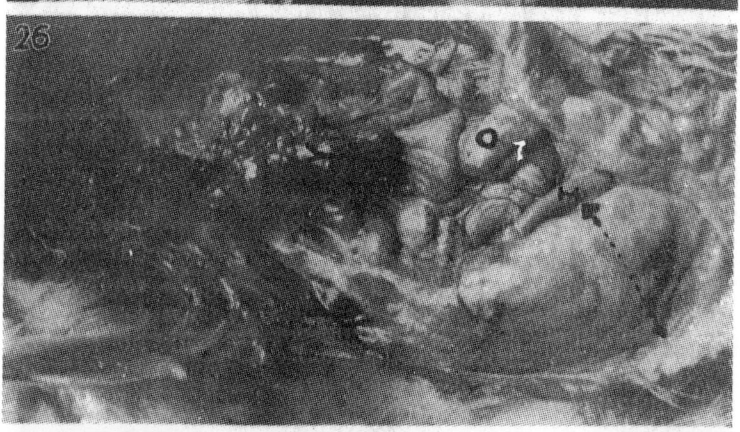
26

27

will have plenty of trouble and heavy losses. So do not treat colds lightly, and remember that *prevention is better than a cure every time.* Prevent all draughts in the houses and let the wooden front of each house go up beyond the backs of the birds as they are roosting on the perches, so that the birds are never in a draught, and see that no single pullet can roost near a small hole in the house to get a cold in the eye and spread trouble.

When the pullets are first put into the houses see that they have a good feed in their old quarters before the shift, in case any should fight shy of feeding when in their new surroundings for the first day. And for several days the drinking-water should be disinfected, and onions included in the diet as a tonic. I know of no better tonic than raw minced onions, whether for adults or young chicks, but, while they may in excess flavour the eggs if fed to laying stock, they can be given to pullets coming into lay and hens in the moult which are not producing eggs. I would like to see onions fed by poultry-keepers in greater volume to all classes of stock, although if given in excess I have an idea they can force the pullets and make them lay early.

Next comes the commencement of laying, and if you place pot or dummy eggs in the nests see that they are removed directly a few pullets start, as these will teach the others. If pot eggs are left in the nests all the year, the chances are that eggs when laid will fall against the china ones and get broken, which may lead to the abominable vice of egg-eating. Keep an eye on egg-eating, because it undoubtedly causes more losses in average poultry-houses than many believe. Every now and then a few marked eggs should be placed in the open nests to see if they are eaten, just as a check, and especially where one is not satisfied with the laying of any pen or birds. Have a depth of sawdust or fine peat-moss at the base of each nesting-box, and put the nesting material proper on top as a precaution against hens and pullets of the fussy kind, which scrape away all the litter before they lay, and reveal the hard bottoms of the nesting-sections for the eggs to drop on and break. Egg-eating can often spread from the eating of the first broken or soft egg. Have all nests well off the ground and not in too light a place.

When pullets begin to lay keep an eye on any which are sitting on the floor of the house, and put them into the nests, which will get them used to the boxes. Again, when raking over the litter, look out for any eggs therein so that they can be removed and not be left there to get broken, and, maybe, to start egg-eating. Also remember that while pullets need to be built up for laying, you can overbuild them so that they are really fat, the pelvis bones being as thick as one's thumb and the abdomen being blown out with internal fat. Such over-condition is dangerous, because the pullets may lay soft eggs and will often continue to produce such eggs without a shelled one, thus being ruined. I never like to have birds with very short breast-bones in a laying house in case they start laying soft eggs and egg-eating sets in.

While the pullets are settling down, be with them as much as possible and get them used to those who will have to look after them. Let them feed from the bucket, and handle a few that come near; do not merely throw down the grain and depart. Also let all birds have a dose of Glauber's salts once weekly from the start for a month—say a teaspoonful to every nine adults—in the hot water over the mash. This will help egg-laying and digestion and should be continued now and then throughout the season, while the drinking-water can, as a general rule, be disinfected once weekly to advantage to freshen up the system and to keep down gases. Liberal feeding is imperative during all the time of settling down, and one must look out for increased appetites, which commence directly egg-laying falls due. Increase the rations just as the birds eat heartily and let them have all the mash they will eat eagerly each evening. I would here impress on readers the importance of educating growing pullets to the foods and system of feeding to be practised when they are matured and in the laying sheds. Often I have had pullets which would not eat wet mash because they had been put on grain only—no doubt to keep them back from laying in the closing stages. To lay well, the pullets must enjoy all they eat, and to get fleshed they must eat plenty as maturity is reached, so bring them up the way they are to go. If they are to have wet mash, see that they get used to it, the same applying to dry mash. Pullets will refuse dry mash if they have not been educated to it, and where pullets are sent to the Laying Tests, obtain details as to how they will be fed and bring them up to the same foods and systems.

When you buy pullets at maturity be sure and ask the owner how they have been fed, so that they can be kept on the same system and foods until settled down into their new environment, when they can be gradually placed on any new methods the purchaser may have in mind. Breeders who sell stock may, in these days, well bring the pullets up to both wet and dry mash, so that they will take more readily to the system of feeding likely to be adopted by the purchaser at any time. Some owners adopt both methods and swear by the combination.

Before the winter season sets in, the poultry-keeper should be ready with an isolation place, so that any ailing bird can be isolated directly it ails. No matter whether six or six hundred fowls are kept, *an isolation hospital is very important* as a safeguard against heavy losses among the stock. It is asking for trouble and taking risks to leave a bird with even a common cold in a flock of maturing pullets just because no isolation place is handy. Plan for this imperative accommodation before trouble starts; map out the schedules ahead of losses. Isolate any invalid at once, and *always be ready to kill one bird to save many.*

Having built up the pullets for laying, the eggs to be produced can draw from the extra flesh and fat in the system and not on the latter itself. The fat will gradually decrease as laying continues through the winter until the abdomen will be pliable, together with the pelvis bones, while the breastbone will be prominently felt in February after 80 eggs or so have been produced. One grades that way in February for the likely winter producers. If the pullets are thin when laying starts, then egg-production will suck the system, and the birds will have to stop and build themselves up again. They can easily go into a moult as a result of loss of condition, or may get colds, etc. Those pullets among the early broods which come into production in late summer may be plump when commencing egg-laying, but about September, when they have been laying heavily, they will become thinned and therefore fall a prey to colds and the partial moult.

If I am picking pullets for Laying Tests, I do not study pedigrees but just the individual birds I examine, and I always handle all pullets in order to select every " plum " that is on the place. Where the breeder just breeds from those which lay the most eggs, and backs the high individual producer, I find my best stock among his or her general pens; but where the man breeds from mediums, and aims at flock-average, I can find the best birds from his special pens. Personally I prefer the pullets from the general pens sired by the best cockerel and out of nice medium dams. If I want cockerels, I go up to the " extra special " matings, where very high records reign supreme. Final selection for Tests should be left till the last moment, so that one can pick the desired number of pullets timed just ready to lay a few days after arrival at the competitions. I never send pullets in lay if I can help it, as that is too risky seeing that should they stop production and rest or moult partially I lose thirty to fifty eggs per bird, and one cannot catch up, especially where values decide the placings. Many have luck and send pullets in lay which do not stop, but such is the exception to the rule, and

I would rather have strict rules than otherwise, especially as I have given you my methods of telling roughly when a pullet will lay her first egg. I allow for the pullets to arrive at the Test, pass the examiners for health, and get into their quarters to settle down and then to think of laying.

As far as possible, I prefer to advise competitors to send pullets which will be breeders after they have done the laying. I see no ultimate value in dispatching high fliers which, when they return, will let the strain down from the breeding viewpoint, and never be good breeding hens in their second season. Often a breeder will send on pedigree pullets that are too fine or too well bred—all capability and no capacity, and handling like a "rabbit"—thinking that the Test will be won by birds which each lay 280 eggs. Some of these pullets go under because they are too fine, but one or two on the border-line win through and lay 260 or so eggs—mainly small perhaps. Then the owner sympathises with himself at having got one or two champion official sprinters to compensate him for his entry fee and loss of the competition. Some do lay well solely because what they lack in bone is put in by feeding, the manager adopting clever feeding methods. In the hands of ordinary poultry-keepers they would be bad producers, always being out of sorts owing to superfineness, therefore of undesirable type.

I wish all to bear in mind that they are picking a pen and not one or two odd layers; importance must be attached tc flock-average. Every bird must lay up to sample to help, and not drag down, each of the others. Having picked one gem go all out to pick every other like her; then you will find that a pen of good layers will win and come home and prove decent breeders. Quality or breeding or pedigree is easy to tell by handling, seeing that it is the sheep and the rabbit over again. If I take hold of a pullet on the point of lay and she feels like a silver-grey rabbit, then she is well bred from a high dam, and herself has texture or laying capability very high. If the next handles as if her feathers remind me that I am taking hold of sheep's wool, then I have a low pedigree and inferior laying qualities. *A waxy or "rabbit" feel about a bird will always denote high laying ability or capability*, but you must not let that deceive you, because you find it also on the super-bred pullets which have flat chests, cut-away abdomens, little or no capacity generally or stamina, denoting that you have gone a stage too far in pedigree or quality. It is a merit point indeed when you can find the "rabbit" feel about a pullet which is nice in frame and high up in capacity, because she will lay well and then be a sound breeding hen on her return.

Supposing I am out to select five pullets for a Test, I study very closely the date the competition starts, the method of scoring, and the date the birds are to be dispatched. Laying Tests do not study you, and some seem mainly interested in taking your entry fee. It is common for you to be told when the pullets are just ready to be dispatched that the date has been postponed and you must send in the birds a week later. During that week the selected pullets start to lay. Again, you may never know till the last moment whether you have been balloted in or out. Having made your entry, you just pick out the pullets and then may find that you are in, or may hear that you have been balloted out, only to be notified later by wire that a pen has been rejected and that you are reserve, so can send a pen per return; other pens have long since settled down. As a rule, Tests go by ballot, and you send your fee and entry form which you obtain on application from the competition concerned when you see the schedule advertised in the poultry papers as ready. You may enter many Tests and many breeds and have about £40 out in entry fees without knowing for ages whether you are balloted in or out. Most Tests ballot for positions, and if you are not lucky you get your fee returned, which is often the only signal you have been unsuccessful in the ballot. Then when the Test starts, you see that Mrs. Brown has got a pen accepted in four different breeds at the same Test and you have been balloted out in them all. Some of my students have been out of the lot very often except for an odd pen. Often a winner one year is balloted out the next.*

Tests should be run on quite different lines and the sooner this happens the better. All Tests should be amalgamated and should adopt the same system of scoring. And when one breeder has got a pen accepted he should not be balloted in for another until all the rejected names have been accommodated in other breeds. In some Tests there is a champion section open to those who have won so many medals in competitions with the same breed. I always admire the sporting spirit of Miss N. H. Bell, who enters only the champion section and wins there. I consider that there should be eliminating trials; those who are entering for the first time having a preliminary section; those who have never been placed but have entered several times going into a middle section; and all who have been placed and won a certificate or medal into a championship section. Winners in the latter then claim a supreme championship strain. Another plan would be to award prizes solely on the performances of the

* *Since writing this, the N.U.P.S. competition officials have decided to pass in previous winners.*—AUTHOR.

pen, and not according to how they finish. For instance, I send a pen of pullets which finish second, but at the end of the Test they are all in lay while the top pen, which has beaten mine by four eggs, are all out of lay when sent home. There should be a system of scoring points for each bird which goes home in full lay and a deduction for each which is out of lay. Again,* breed points are not counted, and you can send Buff Rock-Rhode Island Red crosses in either pure Rhode Island Red or Buff Rock sections—a thing that was done very extensively some years ago. You can send White Leghorn-White Wyandotte crosses as pures, or Black Leghorn-Anconas as pures, to either of the two breeds. All typical of the breed should have merit points awarded when they arrive, and should again be judged when they are returned. Then awards of red, white and blue ribbons would be worth having, and would be fair awards, doing away with the lottery side of Tests, where, if one bird dies, all is over for another year, and the public, missing your name in the winning list, declare you have lost your utility strain, and rumour blames a serious outbreak of roup or a building up again of the strain for stamina or size of egg. So many red, white and blue ribbons would entitle you to a supreme championship mauve ribbon of the Tests. If our laying trials were approached in a broad-minded way, we could get some good from them and fair tests of strain instead of the lottery of to-day. Some day we shall see my ideals accomplished, and Tests will call for something more than five or six specially chosen pullets—plums of the flock.

Supposing we have the details of the many competitions we have entered before us, we will study very closely the rules. Every Test is somewhat different, and, while some have five pullets to the pen, another may have six, and a third may have seven, six birds taking part with the seventh as a reserve in case of death. In an American Test you enter as many reserves as you like, and if a bird dies another takes its place and scores on from where the other left off; several birds can thus make up the 250 eggs in a year credited to one pullet instead of the three or more according to deaths and the number of reserves brought forward to take their places. As a rule, the reserve bird does not count unless another dies. In some Tests the reserve does not count while the competition is running, but directly it is over the best six scores of the seven pullets count for the final placings, the reserve bird's total coming in if it is higher than another pullet's among the six. In one Test, when a bird dies her record to date

* *For the first time, a pen has this autumn been returned from a competition as untypical of the breed. This has come to my notice since writing the above.*—AUTHOR.

dies out with her, and many refuse to have reserve pullets, saying they prefer to have a trial for stamina as well as egg-laying, although one year at a Test a fox took a few competing pullets from the broody-coops, which was rather against the stamina idea. Any five pullets will lay less than six pullets, just as a team of eight footballers cannot beat a full eleven when every unit is a good one. Competition is out of the question to be in any way fair, and reserves should count and be allowed in all trials. Credit here to Harper Adams.

One should begin by having several baskets handy, each properly labelled so that every pullet likely to suit a given Test can be placed in the baskets reserved for that competition. Each bird can be rung with a special coloured ring, say blue for Northern, red for Bentley, and so on. Having taken out of the flock those pullets which are found to suit the requirements of the individual Test, they can again be run through and the best picked out until there are many more left in than the number required. I should make the first selection at four months or so old, and do the general grading of the pullets throughout the farm at the same time. All pullets would be driven into the catching-box and be handled carefully until all which we desired were placed in the baskets.

The pullets aimed for will be those which, *when in full lay*, will measure five fingers between pelvis bones and end of breast-bone in heavy breeds and four fingers (to five) in light breeds, with a three-finger space between the pelvis bones. That is our ideal in the pullet at a later date when she is doing her duty at the trials. She will also have a very wide back, full abdomen, bold front, wide pelvis arch, great width between legs at all times when standing or walking, neat bone, shanks and beak, and just that frame to help size of egg and stamina and to make her a breeding hen in the second year. In addition *she must have as high capability as we can get.* She will be faultless as regards health, construction, etc.

We will reject any pullets which, on handling, reveal signs that denote a likelihood of laying too early. Many pullets may, for instance, on handling, be found already to have filling abdomens and increasing vents for size, and these will not be placed in the baskets. They will be rung especially as likely to lay before mid-September, and can be removed later to their special laying sheds which have been reserved for all pullets laying before mid-September. Being rung with special rings, such pullets can readily be collected up for removal, but careful records of all such markings should be taken in note-books and never on loose sheets of paper. They should also be written down in ink or indelible pencil.

Supposing we are handling pullets which are set in bone, with pelvis bones and breast-bone rigid, but with vents very small and abdomens empty. We can regard such pullets as six to eight weeks, approximately, out of lay. All we have to remember is that laying is responsible for the breast-bone widening out one finger and dropping, so that a pullet nearly laying and handling four fingers at abdomen will go five fingers in lay. In like manner the pelvis bones widen out, and a two-finger space between the pelvis bones when near lay will be likely to mean three fingers when in production in the majority of cases.

It will be well, then, to select pullets in heavy breeds which, when set in bone, will measure about three fingers or more at abdomen and nearly three in light breeds. At the same time one will concentrate on the right type of breast-bone, namely, straight and supporting to the abdomen, because the breast-bone will have to drop or swing down when laying starts. Often a pullet with small abdominal measurements increases, upon laying, beyond expectation, but such are exceptions which do not concern hand-graders. We have a clear-cut course as to what we want and do not depart therefrom, consistency in handling and balancing the birds being very desirable. When being graded early, the end of the breast-bone will be almost level with the pelvis bone end, *i.e.*, almost in a direct line below.

Next we want a good type of pelvis arch and will see that the pelvis bones are straight and not too coarse in the bone, and that there is ample space between the two. One may be approaching two fingers between, or one-and-a-half to two. But one cannot advise in cold type what is the ideal, because experience in handling alone tells, and you must handle plenty of pullets and keep detailed notes at different stages of growth. Then one will become efficient, as naturally the likely date of laying the first egg will decide whether a bird is handling well for measurements. I am after a wide pelvis arch (which will be present where there is ample space between the pelvis bones) because with it, when checked up, go width of back, space between legs, and later a capacious abdomen. You generally obtain a narrow back, narrow width between legs, and small abdomen where, in the finished article, the pelvis arch is narrow, *i.e.*, when in full lay.

Next, one will bear in mind the amount of bone in the bird, and this will be decided by the beak, the shank, and the stand of the bird. We will grade out those birds with long, thin beaks, and shanks like pins and needles, also those with very coarse (long and deep) beaks and shanks, aiming at the ideal pullet with a short and deep beak and nice medium shanks. The scales will be tight and small, giving a rounded appearance to the shank, while the sinews at back of shank will be soft. *We will always select pullets which handle plump, which are of the heart shape* when they come into the hand, and which are compact. Capability will be judged by the bold and kindly eye, placed high in the skull, so that each eye is prominent from back and front of bird. The pupil will be large and the iris bright (red preferred in most breeds, or a bright yellow). There will be a look of contentment about the eye. The handling of the feather at thighs and back will denote silk or wax like the rabbit and not the wool of the sheep. The feather will be sparse and tight. About the pelvis bones there will be a healthy feel as if they were fleshing up, and the same applies to the breast-bone. Keep to the heart-shaped and roomy body; avoid narrow bodies for width.

Having placed in the baskets all the pullets that interest us (the others being rung and graded and put down in the run), they should be taken with the baskets into the house, the perches being removed to leave the drop-board clear. One should now put the pullets through their paces on the board so that they can be watched for carriage. Never be let into the trap with handling alone, but after handling make full use of observation and see each bird walking. When I am judging, I always have a final look at my birds which I have out for quality, and it is then that I ascertain just whether or not there are any defects. I remember at a classic Show a judge gave a pullet a prize and "best in show," and, as a test of grading, I asked several to look at her and tell me her defect, but none could get it right, so I gave one of them my judging-stick and asked him to stir her up, when she turned one of her toes under and round to walk. The judge must have missed the serious defect solely because he went by the bird when in handling. I have found an extra toe on a bird before now, many birds defective in eyes and almost blind, others with extra toes under the thighs, and one with two breast-bones.

Taking a pullet out from the basket, let her walk along the drop-board. See that she has a well-developed front and *walks on her pads and not on her toes*, as with cut-away front and abdomen. Let her be well balanced and not too tall. If she is too tall she will be knock-kneed, showing too great a length and excess of bone, and, with age, that will increase. See that she stands wide between the legs. Also see that she walks with straight and not tilted back near parson's nose, which will throw her forward and make her unbalanced. This is a deformity I have often noticed missed by a judge who has placed his cards solely by handling. Balance or symmetry is very imperative with utility stock of all ages and both sexes. Test also for temperaments when the pullets are being finally examined on the board. Take note of those which talk to you and are friendly, and omit any which refuse to be handled and struggle violently, all the time refusing to settle. Also reject any which ruffle their neck-hackles in fright, or show fright in the eye, or are high-strung. Pass out, too, those which are bad-tempered and which peck at you. I remember lecturing at Street Court, Herefordshire, in the open, and asked for a good layer to be fetched from the run, and immediately lectured about her, taking *docility* as my text. Then I asked the manageress to fetch me a bird which ruffled her feathers when caught, and directly this bird came along and I attempted to take her, she immediately gave my coat several hard pecks and ruffled her hackle. Do not mistake this with the male bird that shows fight when being handled, as that is breeding power and desirable in the best males. The " funk " I have in mind will, when you pretend to hit it, dive its head and face into the socket of your arm out of the way and will raise its neck-hackle and squeal, especially when lifted up by the wings at the junction near the shoulders.

The selected pullets that grade in will next be taken to a small house and run, where they will be conditioned for dispatch to the various Tests when the time comes around. I prefer early isolation, because one can manage them in an ideal and special way, and if they threaten to develop too quickly they can have another shift of quarters. One should place in the house more pullets than are required, so that any which come on to lay can immediately be withdrawn. One must prevent them from laying, and no nests should be provided, while a keen lookout must be kept for any eggs in the litter. Don't be like one breeder I know who asked me why his pullets started so badly at a Test. I suggested at the time that probably they had laid before dispatch, but he assured me that such was not the case. Several months after, he quietly admitted that he had just found twenty eggs under the poultry-house occupied by the birds. One must watch the pullets carefully to make sure that they do not lay astray, and one might handle the selected pullets often to see how they are progressing at abdomen, on the lines mentioned, to ascertain the development of the ovary. The sending away of pullets in a bad season as the Tests fall due naturally tends to frighten them a little and may make them sweat, so that the change is quite sufficient to put them off lay and into a partial moult. The latter is detrimental to a win, seeing that

Tests are won through a good start when the pistol goes off and a splendid finish. Send pullets therefore that are not in lay, and always the friendly ones. During the time that they are being prepared for the Tests they should be handled often to make them docile, while they can be placed in a trap-nest and allowed to walk out again after the slide has dropped. This will get them used to the routine that will be meted out to them when they get to the trials.

I am very careful, too, to bring the maturing pullets up accustomed to the system of feeding that will be practised at the selected Test. By eating well from the time of arrival, the keen appetites will enable them to fight against colds which might follow the change of quarters. One can always ascertain the methods of feeding that the selected Test will adopt, and, as a rule, each uses a different system. The best plan is to bring the pullets up to both wet and dry mash, and then they will do well on either method of feeding.

The time will come when the pullets have to be finally selected and dispatched to the trials. The final overhauling of the pullets must be very carefully executed. You will begin to think that there is plenty to do when an entry is made in a laying competition, and I can assure you there is, especially if you wish to win or do well. *You will only win by luck if you rely upon pedigree, and just send pullets as you take them unselected from the run or house.* I am always well up in the many Tests at home and abroad with my own selections, but for this I have to thank my knowledge of handling and grading. I ask you to bear in mind the pullets you should send. Always pick them to the minute so that they will begin to lay a week or so after arrival on the trial ground. *In Wyandottes and heavy breeds I have, in my finally-selected birds, pullets which measure four fingers from end of breast-bone to pelvis bones with two fingers between the pelvis bones, the vent being large and moist and the abdomen half full. In Leghorns and such light breeds I am satisfied with pullets that handle three fingers at abdomen with the same conditions as mentioned above; if the birds go a little over the three fingers then I support this good point.*

It is compulsory to have pullets with straight breast-bones, and the latter in full support of the abdomen. You must, in handling, know that when in full lay the pullet will handle as you expect her to, and here practice makes perfect. Pay particular care with each pullet to see that everyone acts as a separate unit, and does well in the laying line. If you hurry matters you will only include a pullet not up to sample which will let you down. It is the level production of the pen that wins the Test.

CHAPTER XVII.

DISPATCHING THE PULLETS TO THE LAYING TESTS.

THE actual dispatch of the selected pullets to the Laying Test always reminds me of the fancier who gives his birds a final touch up before the judge goes round the pens to make the awards. I am always very particular regarding the dispatch part of the "lottery." One must be careful not to let a pullet in with a deformity of any kind. Next see that each pullet handles plump and well fleshed at abdomen, pelvis bones, and on both sides of breast-bone. In other words, each pullet will handle in the best of health, which is essential, because if the pullets are well built up and in ripe condition they will readily fight successfully any colds, should the same come along during the settling-down period at the Test. Otherwise, should the birds handle out of condition on arrival, colds would soon affect them and their laying suffer in consequence.

Once one masters the handling of pullets on the point of lay and gets fixed the healthy weight of well-conditioned pullets, it comes natural to the grader to say if a pullet is handling fit or if she lacks tone. If you hand me a bird that weighs like a cart-horse, then she is too heavy; if she weighs like a Bantam, then she is out of tone. When pullets handle light, as if they were but feather and bone, something serious is wrong with them. A cold may be coming along or they may have had a cold that the unobservant owner has missed. To send pullets that handle light to a Test is serious also from the view-point that they will have to waste three months without production while they build themselves up. Do you think that pullets can win a Test under such a drawback? *Yet when handling pullets which have arrived from all parts of the country for a laying competition, I always have to comment upon the very large number of birds that arrive completely out of condition and which I know will take months to get fit.* Such is not fair to the pen or the owner, the breed or the Test manager. I have often hinted that such and such pullets would not lay for four months, and once, when handling the pullets that had arrived for a Test in October, I said that a certain pullet would not lay her first egg till the following March, and it was April before she gave her first product.

Again, I have always had to complain of the odd types of pullets sent, a competitor so often sending a tip-top pen with the exception of a "dud" or a deformed bird. My notes on pens upon arrival are very interesting. Remember my advice and send a level lot.

With the pullets sent in good plump condition, the chances are that lice will make their presence at abdomen felt. They always do on fat or thin birds. With pullets ready to lay and fat, I adopt the general rule of plucking the fluffy feathers from the lip of vent and also from the parts under the vent, which in the ordinary way form the nesting-ground for lice. You will find the parts I mention if you examine a fat hen, namely, where the lice eggs are. I also give the pullets a thorough dusting with Izal powder before dispatch. With pullets plump and well fleshed, lice will atack them at abdomen, and yet they will all disappear when the birds get pliable at abdomen and in good condition through laying. With pullets fat and attacked with parasites, I find that the irritation has a tendency to make them lay soft eggs and to go wrong in the laying organs. As a safeguard, then, the plucking of the fluff in the parts mentioned will be helpful, and I always give pullets in the laying sheds a good dusting with Izal, also hens that are coming into lay after the moult. *In fact dust all poultry when they happen to be plump or overfat, male birds included.* Birds being dispatched to clients can be dusted to advantage and should be overhauled for deformities, etc., in the same way as pullets being sent to Tests or kept for laying at home or trap-nesting.

Send to the laying competitions pullets which will be good breeding hens upon their return. The pullets are returned from the Tests at the end of the year, and the owner is supplied with the official record of each bird. True, one has to wait ages for such data from some trials, but I am hoping that soon Tests will study a little more those who pay the piper. The competitor pays the stipulated fee and the Test management keeps the eggs, but fees are returned if the owner is balloted out of the trials. I mention this because so many seem to think that they get the eggs laid by their birds, while others think they lose their competing birds. If they drop a line to some Tests to enquire procedure, then they are kept waiting ages. I am aware of all the drawbacks and handicaps which face the average Test, and some day, with Government help, all may be put right. But I would like to see competitors treated better by some competitions. Only to-day, when judging at a Show, a small man told me that he would never enter again, as he could not get to know where he had finished at a certain Test, although the birds had arrived home. Some of my students still await their medal and prizes won at a Test ages ago.

Now for a few words with would-be competitors. Please help the Test manager by sending only healthy birds. Often a manager is a poultry-breeder with his own flock of birds handy, and any outbreak of roup in the Test might spread to his own flocks and mean ruin to him. This applies to many who run small competitions for societies, and their number is increasing each year. Again, it is not fair to other competitors to send pullets which are unhealthy, for colds, etc., to spread to all and sundry. See that colds are kept away from the selected pullets at all costs, and keep them away from the other pullets on your farm by the special precautions and careful management already advised.

Apart from seeing that the birds handle plump and healthy in weight, examine the nostrils to see that they are perfectly dry and not watery. See also that the mouth is clean and the breath sweet. Never send a pullet with running nostrils, watery eyes, or with cheesy matter at back of throat or under tongue. Every year examination of the birds on arrival will get more stringent, so that you will be well advised to take the precautions advised. At some Tests replacement of rejected birds is not allowed, while at others only once, and providing you can guarantee to send birds from a flock that is healthy, and with which the other birds were not running. I have often been to a farm to select the pullets for the Tests only to find colds rampant, while I have on many occasions located diphtheritic roup in an odd bird or two, the owner being very surprised.

The main thing is to keep the pullets clear of colds during the finishing-off period when they are on their own. It is useless to find things are wrong at the last moment. Often birds with blocked nostrils will be found, telling me that they have colds; black or dark nostrils and light handling will tell me that they have had colds. The treatment is to remove immediately the matter from the nostrils so that the birds can breathe properly without using their mouths only. Ducks do not have roup, and I presume it is because they always have their nostrils dipped in water, thus keeping the passages clear. If pullets are rejected for health, do not run to a " vet " to get a statement that the birds are healthy and fit. Do not write to the Press and complain. I remember a Light Sussex pullet which was rejected from a single-pullet Test which was taken to about a dozen notabilities, also to a " vet," and all declared that the manager had made a serious mistake. The pullet then came into my hands and I satisfied the owner that she was bronchial, yet the others had missed it. Do not think it is clever to return to the Test the same pullets that have been rejected, because I have known this done with resultant colds and trouble. I have traced this personally. Be sportsmanlike and look at things from the manager's point of view, bearing in mind that the inclusion of just one "wrong" bird may mean a serious outbreak in the flock, especially if the Test is a large-flock one with all birds in one laying house.

When the time comes for dispatching the pullets, remember that the birds can easily get colds on the journey if badly packed. From my experience in examining pullets which arrive for competition, I can say that too many use open crates or baskets; it is this that is responsible for the majority of rejections. Do not use open crates, boxes or hampers wherein the birds can get in a draught at a railway station. Do not use closed-in boxes which may cause the birds to sweat and chill. On the other hand, use proper hampers which will keep the birds cool and yet protected from all draughts. Send them off by passenger train by the shortest route, and if you happen to be near the trial grounds, take them over personally and save a railway journey. If you will take all the precautions mentioned, you will have a fair chance of seeing the pullets accepted, and can hope to be well up in the final list of results. And I hope you will win. I suggest, too, that you take your loss or win in a sporting way, knowing that Tests have a lottery side. You may have bad luck this year and good luck next. I often pick several lots of pullets in the same section, and frequently get two pens placed in the first three. I tell each student that I hope he will win, but I know that two cannot take first position, and you should always understand the same. And if your pullets fail to start production the first month or so, do not blame the Test management and ask for the return of the birds because your own at home are laying well. Do not compare home production with that at any Test—the conditions are vastly different. The very fact of sending maturing pullets on a journey to a Test means a risk in upsetting them.

CHAPTER XVIII.

GRADING OUT DEFORMITIES.

JUDGING so many Utility Shows as I do, also grading so many flocks, I can honestly say I am appalled at the deformities met with. When I have to withhold prizes because of the number of birds in a class that are deformed, the matter is urgent. My judging-books are very instructive and interesting on this point; frequently I find more deformed birds in a class than healthy ones. But one need not be surprised, seeing that in the past the system of breeding has too often been just the mating up of the layers of the greatest number of eggs. Such birds have been deformed in some way, and close breeding and the mating up for increased egg-returns have seen the weakness worked properly into a strain. Remember that all the bad points go along with the good ones when mating. Remember, too, that every bird should be minutely overhauled before it is placed in a breeding pen. My system of hand-grading is intended to teach breeders to be observant, and deformed birds will not be tolerated before long; in that direction handling can help to breed better poultry. To-day, the stock shown under me are a vast improvement to those a few seasons back.

When the pullets are being overhauled for dispatch to the laying competition, examine the breast of each bird to see that there is no tumour or blister present, caused by roosting on sharp or narrow boards or perches and the friction set up. Next take each thigh to see that the bone and the flesh are normal. Many a pullet dies the first month of the Test because the final overhauling has been rushed. Sometimes one thigh will appear withered up, or the bone may be enlarged. Next examine the pads of the feet for bumble foot, growths or corns. Flick the hand towards each eye to ascertain that perfect vision is present in each. Throw out all pullets which have masculine features; in like manner discard all males with effeminate ways and appearances; take no risks with these.

Pass the hand down the back to make certain that it is as flat as a billiard table. Rounded backs are too common in utility stock of both sexes and denote weaknesses. A rounded back will often cause the tail to swing over to one side and become "wry." Or it may result in a "squirrel" tail, *i.e.*, carried upright.

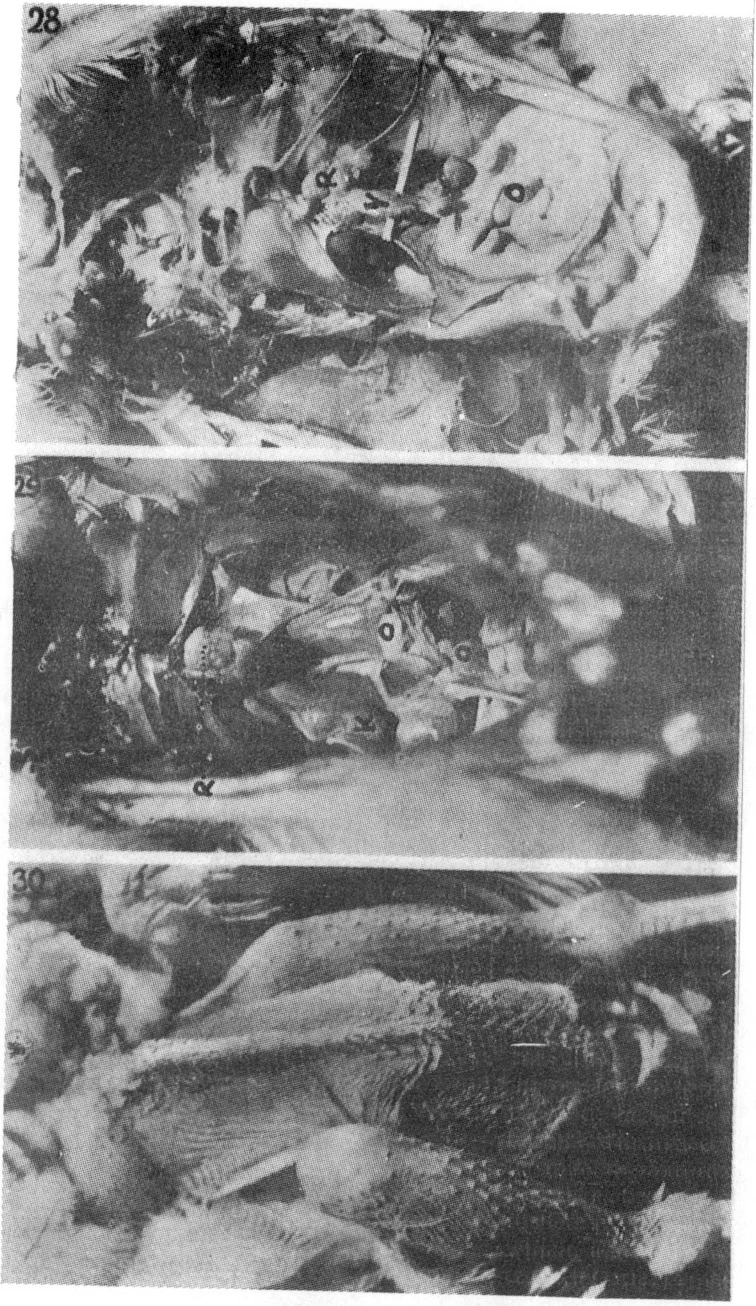
28
30

Often the back appears straight and yet the tail is carried weakly, and I appeal for the use of those birds which have good strong and natural tail-carriage. In the past it has been the considered idea of most persons that high tail-carriage meant more eggs, but often it is due to weak backs. By all means let us have well-carried tails and at a high angle, but not squirrel-fashioned or wry.

To test for wry tails the bird should be placed in a training pen and be watched. When a bird is alert and on the move, he or she can control the muscles, and the tail will be carried erect. When resting, with the muscles relaxed, the tail, if weak, will flop to one side and remain there. The way I test for this weakness, when it is suspected, is by holding the bird in the hands, as in Plate 49. By jumping the bird up a little, she will move her muscles and then relax them. When relaxed, see that the tail remains straight out, as in the Plate named, and does not settle down to one side. I have often had a male with sickles and coverts all growing towards the one side, yet the deformity of the bird has been missed solely because owners are so seldom looking for defects. If you become automatic in mating, then you do not use your powers of observation; no wonder deformities get established in a strain. The owner will often dismiss my suggestion that his bird is wry-tailed because, when disturbed and made to walk in the show-pen or in the run, it carries its tail well. The weakness should, as mentioned, be looked for when the bird is resting with muscles relaxed.

With squirrel-tail deformities the tail is carried upright into the air, while the sickle-feathers often bend round and nearly touch the back of the male's neck. In the male, select the bird which has a graceful carriage of tail, as in Plates 45 and 59. In the former we have a White Wyandotte male bred by Mr. Leslie Williams and successfully exhibited in utility classes. It depicts quite a good pattern for breed points as well as tail-carriage. The bird is deep in body, well curved in most parts, and with rising "bushy" tail. There is a complete absence of long, streaming sickle-feathers to tail such as we find with Leghorn-Wyandotte crosses, which are also accompanied with white in lobe, much black in plumage, and long, shallow bodies, not forgetting length of leg. Here we see a high tail-carriage and yet one with a graceful sweep.

The Australorp cockerel in Plate 59 has, like the White Wyandotte, also won under me, including First at Leicester Club Show as a cockerel, and First at Shrewsbury Club Show as a cock. Note the high tail-carriage that permits of a sweep in the tail. The sickle-feathers leave their base and finish with a half circle in my ideal. In the bird with squirrel tail, the sickles often point straight up into the sky, just bend round at the top or even make the full half circle, but towards the neck of the bird instead of out at a decent angle.

In the female, while aiming at high tail-carriage, I detest the tail which goes straight up to the sky in proper squirrel fashion. In Plate 58 we have a well-carried tail and yet one which is fairly high. *Aim at a nice sweep of the tail in the hens and pullets bred from, and avoid the upright right-angle tail, where the muscles seem so contracted that the tail and parson's nose stick up like a paint-brush.* One has but to handle thousands of birds to find how deformities of the back go with wry and squirrel tails. Often when there is a weakness here, the feathers at the base of the parson's nose along the back become worn, and any wearing at the side of the bird prompts one to look for a tilted back towards that side. Examine the tail feathers of each bird, whether male or female, to see that each feather follows in sequence, because if they are growing at different angles a deformity can be suspected. A bird with a deformed back is unbalanced and only handled with difficulty.

The danger of a rounded back towards the parson's nose is manifested in the progeny, seeing that ninety per cent. of the sons will follow the deformed sire and a small percentage of the pullets. I was grading at Wrexham with the object of picking out a few birds to lecture upon, and found nearly every male with a rounded or deformed back, and several were cripples in the pullets that were sisters from the same brood. The eggs were bought at a good price from a breeder, and told me that the sire used by him was deformed. Again, I recently graded a flock of Rhode Island Red cockerels and nearly every bird had to be discarded for deformities in back, the same fault being located in the sire used in that particular pen. Another recent case concerned White Leghorn cockerels, almost every one of which was deformed, as was the sire, although he cost a high figure. A correspondent wrote in answer to one of my articles that the percentage was higher than the eighty to ninety I had claimed, as he had had a hundred per cent. deformed in the males and was able to locate a rounded back and wry tail in the sire after I had explained what to look for. Why, then, allow any deformity to get a hold on the strain just for a little care in handling for deformities? I have before now had to pass aside a whole flock of White Wyandottes because of deformities, and know of a strain of Light Sussex which I could nominate from the progeny by the deformities. Again, I once turned down a Rhode Island Red male for a bad back, but the owner left it in his leading pen, the progeny answering to deformities whether out of eggs and chicks sold or from the breeder himself. I know it has been an expensive lesson to him, yet I warned him when handling his males at the head of the pens during the mating-up time. I always have a strict rule, namely, *when in doubt leave out*.

Deformities soon come with close breeding, and where in exhibition there has been much close breeding for certain points like colour, I know I shall find plenty of deformities. In like manner, breeding for heavy egg-laying tends with close breeding and near matings to release in the progeny deformities of all kinds. The systems of breeding adopted by many in order to breed layers are various, but some have no practical use whatever. While one should keep to near relationship as regards a strain, one must always see that every bird used is vigorous to the letter. To get a 285-egg hen and to use her sons in every mating for several years is just asking for weaknesses in the progeny. In other words, one fails by attempting to over use one high layer; yet so many have slipped at this hurdle. Heavy laying is a strain on the system, and you must see that every hen and male used in every breeding pen answers to vigour, health and perfect development. In the past, a hen has laid extra well but has been deformed in some way. Her progeny have been used to maintain and increase egg-production, and at the same time deformities have been carried along as well.

Some have an idea that high production is obtained only by using the blood of a high producer, and there is a general employment for years of this individual bird's sons and daughters, grandsons and granddaughters. The breeder often thinks that he must never introduce any fresh blood or his strain will be ruined. Breeding is a weak link indeed, but common-sense rules will win out. Always be building and maintaining, and at regular intervals bring in a few fresh-blooded females, the best of which mate to your own males. I think the best plan is to rely upon your own males as far as possible, but to bring in the fresh blood with females of a different strain. Utility breeding would be greatly facilitated if each breeder would specialise and keep one breed and variety. Then he would be able to work on many strains, and thus keep up stamina by having a wider field for selections. *To-day, most seem to think that one is not a breeder unless forty varieties are kept*, which prevents one from making full use of matings and available room, or of recording in a proper detailed way. Work with but one variety can be done very thoroughly and in all directions. If it comes to advertising, one can put all the money into large announcements which the readers can see, instead of into several smaller ones.

If you start with several breeds, at your earliest decide which is to be specialised in, preferably one that is in greatest demand with you.

Here I have not the space to go fully into breeding, but I wish to touch lightly upon certain matters of importance. In utility breeding there is safety in having many distinct lines and strains, so that vigour can be kept up. If you attempt, as many do, to establish a thousand head of breeding stock in a given number of years from one pen of eight hens and a male, proudly to boast that you have never bought from an outside source a bit of fresh blood, then you will surely court trouble. I think that, while there will be matings of one's own strain, there should be plenty of half own and half another's, and three-quarters own and a quarter another's. Supposing I have a cockerel from one of my own strain high-producing hens, I would include matings of her sons to hens boasting of half of my own blood and half of another breeder's. The progeny would then carry three-quarters my own and one-quarter the new blood. It is quite simple to build in this way, seeing that we buy in a few females from a reliable strain and mate them to one of the best males of own breeding. The resulting progeny will be half and half blood, and to them we will mate own males again to get the three-quarters own blood. I always prefer to introduce fresh blood on the female's side and not on the male's. I prefer to have own males as far as possible for the main matings, so that they will carry a long line behind them, to my certain knowledge, in high laying, vigour and freedom from deformities. If you buy in males, the progeny should be tested before you go further. One should remember, too, that *buying in from a client is a sound policy so long as you are sure the quality of the stock has been kept up*. If you sell Mr. Jones a cockerel of your own breeding and he mates him to hens of proved quality of another breeder's strain, and you like the progeny bred from the mating, then you can buy in a few progeny-pullets to be mated to your own full-blooded males. These progeny-pullets available from your reliable client will be half and half blooded. Breeding on common-sense lines should be studied very closely by every breeder, and, above all, I appeal for sound, common-sense matings with vigour assured in every single bird used in the pens, whether male or female. Never let in a weakness or a deformity, not even if the hen has laid 290 eggs in twelve months.

Every bird, then, will be handled to see that the back is flat and does not round off to the tail, as tested by passing the finger down the trunk. In the case of a rounded back, the trunk will also be rounded to the pelvis bones, thereby causing the latter to be much lower than they should be. Consequently one can have what should be a four-finger bird which only measures two fingers from the end of breast-bone to pelvis bones, because the rounded trunk knocks off so much space. I constantly handle birds wherein the trunk is so rounded that the pelvis bones are in a straight line with the breast-bone, itself meaning no depth of body. Cut all these deformities out by careful grading, not only of pullets sent to Tests but of males and females earmarked for breeding as well as for sale. Do not introduce any deformities into a client's yard.

Another deformity to look for is the back that is higher on one side than the other, or which rises badly from neck to tail. Also see that there is no raised hump on any part of the back. I would warn hand-graders that deformities usually set with the bone, so that one should also grade at maturity to make sure all is well. Breeding stock should have a final examination before mating is completed and eggs incubated. I am always keen on watching each bird run and walk, as deformities can then be located very readily. After handling a bird I make my notes and then let it run, even giving it a send-off by chasing it. If the bird is sound it will be well balanced in running, and symmetry will be depicted to a marked degree. The male or female will just run off on its pads with graceful action. If it runs in a stilted way, or with tail badly carried, I cancel my grading and then have the bird caught again and passed out. Sometimes young males run like camels, with the back much higher at parson's nose than at shoulders, and these I send to table very quickly.

Pullets sent to a Test will also be handled at vent to make sure that there is no ill-smelling growth inside, as in cloacitis and vent disorders. I remember when examining the birds sent to a competition, I was always behind, as I gave every bird a thorough overhauling. But I was rewarded by finding two cases of vent growths. One cannot be too careful with vent diseases, and when handling birds from the trap-nests, the operator should examine the vents now and then to see that nothing is amiss. Always examine at vent any bird which, when handled, seems to offer an ill smell. If birds suffering from vent diseases are mated up, the male will carry the disease, and he should be examined before mating for any disorder. Just now I am in touch with a breeder who has every duck down with vent trouble. Often when a bird is wet at abdominal fluff, isolate her and bathe the parts, thus making sure that there is no vent trouble to be spread to the flock.

It only needs a slight tear or scratch at vent or in passage to set up serious disorders, bacteria getting in. Every pullet, when laying its first egg, runs the risk of a tear and vent disorders, as do pullets that are constipated, or through sharp grit passing down the intestine. Watch all pullets very closely when they begin to lay. Stretch the vent wide (first finger above and second finger below), so that the interior of the passage is revealed. Often it starts by a cheesy growth in vent, which is the exudation from the walls of the channel. Such should be carefully removed and dressed with antiseptic ointment. If neglected, a fluid may commence to show itself and cloacitis follow in a virulent form, spreading rapidly through the flock by means of the droppings or the male bird.

Often canker of the vent is the sign noticed, appearing as ill-smelling white crusts around vent. Attend to this very promptly by anointing with any good antiseptic ointment, or serious trouble will result. Wear gloves when attending to such birds, and, apart from examining birds at vent now and then, always examine any bird appearing mopish or wet at abdomen. Blood on every pullet's first egg denotes the natural rupture.

My score-card is the only one calling for examination of the vent, and yet I get birds with such disorders at nearly every Show. Male birds affected seem to waste away and to have wet abdominal fluff. Treat this problem as a very serious—and yet all too common—one.

Often diarrhœa is set up by the abdomen being dragged in long, wet grass. When mating, therefore, see that you discard birds of either sex which are too short in the leg; it is unwise to breed from dumpy birds or to send them to Tests for the reason stated. Any looseness may readily end in the laying of soft eggs through the irritation set up, or in ruining the bird's egg-organs. Dropped abdomens and excessive thigh and abdominal fluff will have the same effect when coming into contact with long, wet grass. Apart from diarrhœa, chill on the liver and other disorders may accompany such undesired types. Wet abdominal fluff may often mean in hens undermating, and in males overmating, due to exciting the organs.

Place the ear to the back of each pullet to make sure that there is no wheezing present, or bronchial disorder, when the air will be heard passing into the lungs as if through a bicycle pump. Again, apply this test to every male at mating-up time, because if you should use a bronchial cockerel or cock, he may fail you just as fertile eggs are wanted. Take no risks in this direction. Should any males develop rattling or bronchial disorders, grade them out from retention as breeders no matter what their breeding and pedigree may be. But it is for you to take care of the males so that such troubles do not follow bad housing. The same applies to "hearty" birds which, when handled, go mauve in the face, denoting heart troubles. They will be noticed to breathe through the mouth as if exhausted, and may be suffering from fatty hearts. This is common to heavy birds or those which have been allowed to get overfat when resting. Consequently the utility male is the active bird of compact weight which will last many years. One must not ruin good males by fattening them or allowing them to lose condition. Often when judging at Shows, I have to put "hearty" males back into the pens very carefully in case they should have a fit and die. If the comb is black, liver disorder may be suspected; if it is yellow, then jaundice on the liver. Observation will come to those who hand-grade their stocks.

One should keep an eye open for a badly-carried head as likely to denote brain trouble. A bird will often walk away and continue to turn its head round to look at you. Peculiar carriage of head would see a bird graded out, whether for Tests, breeding, or sale. A peck on the head may be the cause during the growing stages, and I am always against having food-troughs or dry-mash hoppers too low, which is likely to cause the blood to run up to the brain. Defective sight will also make a bird run or walk with peculiar head-carriage, because the bird always endeavours to be looking back with the eye out of which it can see best. Often at a Show when judging, I locate birds with defective eyes because they insist upon standing in one way when gently persuaded to move round by the judging-stick. I may add that at most Shows I have several blind birds shown under me, showing that hand-grading has not been practised by the owners. I have had shown under me, just before the breeding season, males almost blind in both eyes, so that they would have been mated up in due course had I not written the owners. I suggest that every male be examined at eyes to see that vision is of the best.

I always grade out any birds with odd eyes, and aim at having an eye that one can read. I think the eye tells stamina —or the absence of it. An eye which is dull never attracts me, especially in the male. Keep the pupil large and defined, as therein lies stamina. I often find males with scarcely a pupil at all, and certainly never hesitate to grade out such birds with pupils like pin-pricks. Large pupils should be the order, together with a bright iris.

Next examine the nostrils to see that the comb does not grow over and close the openings, causing the bird to breathe through the mouth. Test this with males and females.

Another form of defective tail-carriage is on the lines of the rumpless type. Here, there seems to be little or no parson's nose, and this is found by handling. In another case, the tail grows down through a weak and deformed parson's nose. I often come across birds where the tail drops and acts as a door by closing the vent, the droppings being unable to escape, and the owner never guessing the actual cause. In ordinary cases, one would cut the tail and fluff to help matters and just keep the bird at home for laying. When a bird carries her tail down, do not think ovarian disorders *must* be the cause. Test for deformity.

Then there are deformities of shanks, where the ankle is turned or the thigh put out, so that one leg appears as if it had been broken at some time. Look over the Test pullets for such and for crooked or broken toes. To test for deformed toes or ankles, support the bird under the left arm, with its shanks straight out at back. Place one thumb on each hock at the back, and one finger on each ankle underneath, and compel muscles to tighten. Curved or deformed toes or ankles will soon be located. You cannot bend the toes straight.

Coming to the beak, see that it closes properly and that one section is not longer than the other, causing the bird inconvenience in eating. Send to Tests pullets with perfectly-formed beaks. I often meet with birds so deformed in beaks that they cannot pick up grain and so must live on wet mash. Yet these birds are often found in the trap-nesting section, causing a waste of time, labour and expense.

Finally we come to crooked breast-bones, which will raise a controversial matter. At Shows I have always just penalised such birds, and not disqualified them unless they have been very bad. I know that most of the best laying hens have crooked breast-bones, and even if they are sound as pullets the defect can arise the second season. I rather put it down to utility birds having to be medium in bone and to heavy laying at a time when the birds are a little underfed, causing the bone to go. The males suffer in the same direction through being bred rather fine in the bone. At any rate, while we breed heavy layers, we shall always be with this defect, and my advice is rather to control it as much as possible. I have mated up males with crooked breast-bones in order to see if such were carried on in all cases to the progeny, but have found the latter quite sound in breast-bone. On the other hand, I have found it handed on when badly in the strain and accompanied with other defects, such as rounded backs and wry or squirrel tails.

We should divide crooked breast-bones into grades. Personally I never hesitate to grade out any bird with crooked breast-bone if it is accompanied with the least deformity in back or elsewhere. Secondly, I grade out any birds with very bad "crooks." Often a bird will have a breast-bone which cuts deep into the heart or allows the abdomen to sag, and such should be looked upon as undesirable. Dented breast-bones or those slightly crooked, if unaccompanied by any other defect, would not be a disqualification from the breeding pens. At the same time, when you have plenty of good birds without this defect, use them, and when you use a defective bird let it have a slight failing only which is not likely to be a serious deformity and to get into the strain. There is safety in a long breast-bone that supports the abdomen as already mentioned, seeing that it is an insurance policy should the bone become crooked; if the latter happens, the abdomen will still be supported instead of sagging down. Aim always at the ideal and most perfect bird when mating, and remember that there is a next best with faults; let the latter, however, be few and not too serious. Sometimes a male is exhibited under me with a breast-bone like a figure eight, and I never hesitate to grade out such a bird. Often a crooked breast-bone will seriously affect hens in lay through the bone going right into the heart or internal organs, altering the latter's course. As a result, the gizzard is often pressed down into the abdomen. Always handle each pullet sent to a Test to see that the gizzard is in the right position, and it is wise not to send any birds with crooked breast-bones in case they should become worse. One should discard any bird wherein one feels at abdomen an enlarged or scrofulous liver. It feels to the touch like loose grain.

Sometimes breast-bones are turned aside by perching, and in the perfect bird one should aim at having the breast-bone centrally carried. When handling a defective pullet or male at abdomen, the end of the breast-bone will appear to be out of place where it touches the hand; that is when measuring the abdominal space from pelvis to end of breast-bone. I should not send such pullets to the Test, and should not breed from them if they have any other deformity. Aim at symmetry in all parts, whether it be back, breast, breast-bone, tail, or shanks. In short, get the well-balanced bird every time.

Handling a bird at abdomen will often give the operator the idea of looking for deformities. For instance, when the hand is placed between end of breast-bone and pelvis bones, as in Plate 50, the hand-grader might have made up his mind that the bird being handled should, on its size, etc., measure five fingers, but actual test shows but three fingers. Such will suggest to him that the back may be deformed or rounded, thereby cutting away capacity at abdomen. Again, during the same measurement, the end of breast-bone may be a little to the right or left of the centre part of hand where a normally-carried keel should touch. The keel should be touching the hand at a point midway between the pelvis bones. Often, too, one will detect a tilted back by the badly-carried pelvic arch, one side being higher than the other, and tossing the tail, parson's nose, and pelvic arch to one side. In that case, one pelvis bone will be higher than the other.

As a final check up, one will send pullets in Leghorns, Anconas, etc., with combs quite clear of eye as they fall. The comb should have a nice base, to throw the whole well out and clear of the eye. If the comb is but a trifle clear, and should it be on the coarse or heavy side, it may affect the eye directly the bird is out of tone and the comb limp and drooping.

CHAPTER XIX.

SELECTING THE MALE BIRDS.

THE selection of the male bird is not always given the attention it deserves. Often one uses a male just because it happens to be handy or the last left from a batch of chickens. I suggest that one should be looking for likely " merit " males during the growing stages, and the progressive breeder of utility poultry will go all out to master my selection methods as applied to males. Then he will be able to send to table early all that are not likely to make good stock birds, and he can devote all his attention to those which are promising.

The cockerel which so many seem to like is the perky finished article which, maturing in a very short time, is early ready for mating and fully furnished. But in size he is what he handles—a mere Bantam—not worthy of a place in a roost for sleeping, and certainly no stock-getter. One might mate one hen to such a Bantam cockerel, but he would be lost with a full pen of eight to twelve good-framed hens.

When a pullet lays her first egg, she may be said to be fully furnished and to have finished her growth. Poultry-breeders seem unable to tell when a cockerel is finished, so I will deal with that part of my handling system. Imagine a cockerel with saddle hackles and sickles up, and he is finished; the saddle hackles falling over the back near the tail will be level at their bases, and the two main sickle feathers will have finished their swing in the half circle. I constantly have young cockerels shown under me with sickles and saddle hackles up, and when I tell the owner that I have passed the bird for being too small, I am reminded that the little bird is only five or six months old. Dismiss age of the males and go by handling and observation. Imagine a very coarse male too heavy for utility purposes, and watch the bird grow. He will run up on leg in growing bone, and will be very angular. At the end, he will grow feather and tail and sickles, and finally become furnished. But the heavier the bird, the longer will he take to mature. With the Bantam, everything is complete together, because of the absence of bone, and he is fully furnished in a few months. We should aim at the medium bird which matures early and yet has made that size which we consider ideal—neither a giant nor a Bantam, but a compact and active bird of medium size.

The abdominal capacity of the male will not naturally equal that of the female, although I often have males which have female measurements and yet handle compact and are of fine texture. It is well to bear in mind that *we desire high texture —or my capability part of card—in all males*, and, while one uses the card as it stands, one needs to read the scores a little different to those for hens and pullets. It has become general to say that a bird handles well when capacity alone is meant, instead of calling capacity size of frame and referring to my capability side as a token of good or bad handling. After all, that is the main part of handling in utility where number of eggs is concerned. Thus we should say of a pullet which is down in capacity and high in capability that she handles well (for the latter) but is too small. And of another that is well up in capacity but low in capability, that it handles badly and is too big.

I have had many males which have measured at abdomen like hens, and naturally I have seen them used as stock birds, and they have bred some splendid stock, being, of course, high up in capability. But bear in mind that *capacity is not everything*, just as it is not in the females, because we must have capability high, and more so, in the male bird. In the ordinary way, for males I like to aim at about a four-finger distance from end of breast-bone to pelvis in heavy breeds, like Wyandottes, Rhode Island Reds, etc., and about a two-finger space between the pelvis bones. In light-breed males, such as Leghorns and Anconas, I aim at an ideal of three fingers between the end of breast-bone and pelvis, and approaching two fingers between the pelvis bones. Remember that my methods of handling hold good with males, and I want the breast-bone straight and in support of abdomen as my desired type. If one uses excessively short breast-boned birds, one will naturally breed such a type.

At this stage, let me explain that it is easy to get capacity measurements on the largest males, but I have never said that capacity sets my standard for utility poultry. True, the majority have always gone by capacity, but my card stands for my teachings and methods. In fact, not till I hit home capability did hand-graders understand the proper way to select utility birds. I was the first to impress upon all the value of capability and what it stood for, and you will see that on my card capability has always dominated capacity in the number of marks allotted, namely, with size, 80 to 70. At one time, capacity was considered the real thing by most, and then they followed my methods and learned the value of capability. But of late, many have gone the other way and are neglecting capacity in order to get high capability. Keep to my card and you will not go very far wrong, and do not have any pet fads that are not based on practical handling, not of six months' experience but of year after year and constant handling, borne out by trap-nested records.

When getting a bird to measure well on the lines I have advocated, you will always keep capability high and over the fifty of the seventy marks. Hold that capability and you will not have got too large a male. When you have exceeded capacity at a point, and incidentally gone above the ideal weights, you will have gone down in capability, getting coarseness with the extra frame. Well, I have shown exactly what I mean by capability, and how vitally important it is, so that you will bear it in mind every time you are grading stock. Abnormal abdominal capacity that exhausts the bird is useless to either hen or male. In the first place, when you exceed capacity at abdomen and get a coarse, excessively big bird, you lose breeding power as regards fertility, which also follows short breast-bones. Recently I was informed by a breeder of Columbian Wyandottes at Moore Show that he had a male which had won under most utility judges but not under me. The bird measured six fingers and was short in breast-bone, which accounted for the extra measurements. Fertility was always bad from him, and in the end he died suddenly from abdominal dropsy. Utility pullets and hens that are neatly "packed" at abdomen, with the latter in nice control, are the ones that lay well and for many years (longevity), and which last as breeders, giving little trouble from infertility. Again, the male with controlled abdomen is the breeder and stock-getter. When, as I often see in utility classes, hens win because they have abdomens like footballs, I think what would happen to them were they scored up on my capability side and challenged to get over fifty of the seventy marks which stand for egg-laying to numbers. Excessive *abdominal* capacity is not an ideal, but *general* capacity is, so read my card closely for what I mean by *general* capacity, and do not listen to critics who refer loosely to capacity and yet mean only abdominal capacity.

Do not listen to the critics, too, who tell you that capacity is nothing at all in a male bird; in fact, just put into practice all you are told about handling and keep strictly to what you prove to be the case. Capacity is naturally of real importance in males, and I will explain why. In the first place, you always have a reminder when you have gone beyond capacity by the decreased capability. But capacity will keep you away from the Bantam, which is just as bad a stamp to utility as the coarse cart-horse type that is all capacity and no capability. You have to bear in mind that capacity runs in the full series as set out on my score-card. If, for instance, the male has a narrow width of pelvic arch (it would measure little between the two bones), he will measure small at abdomen and between the legs and also across the back at shoulders. Hence we want to aim at a wide pelvic arch, four fingers between the legs (wide stand), and a good heart-shaped body, together with a bold frontal or chest. Then we shall keep clear of a Bantam, and at the same time we will aim at having capability or texture very high. You can generally get high texture on a Bantam, but there you have super texture and no capacity, and you get a bird too fine in beak, shank and bone, flat in chest, and a weed really from the view-point of breeding. Aim, therefore, at a male bird that is compact and plump and yet handles fine for texture. You will follow my previous advice regarding other items, as both texture and capacity points are taken in the same way as with females.

One should not overlook the fact that fertility stands before even pedigree, because you must have the right male to fill the eggs and to start the latter off with strong germs. Then will follow rearable chicks. I place great importance in breeding males upon the width and type of the back. For instance, I want a wide neck for the lungs to keep clear of bronchial troubles, etc., and the wide part of back to be right up to the wing sockets, as already illustrated. A wide back is an essential, and here, when you have got this, you will see that the bird stands wide between the legs. You may take Plates 45 and 59 as models. The White Wyandotte male stands wide, has symmetry in a marked degree, nice tail-carriage, yet just the right depth of beak and thickness of shanks, together with the fine texture of tight feathering, texture of face, bold eye, and neat headpiece. The same can be said of the Australorp male. Have the bold eye, large pupil, and just that length of leg which keeps the shanks carried straight down like an Indian Game cock. If the male has too great a length of shank, he will stand knock-kneed, will be inactive, and will grade out. We can test the utility male readily enough when on the run, because he should be graceful, easy in movement, and an aristocrat. There should be nothing angular about him, and certainly a male with too long shanks would not answer the test. I remember lecturing at Bulwell on the right type of male to use, and suggested that if you put two cockerels into a run and chased them, the better one for our purposes would be the bird caught last. The faster you run and the faster our ideal goes, stepping out like a trotting horse. The undesirable one would jump with both feet after going a distance, and would easily be caught. A breeder then followed my advice accidentally. He had ordered a White

Leghorn male from Mr. R. Rodwell and, opening the basket, out went the male and down Bulwell streets, chased by several who happened to be handy; the faster they ran the more he stepped it out.

I also want a good-tempered cockerel which talks when taken in hand and when feeding with the hens. As you take him in hand he has plenty to say and can be handled readily. Often he will crow when in the hand and will be quite friendly, yet a little "ginger." If you handled his hens in the breeding season, he would soon have a lot more to say to you, and would get into action quickly. I am not interested in the male which arches his neck-hackle and pushes his head out of the way when being handled, as if afraid his last moment were up. I like a male to be masculine. The good-tempered, talkative males are usually high in capability, and you just have to check them up for capacity to make sure they are not too small. I am often told that training makes the male docile and friendly, but you will never make a "windy" male answer my ideal. I remember handling some cockerels of a very young age for a breeder, and she said that handling would make them all docile. So by observation I picked up two youngsters and asked her to fondle them and improve their tempers, but I knew she would not succeed because I had picked two which arched their neck-hackles in fright and looked the part in their eyes, showing a wild appearance. Docility is quality!

As regards grading the males, always pick the best for breed points, so long as every bird first answers the type you require. I never want you to go for breed characters alone, but just to breed utility with beauty. Take the question of breeding white-plumaged progeny and you will need to know a little about breeding for colour. If you use yellow-topped White Wyandotte or Leghorn males, you will not breed stay-white progeny. I should always select males that are bred-white, and the progeny will be white, even keeping their colour well when exposed to all weathers. Many think that white fowls will keep their purity of colour if shaded, but you must learn that it is in the breeding and mating. Use white silvery-topped males and you will have made a good step forward to breeding progeny of beauty and utility. Shading is dangerous, because you need to let growing stock run out to get bleached by the rain and wind. If you should confine young stock in white breeds, you will find that the shading will keep the sap in, and they will be yellow. A bred-white bird will keep his or her colour. The same applies to the females, and if they have yellow neck-hackles, then they will be likely to breed yellow-topped progeny. I should select females which have silvery necks and tops. It is just as easy to breed yellow Wyandottes as it is to obtain by selective matings really nice-coloured White Wyandottes.

Always work to the principle that the male influences externals. If he has feathered legs, a bad-coloured top, plain neck-hackle (as in Light Sussex), an excessive comb, or the like, he will pass the weakness on to the progeny. Always have external points strong in the male. Again, when breeding Light Sussex, it is possible by lack of selection to breed them with white necks and no hackle-striping at all. Hence they become far removed from Light Sussex. I suggest that the breed points be carefully mastered, also the way to breed them, so that they can be added to first-class utility. Selective mating is what is wanted, and an admission that beauty can go with utility. A black-legged Black Leghorn will lay as many eggs as a yellow-legged specimen, but the former can be a Minorca, a Bresse, or anything else, whereas the latter is typical for breed points. You cannot, however, breed good specimens for purity if you argue that it does not matter and persist in mating up birds far removed from typical representatives. If you do this, you should, as a stock breeder, just call them black layers. To get two more eggs, you should not mate up a Black Leghorn male to dark Ancona pullets and call the progeny Black Leghorns. Nor should you cross White Wyandottes and White Leghorns, Buff Rocks and Rhode Island Reds and call the stock pure. Don't listen to the critic who says I favour fancy types—my ideal is, always has been, and always will be, utility first, with beauty added.

Faults can readily get into a strain if allowed to pass, and it is a pity that so much black is allowed to get into White Wyandottes. It is, in a way, one result of crossing with the Leghorn, and crossing will always bring out odd colourings. You will have males that are good in breed points and others that are badly mismarked. *All I ask you to do is to carry on with the breeding of utility stock, but use the purest stock birds in order to wipe out faulty plumage.*

Here I would also like to touch upon head points. I have known many strains of coarse-headed stock resulting from the use of bad-headed males. I remember taking three years to get neat heads on a certain strain of White Wyandottes where this was a failing. If you use males with coarse, excessive combs or surly features, although in other points texture may be good, you will ask for such faults to be carried on into the strain. I can never understand why certain breeders use males in Leghorns with combs that are quite out of proportion to the size of the bird carrying the same. I remember taking four years to rid a strain of White Leghorns of beefy combs and

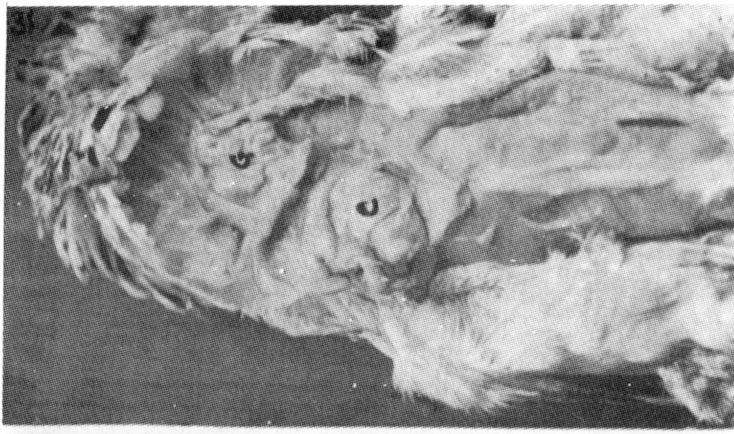
31

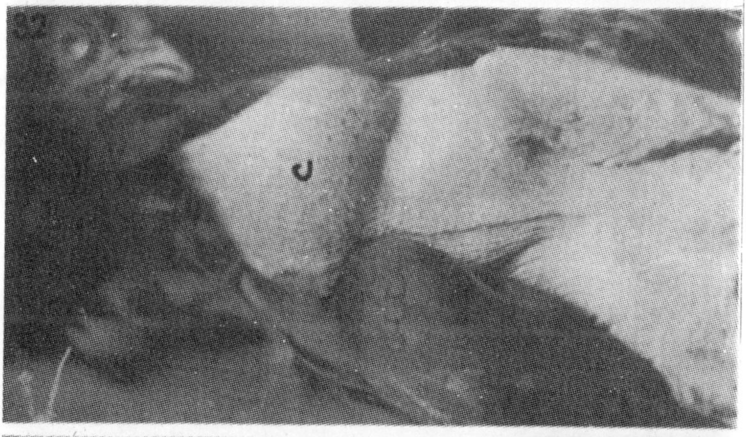
32

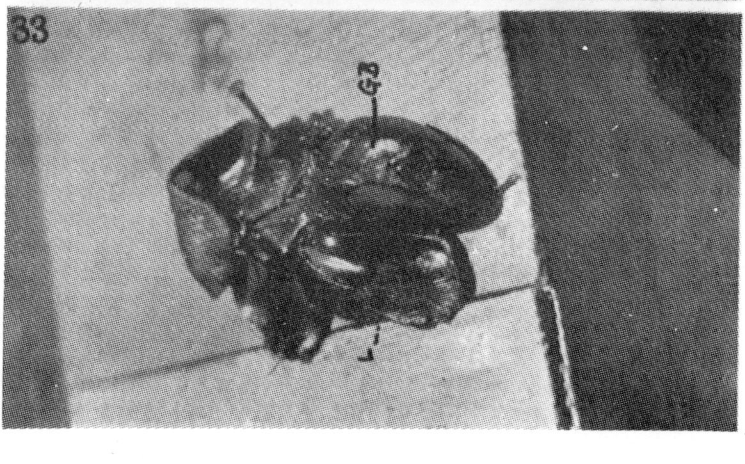
33

wattles introduced by one original male. True, he was the son of a tip-top hen, but I did not lose a single egg in getting those top-pieces down a bit by selective matings. Utility calls in Wyandottes for a neat comb that is well out of the vision, just as it does in White Leghorn females. It is against sound arguments to have a beefy comb in any breed where the comb falls over the side, as in Leghorns, Minorcas, Anconas, etc. The breeding for a certain practical or useful comb in these breeds should be the aim of all utility breeders, for the following reasons. The model head of such a hen should see the comb neatly swung over one side of the face and clear of the eye. As one looks under the comb from front and back, the eye should be clearly seen; vision should not be affected in any way. The comb can fall over into the eye for two reasons—because it is too beefy or because it is too thin and "papery." Either is objectionable. It is the base of the comb that will swing the whole nicely over and out if it is good at the root. I should therefore aim at having the comb, in both male and female, firm at the base where it joins the skull. Take no notice of the suggestion that the bigger the comb and the more eggs the birds lay, or the less they go broody—Keep to my original point that a hen has two eyes to see with every day of the year and not for the months she is moulting. Breed for such ideal combs as far as is possible. *If you have a male with a comb as thin as paper, grade him out of the stock pens.* Do the same with females. Do not have mean combs by any means, but let each be symmetrical, and let its size be governed by the body of the individual bird. A mean comb is often accompanied with a lean, long beak, denoting weakness. Go for the happy medium in all things utility. Keep an eye, too, on types of combs in the male, because any fault will be an external and will be handed down. If you use a male with a bad side-sprig on comb, or with a comb like a Redcap's or like a Hamburgh's, then I will be able to come along and pick out his sons and grandsons for you. Have as perfect males as you can get, and even if your original birds have faults, just carry on, but insist on improving things bit by bit, never being an extremist either way. I like a comb in Leghorns and such males which is approximately one-and-a-half times the depth of the face.

I have always been greatly interested in ducks, and, by setting white-shelled eggs and using drakes from same, I have got strains of white-egg layers in breeds where they usually lay green eggs. I consider the offering of green duck eggs to the public to be asking for trouble and spoiling the market for the duck eggs. As a result, by gradually working for an approved aim (it has ever been my original objective; *vide* my books), I have got what I wanted but have never sacrificed an egg in totals. I can breed layers as well as obtain an objective which I mate for. Miss Barbara Raye, one of my students, will tell you that when she started with ducks many years ago I advised her to keep to the white eggs and she has now an established white-egg strain in Buff Orpingtons, although most of the original lot laid green-shelled eggs. And, as I write, she is top at Bentley Duck Test in Buff Orpington ducks, and won the Bentley Duck Test (just concluded) with ducks bred to and picked by my system of handling.

I would ask readers to make use of training pens so that several males can be placed in the show-pens (same pattern as used at Shows) in an odd outhouse where they can be properly observed at selection time, when the final grading takes place and the pens are mated up. Pay special attention to the selection of the males; keep always in hand several stock males of tip-top quality and take care of them. The stock males will do much to help you along in the breeding and maintenance of heavy-laying strains.

When grading males, you will keep in mind your objectives, and should find little difficulty, after experience, in grading out at any age those young males which are not promising. The males which will never interest me are those which are Bantams and which, when fully furnished, have pelvis bones close together with the pelvic arch about half an inch wide. I can find you plenty of such males where the pelvis bones touch, and such birds have narrow backs and frames. I could find better measurements on a utility White Wyandotte Bantam. I am very keen on a broad back in a male as denoting strength, stamina, and fertility, and pass the hint on to you. I am sure, therefore, that you need not bother with those males which at the maturing stage have narrow backs and are half the weight they should be. Keep to the nice type of wide pelvic arch with some space between and with the pelvis bones straight and medium to fine in texture. Be no extremist.

The best males will often be found among those that are not looking at their best when others (the Bantams) of the same age and from the same brood are ready for mating and fully furnished in feather. One generally keeps the latter because they look saucy and pretty to the eye. But in hand-grading, the hand and eye are the two instruments to be used. A badly-formed pelvic arch so crooked as to resemble a cow's horns is undesirable in type, and I have always traced any failing in the male's pelvis to his sons, like any merit point. I would just add that a medium male may have to be your best until you get another, so long as you are on the lookout for improvement. *You do not discard your best male because he is not perfect.* Every bird other than a "dud" has some value if it is mated the correct way.

I like a male which is well above the heads of the hens, and yet one not leggy and coarse. When grading, you will easily locate the latter brand because they will have narrow bodies and abdomens which, when spanned, are after the triangle shape instead of the heart. In other words, they will be empty, sucked by the excessive bone, which will be the case if feather is in excess. Always keep a check on feather because it so easily gets into a strain. Avoid males that have excessive baggy fluff at abdomen, thighs, and cushion, because if such gets on to the females, then egg-laying is sapped at abdomen. Again, infertility may result. One can see at an early age which males will turn out to be "elephants" for size and they can go early to table. In young males, one will allow for a little length of shank, because when the bird settles down this will be put right. Observation and experience must be the key-notes.

The successful hand-grader and the successful breeder is the one who can "read" a male or a hen for all that. Handling in numbers—and constantly—teaches one that power or happy knack of selection. One becomes efficient in selecting males that will prove splendid stock-getters without exactly being able to describe the points of guidance. It is the practised eye as well as the hand which will make hand-grading sound and enable the reader to get the best results through clever execution of the system. Hand-grading is just as much a science as feeding, incubation, and any other section of poultry-keeping, even if my critics speak and write to the contrary.

CHAPTER XX.

SIZE OF EGG AND INFLUENCES THEREON.

SIZE of egg is a very interesting and important problem with utility poultry. One naturally wishes to ascertain the likely size of egg a pullet will lay before she has commenced production, especially where pullets have to be sent to Tests where eggs are scored by weights. As far as handling goes, I have made a very deep study of this part of selection, and give my views. When the pullets are on the point of lay and are being selected for Tests or for trap-nesting, I go by the following measurements. In the first place, the back must be wide if the yolks in the ovary are to be allowed to mature up to standard size. And the wide part of back should go right up to wing junctions, as explained. Then I lay great stress on the distance from parson's nose socket, *i.e.*, under the tail to the pelvis bones, aiming at great depth there approaching two fingers clear in the space mentioned horizontally. Then the pelvic arch must be so wide as to allow the three fingers to go between the pelvis bones when the laying stage is reached. Finally, the size of vent when stretched comes in for circumference of egg, and the three measurements go hand in hand. With them, of course, I aim at the long five-inch back. The pullets which lay small eggs mainly fail in being shallow from parson's nose socket to pelvis, narrow between pelvis bones, small in size of vent and width of back, also in width of pelvic arch. It is a matter of selection by experience, because one is unable to control or arrive at the length of the egg. Where one is defeated, however, in my above-mentioned original system, the egg is often narrow but long and not a good type of hatching egg. I rely solely upon these points when selecting for size of egg, and the fact that last year I picked dozens of pullets for one breeder, and for the six winter months had only four eggs per bird second grade out of an average of nearly a hundred for each pullet for the winter months, shows we can rely upon my methods—failing any others that are better or more accurate. One pullet I picked laid over 200 eggs in the single-pen Test at Harper Adams, and all were first grade except the first.

Appetites have a say in size of egg; the bird which plays about with her food will not be the layer of large eggs. Again, underfeeding makes for small eggs, as does the use of cheap foods or an insufficient supply of green food, drinking-water, etc. Heavy laying day after day without a break will affect size of egg. At some Tests they get large eggs from pullets while their sisters at another competition produce small products.

Here I would appeal to readers to keep up the size of egg as far as possible by breeding and rearing. A duck lays an egg weighing one-and-a-half ounces or more above standard weight, and one could breed ducks for numbers for many years and not get down to small eggs. With pullets, however, the standard is two ounces, and here the breeder has little to play with. I know that you can set two-ounce eggs for years and then get pullets producing eggs that do not scale the standard two ounces till they are second-year hens. I suggest that by setting the 2¼-oz. egg we have just that odd ¼-oz. to play with, and are more likely that way to keep up size of egg. Keep sending into the strain hens which lay 2½-oz. eggs and plenty of them, so that the size of egg can be balanced. We have but to admit that in breeding heavy layers we lose size of frame and also size of egg and we shall know the full value of the medium but profitable layer of large eggs.

Pullets are made or marred for size of egg in the rearing. I contend that the pullets of high texture tend to lay too early, and, if they begin to produce eggs before they are fully ready, the yolks are immatured, so to speak, and the pullets are ruined as regards size of egg. Keep the pullets growing so that they will be nice for frame and show a tendency to lay large eggs. But above all, breed from the medium hen with a breeding frame and she will help you with the progeny more than anything. Mate up textured males of small capacity to textured hens of small frames and you will breed Bantams quickly enough and get 1½-oz. eggs, and you might as well keep Columbian Wyandotte Bantams, as they will lay eggs as large.

Avoid late hatching at all costs; hatch early. One may do better with late-February-hatched Wyandotte or heavy-breed pullets over March ones if the owner knows how to hold them back a little and to build them up for October eggs. Local conditions alter hatching seasons, of course.

Overfat stock will lay small eggs, as will hens which are closing down the laying organs ready for a broody-rest or the moult. Hens that are laying and moulting may also produce smaller eggs. One should not take the first egg laid by the pullet as her real standard for weight, but rather wait until she has got well into her stride. Perhaps the twentieth egg or so can be taken as a rough guide, and if a pullet starts off with a 1½-oz. egg and does not increase to 2-oz. within a given time, she should be discarded. I like the pullet which starts off with a $1\frac{3}{4}$-oz. egg and reaches the 2-oz. standard product after laying well for a month or two. If a pullet gets to 2-oz. eggs quickly, that should be regarded as a merit point. Pullets laying $2\frac{1}{4}$-oz. eggs in their pullet year, a few months after commencing to lay, should be used, at all costs, as breeders if they produce 160 or upwards for the first twelve months. Here quality of egg will make up for the quantity. *In breeding, keep up the body size of your birds, at the same time seeing that they handle well for texture or capability.* Then they will lay well, and large eggs. Once lose capacity and you will lose size of egg. Some small-looking birds lay large eggs, but they handle big for frame; exceptions will always turn up. We must keep to rules in hand-grading, and not exceptions.

Where I have no trap-nested records of hens I wish to breed from, I go by the size of vent, width of pelvic arch, space between the pelvis bones, and the distance from tail socket to pelvis bones. When the vent is very large, we know the egg has the good circumference we want. These measurements should be taken with the birds in full lay. Even if some birds are trap-nested, the breeder should at the right times handle all his other pullets which have been run on for laying and untrapped. This plan I advise so that he can make use of any extra good hens which he has missed. The fact that I have no trap-nest record of a hen does not make me shun her if I am pleased with her. I remember a Light Sussex cockerel being shown under me as a chicken and I gave him "best in Show." Later, when he was grown up, I gave him a medal and a first prize. A year after, when a cock, I gave him first and "best male in Show," and those were the only times he was shown under me. As a matter of fact, the breeder did not know what pedigree he had, as he had jumped the pedigree tray when a day old (started vigorous!), and was dubious of using him, although he knew that the bird was hatched from his very best pen. Well, I don't think I would have passed over such a bird because I did not know his individual breeding. I would have soon found him the right hens and bred some stock worth having and owning. The moral is that one can go to the extreme in pedigree work and miss valuable birds. To me this is not sound. Every bird has a use! Some can go to the table, others can make nice breeders, while a few may be usable because of their large eggs, and so on. Often a hen may be a valuable stock-getter when mated to a certain cockerel because, while she herself is nothing great, the special mating may lead to the right progeny.

The size of frame will, in my mind, always be associated with size of egg. Supposing we run through a flock of layers and compare the trap-nest records, we find, as a rule, that *those hens with the largest eggs against them are the birds well up for capacity*. Often they are the heaviest birds, and what they make up in weight of egg they lose in numbers. A Wyandotte hen may lay 140 $2\frac{1}{2}$-oz. eggs and scale 6 lb. Another hen, weighing $5\frac{1}{2}$ lb., may produce 170 $2\frac{1}{4}$-oz. eggs for the twelve months, while a third may yield 250 2-oz. eggs and weigh but $4\frac{3}{4}$ lb. We will always find exceptions and discover a pullet that is small which lays large eggs. But keep to general rules and you will find that the above is accurate on the whole. When you have frame in excess you have the very large egg and fewer of them, and directly you lose frame and get a more active bird, so do numbers increase and the size of egg decrease. We can go right down to Bantams with $1\frac{1}{4}$-oz. eggs and hardly a frame at all. When you hit upon a large hen which produces a small egg, just check her up for condition, because if she is as fat as butter the egg will be smaller than standard because of the excessive internal fat.

Breeding for number of eggs, then, tends to decrease both size of egg and size of bird. That is Nature's way. Keep the frame up in your hens and use as breeders the medium producers which lay plenty of large eggs. Mate them to the most vigorous male you have, and one that is high in texture, a good crower, and a "ginger-bred." You can easily watch the loss of frame in the high-record hens, because apart from losing frame you will get some very high performers with hardly any chest on them. The next stage is the weed of super texture, with flat chest and cut-away abdomen. Use in the breeding pens those hens and males which have well-rounded and developed fronts. Keep up vigour all round. It is a fight against Nature all the while, and if we are breeding for heavy egg-laying we must put into our breeding-stock what we take out; sound breeding methods hold the necessary keys!

When a pullet has set in bone, I pay careful attention to the width of the pelvic arch by placing the thumb of the right hand on one side and the first finger on the other. I readily get the span of the arch, and if it is wide the frame of the bird will follow as good. The small-framed bird has the narrow arch. Trace the latter bird further and with the abdomen full, the vent extended, and the first egg due in a week, and you will find that with the narrow arch go narrow width of back, small abdomen, small vent, and no width between legs. Handle a very small pullet or two in lay and you will soon see my points. Now take other pullets with wide arches and you will get the other extreme, namely, the wide back, ample space between legs, full abdomen, large vent—and, more often than not, LARGE eggs. I can only give you my guiding principles, and you must handle plenty of pullets and master the items mentioned. I am very successful in picking out the layers of large eggs, but one has to weigh up a bird very carefully. My contention is that the width of pelvic arch corresponds to the width of egg; the depth between tail bone and pelvis to the depth of the egg (being the depth of shelling department); while the size of vent and the fineness of the outer ring point to circumference of the egg. The size of vent refers to the outer circle, seeing that when a hen lays the thin inner skin comes right out, forced by the egg coming through. If the outer rim shows large circumference when stretched to the full, it will allow the inner thin skin to go out to the fullest extent. Again, if the outer rim is soft and silky and fine in texture, it will again allow the inner thin skin full and easy play. If, on the other hand, the outer rim is coarse and hard, then it will interfere with the stretching out to the full of the inner skin. One should naturally note the fineness of the outer rim and vent in general when hand-grading those hens which have laid, because a fine-textured vent stands for easy and quick production, also continuity, just as the latter makes the vent fine in texture. But I have covered both size and fineness of vent on my score-card. A pullet that had been laying continuously for months would not have a dry, coarse vent.

In the male I also want the wide pelvic arch, with which go the wide back, the good abdomen, space between legs and the frame. And I note depth from tail bone to pelvis bone and space between the pelvis bones, also size and fineness of vent when stretched. By so doing, I ensure the necessary frame, the measurements being accurate when the bird has finished growth and is fully furnished with sickles up. I do not discard a male because he is not two fingers between pelvis bones and two fingers from tail bone to pelvis bone, but I want the best space I can get, always with an eye on the wide arch. If we handle a layer of small eggs which has no capacity, we can only get two fingers between the pelvis bones when in full lay and about one finger in the space from tail bone to pelvis bone. You can best learn this part of handling, as other items, by practising on hens and pullets with known records, and with weights of eggs accurately registered against each bird. In short, trap-nest birds of all sorts and sizes in order to learn the efficient execution of my score-card system. Study the low-grade layer for choice, and make full use of hens always in the broody-coops.

Size of egg varies according to the seasons, there being a drop in the summer, when the ground is dry and the grass scorched. In Laying Tests, some pens are devoid of grass, which must mean decreased size in eggs.

Where one overstocks the land with poultry, size of egg will suffer, as will any neglect in providing raw, green food, clean drinking-water, and wet mash.

Poor or even shy feeders and hens that are bullied will not perhaps lay large eggs, so that one sees how environment and special conditions affect size of egg. When pullets start to lay, they are all over the place for size of egg, it being either abnormally large or very small until the bird settles down to steady production. Too much grain feeding, also sameness of feeding, will affect the size of egg.

With Tests commencing in November (too late) and encouraging the dispatch of late-hatched pullets, small eggs run high for numbers. One has often to send pullets that handle for small eggs because no others of the right development are available. Avoid flighty pullets, and rear all stock well. When pullets fail to start in November but grow on to commence in December or January at Tests, one notices an increased size in the eggs, due to later maturity. Span each pullet carefully, placing the thumbs on the back and the hands round the body, for fingers to touch at breast-bone (bird's head towards you), and the good measuring birds tend towards good-sized eggs, with narrow bodies the reverse. I am interested only in the former type and birds which are wide across back near the parson's nose.

Some table breeds will never lay maximum full-sized eggs on feeding adopted at Tests. You cannot feed a Light Sussex or Australorp like a Leghorn.

CHAPTER XXI.

GRADING IN DIFFERENT SEASONS.

IN order to hand-grade successfully one must obtain experience in handling birds at all ages and in all seasons. Practice makes perfect, and everyone must start from the beginning and gradually improve. *You cannot learn handling in two minutes*, if you attempt to, you will be telling everybody that there is nothing in my system.

Remember always your objective, namely, the ideal bird. When a light-breed pullet such as a Leghorn is in full lay, she should be, in our ideal, four fingers from pelvis bone to end of breast-bone, three fingers between pelvis bones, two fingers from tail bone to pelvis bone. A heavy-breed pullet such as a Wyandotte will be five fingers at abdomen instead of four, and will be identical elsewhere. When standing firm, the former might be four fingers between the legs (taken high up) and the latter five, the shanks being straight to show the desired length of bone. Supposing I handle a pullet which is nearly in lay she may measure, in White Wyandottes, four fingers at abdomen and two fingers between pelvis bones, with *almost full abdomen* and large vent. I can safely allow one more finger at abdomen as well as between pelvis bones by way of extra development between then and date of first egg. In other words, the bones will stretch out this odd finger in each of the two measurements I have mentioned. Therefore, if I am score-carding a pullet on the point of lay and wish to find her likely score when in lay, I add one finger in each measurement. If you are score-carding a hen out of lay you must make allowances, and, while it is not an easy matter, you master the allowances with practice. For instance, if I have a pullet on the point of lay and her pelvic arch is narrow, the back will be narrow, also the abdomen, etc. You must study closely the bird in hand. When a male sets in bone and is fully furnished you score-card him with good accuracy, and make no allowances at abdomen as with pullets. Males will, of course, not score so high in capacity as females, although there are exceptions. I often get males which score quite as high as pullets, but they do not as a general rule. Even if they do for total points, they are behind in capacity. When score-carding a pullet not in lay, always mark the card "not in lay" and add "allowances made" if such is done.

When grading young pullets you have to plan some schedule or other. I like to keep to a system of ringing by colours, say red, white and blue, because that is simplest. I keep red rings always for those birds which are low grade. Then when I see a bird with a red ring it makes me suspicious. If a hen is constantly broody she gets a red ring (two, if you like), and she is ready, when required, for the table. At any rate, she will not be bred from. In like manner, young males with red rings go to table early, and throughout hand-grading I find the red-for-danger system of ringing best. Birds that are medium for quality are rung white, while the best have blue rings. I can always have two blue rings on extra special birds just as I can place a white and blue on a bird which is a good one but needs to be handled again later on when more forward. All birds with white rings can later on be gone through again to see whether or not the ring shall be altered up to a blue or down to a red.

The ideal you are after must be kept before you every time you handle a bird. Some judges of utility stock have no definite idea of what they are aiming at, consequently all manner of types get placed. I have one ideal, and a bird is either up to it or below it and grades out. Therefore I am able to keep consistent to myself, so to speak, and as far as it is humanly possible. Repeatedly I am told by exhibitors that they like showing under me because I am so consistent, but any such consistency I put down to knowing what I am after. I want every "handler" to go for an ideal and to keep to it, and then his selections will be strengthened.

If I am grading a flock of pullets on the farm in, say, August, I keep to the following plans. In my experience, pullets laying before mid-September show a tendency to take a partial moult and to tempt others to do the same, whether they have laid, are laying, or on the point. Consequently I take no risks. All pullets laying before mid-September, and those likely to (judged by the development at abdomen, as explained), are graded out into one flock. I then concentrate on those pullets which will commence to lay from, say, mid-September to mid-October, which will be the winter performers. The later ones timed to start from November onwards can be placed in another flock by themselves.

The pullets set apart for trap-nesting will come from the flock due to commence to lay between mid-September and mid-October. Each will be built up for laying with pelvis bones fleshed, also breast-bone and abdomen. As I grade I will have in mind the pullet with the desired measurements, say, in Leghorns three fingers at abdomen and two between pelvis bones when set and coming into lay. As I take the ideal pullet in hand at shoulders I shall note how plump she is, with heart-shaped body which betokens a "good doer." By placing the thumbs together on the back and afterwards near the parson's nose, and the fingers round the bird to touch below at breast-bone, I will get my heart-shaped body throughout. If a bird is triangle-shaped with cut-in sides, I may guess she is angular and that feather or excess of shanks or bone will be sucking away the abdomen; hence the emptiness. Everywhere I go over the bird I will feel that there is something substantial there and rounded as against angular. As I take the bird in hand she will tuck her shanks up neatly below her breast-bone to show how well balanced she is. She will hold her head erect. I can handle her easily, and even place her in any position on my open hand whether on her side or back. The angular bird I shall avoid, and apart from being triangle-shaped in body she will be difficult to handle. When in hand and gently shaken up she will be unable to remain with head up in a symmetrical or balanced fashion. Instead, her head will go down every time she is moved because her legs are too long or she lacks balance. If there is any deformity in back, that will cause the bird to be unbalanced and difficult to handle. The same will be noted if the bird is excessively short in breast-bone, too heavy and cumbersome, or defective in structure, or shallow chested.

The ideal pullet will be a healthy weight and this you can only master by experience in handling. You will drop every bird that is a cart-horse and every one that is a Bantam. The cart-horse will be difficult to handle, being far from compact or balanced. The Bantam will be as light as a feather, without that healthy, plump weight which should be present in all utility stock. For trap-nesting, select the pullet which will make a good hen when she has done the laying. Examine her very carefully to see that she has just that bone which is needed, the beak being short and deep, and the shanks medium, with sinews soft. Let her have a breeding hen's shank, and study the breed, avoiding a Leghorn leg on a Light Sussex and vice versa. Let there be a nice type of straight breast-bone in support of the abdomen, with a good type of arch and pelvis bones, the latter being straight and, in bone, pliable. Let the chest be well developed and prominent, with width between legs, back wide and of the right type at neck, together with a sound walk on pad and not on toes. I pay strict attention to the walk and action of each bird because one detects therefrom symmetry or balance or the lack of it. Let there be good depth at front from shoulders to breast-bone; avoid the cut-away and shallow front.

When grading, I like to have every bird rung with a numbered ring, and as the assistant drives a batch into the catching-box and closes the pop-hole to prevent escape back into the house, another helper stands handy with a note-book. Taking a pullet in my hands I carefully go over her minutely and may decide that her grading is high and warrants a blue ring. This latter is placed on the shank and she is allowed to depart in the run or enclosure. I gently drive her while I take note of her walk or action. If she glides off like a race-horse, with body well balanced and a run on her pads, she is allowed to pass with the grading. If, on the other hand, she walks on her toes and lacks balance or has a side-walk with back high at tail and tilted, or if she shows any deformity, I call back the grading and ring again when the bird has been caught. I am careful, too, before deciding on the right ring, to hold my ear to her back to test the lungs, also to note temperament and heart action. If she breathes peculiarly or is puffed with a little handling she may be back-marked. If she goes blue or purple in the face she may be suspected of having heart or lung trouble; if black or yellow, liver trouble. I always examine the vent when stretched for any cheesy growth in the channel to indicate cloacitis or like disorder. There are many diseases of the vent which are little known solely because there have been too few of us studying closely the "Hen from Within"; but my example may set others at work, and between us we should distribute the helpful information. When examining the pullets on arrival at a Laying Test recently I found a pullet in two pens with cloacitis in its early stage, and the pens were rejected, the owners being advised to kill the birds. Thus I was rewarded for my careful handling at vent of every pullet sent in. Had it gone unnoticed whole flocks might have suffered later. Vent disease may show up in many ways, but one should be on one's guard at all times. Often a bird will have a fluid discharge from the vent with an offensive smell, and such should be noted promptly. Isolate any such bird directly any discharge is located, and, if no change, kill the bird, especially if a male. When mated, such diseases can be readily carried on, and even in ducks. I am just watching a whole flock of ducks which are affected with cloacitis or vent gleet. Often the discharge may be just diarrhœa, but never hesitate to be on the safe side by immediate isolation and observation, as it can spread, if real vent disorder, not only from mating but through the droppings.

Often cloacitis starts with growths in the channel like those in diphtheritic roup, this representing exudation from the walls. I find quite a number of birds with vent disorders at

Shows, and when grading; you will observe that the vent is examined on my card, yet is not by others. I have always included that item on my card, and the fact that I find so many ailing birds shows the unsound plan of those who do not carry out that part of my handling. When handling the birds at the Harper Adams College Conference I passed one bird with serious vent gleet which the other judges neglected to notice. That such a bird could be penned publicly to take part in a test of hand-grading shows that there is a sad need for handling and observation. Imagine a flock going down to the disease at home through the original culprit being unnoticed. Until we have thrashed out the many disorders fowls are subject to, handle carefully at vent and do not hesitate to kill and burn any birds showing signs of growths and discharge when accompanied with an ill smell. The latter often resembles that of strong cheese and that in diphtheritic roup. One may reasonably wonder if there is such an ailment as roup of the vent. When a male is affected he often has the continuous discharge and may waste away in time, losing all flesh, etc. Some may suspect tuberculosis, but be on the safe side and isolate when noted; never be afraid to kill when it means saving the rest.

During the growing stages I bear in mind the same ideal on a reduced scale, namely, the bird with the heart-shaped body, plump, and a "good doer," with balanced walk and body, type in pelvis and breast-bone, deep, short beak and medium bone, wide back and width between legs—together with high capability. I like the symmetrical, bold-fronted and sturdy youngster which seems to own the whole place.

Before I put the coloured rings on, I handle for texture as regards tight silky feather, bone, abdominal flesh, etc., also headpiece. Capability must be high, together with the desired breeding hen's frame. *I like a bird which lays large eggs and plenty of them, looks nice all the time she is laying them, and can be bred from when she has done the laying; a bird, too, I can show any of my friends and tell them she is a Black Leghorn without being contradicted and told she is a Bresse or a crossbred.* THEREIN LIES MY IDEAL OF UTILITY AND BEAUTY!

When I am grading in August or early September and have both eyes on breeder-layers for trap-nesting as well as for future breeders, I will be meeting with plenty of pullets which handle like "rabbits" for texture but lack all frame. For the most part, they will be fine in the bone, with pins-and-needle legs and narrow bodied throughout. I place a red and blue ring on such birds, keeping them together for laying and packing them with food, putting into them by way of food what they lack in stamina. I call them "little gems," which means eggs *ad lib.* but no breeding. I shall meet plenty of cockerels of the same brand, mere Bantams for size with wonderful texture, and I give them short notice to quit. I want you to keep the frame up as well as having all the texture or capability found in these "little gems." Enlarge the edition but hold the high texture. You will, as mentioned, always have a natural check on frame, because when you exceed the ideal you will get the low capability and the "sheep" touch in place of the "rabbit."

When a pullet starts to lay in October she will, as stated, be fully fleshed and built up. She will have a rounded breast as in table condition, the abdomen will be nicely coated with fat, while the pelvis bones will also be nicely coated with soft, meltable fat. The shanks will be full of fat and swollen out to the full. The pigment in the shanks of yellow-legged birds will be noticed. The selected pullets (breeder-layers) are placed in a separate house, and, assuming that the owner is not going to trap-nest them, how best will he be able to pick out the winter producers by my handling system? In the first place he will best leave the pen or flock till February or March, having allowed the pullets to have five or six winter months at production. Then he will remember that continuous laying sucks the system and removes the flesh and fat from the built-up parts, namely, breast-bone, pelvis bones, and abdomen. In like manner, with all yellow-shanked birds it sucks out the fat and makes the legs appear white. In February, then, I shall be handling the pullets and placing rings of the high-class or "blue" type on those which handle pliable at pelvis and abdomen and with breast-bone prominent and yet maintain a nice, healthy weight. At the same time, as a check, the whitish shank will confirm my handling. A red ring can be placed on those which still have fleshed breasts and pelvis bones, also fatty abdomens together with yellow shanks. Birds handling fit for table in February have not been the best winter layers, so that our first ringing will not be difficult.

Those who work on commercial lines will find the yellow-shanked breeds most helpful from the pigment test, carefully checked up with my other handling items. *Generally speaking, it takes six months of continuous production to make the yellow shanks whitish,* although one does not know if it is at the rate of three or five eggs a week. But we are after good layers, and aim at grading out the low-grade performers. After a stop in egg-laying, the pigment soon comes back to the shanks. Many ridicule the pigment test but we can ignore them. You can test it for yourself on trap-nested stock. It is argued that males also lose their pigment, and I agree that many alleged yellow-legged males start with white shanks—a sign of crossing and lack of stamina often like white eyes and small pupils. Feeding and soil have a say in pigmentation of shanks, but do not let that interfere with this part of handling or put you off so long as you use discretion and check as advised.

During the summer you will need to have broody-coops for the pullets being tested, so that each broody-rest is recorded against the individual. Every pullet will carry a numbered ring for recording purposes. As she goes into the coop the date will be recorded, as will the date she is placed back with the flock or lays her first egg. The idea is to have the broody periods recorded against every pullet of the heavy breed or sitting type. This is necessary to help the next grading test.

In August we will have another hand-grading test among those pullets which started to lay the previous October or November, when all birds out of lay and in moult will be marked with a red ring, as not among the best layers. The early moulters are not the best producers, with the proviso I will later mention. We will now note the colour of shank in all yellow-legged birds, because white shanks will stand for the six months of continuous production. If we find a pullet in full lay with white shanks and no broody period against her (or only one early on in the season), we will regard her as worthy of a blue ring. If this makes two blue rings then she may be graded in high.

The moult is a helpful part of hand-grading, although one has to make no hard and fast rule. Many declare that the best layers are the late moulters, and consequently just go into the house in September and ring as good producers all that happen to be in lay at the time. This is risky, because a hen may go broody in June, bring up chicks, and be in lay in September because she cannot help herself, having moulted partially or wholly when with the chicks. Again, some hens have a habit of going broody every month for spells and being in lay in September. That is why I need every broody period recorded against each pullet. If you can tell me of a pullet in lay in September that she has not been broody since March and has white shanks and handles well, then she has been a capital layer. Any exceptions such as birds laying within or going wrong in ovarian organs have been dealt with and will be checked up.

The moult, too, often reflects upon management, because if the birds are allowed to get fat or thin they may, as a flock, moult, and the hand-grader must use discretion. The moult is often timed by the date of hatching, but I am keeping to those pullets which, hatched normally, start to lay around

34
S
G

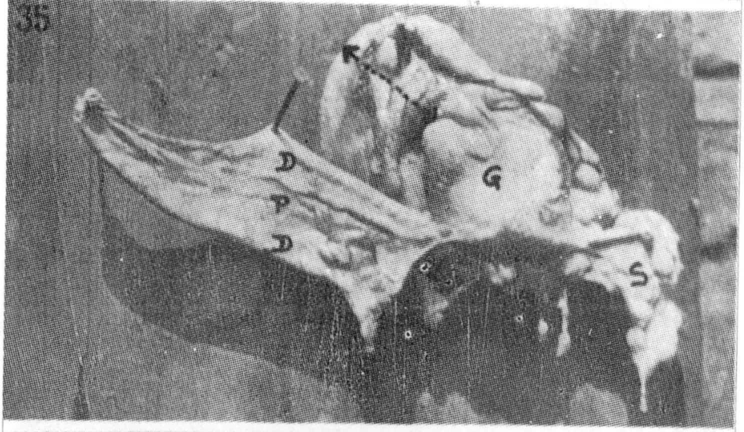
35
D
P
D
G
S

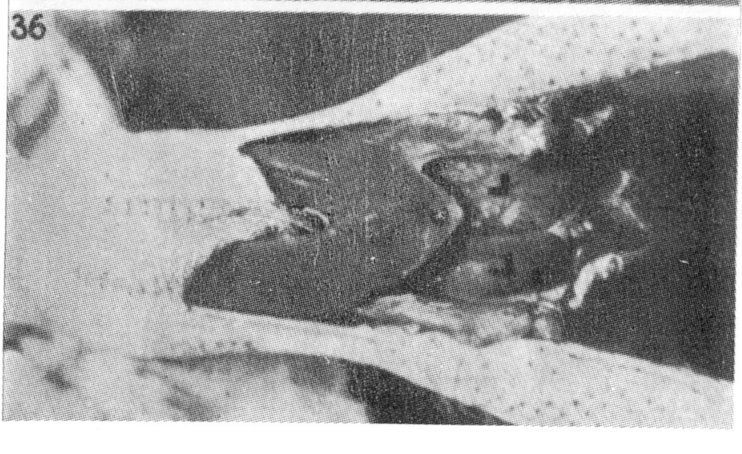
36

October. The late moulters can be noted and the fact recorded. Grade out the birds that have failed the tests and moult them out in the laying sheds so that they can be brought into lay again promptly. Where one is dealing with light breeds, broody periods will not come into the argument. Where, too, the shanks are other than yellow, one must rely upon handling only, together with the moult, as there will be no pigment test to help. If I am handling a pullet in August or September that is still in full lay she will handle well if she has been a continuous producer. The abdomen will be pliable and full, the vent moist and fine in texture, the pelvis bones pliable and almost free of fat, flesh or gristle, while the breast-bone will be prominent and not up to table condition. The fat or fleshy parts will be well out of the shank, showing the sinews at back of leg prominent and pliable. When judging shanks of layers you will remember that before they have laid an egg the shank will be full and must be pictured after they have laid 100 eggs. Often a shank is considered good, whereas a thought for the thinning-down part would have shown the handler that it was too fine in bone.

Hand-grading is made more complete where records can be kept of the date a hen stops laying and moults, and when she comes into lay again. The sooner she is started again the more eggs she will produce. But I like to let the moulting hens come into lay again, leaving the best to be mated up. When they are in the moult and I have marked them with blue rings, I place in the breeding houses all that grade in as breeders. I like a large house where I can put in, say, twenty-five such hens, the building being divided in the centre to take two mated pens with two outer runs. Placing in the twenty-five hens I can grade them again before mating up takes place or before eggs are incubated. I want to be sure of longevity of laying after I have been satisfied of the first year's production. What a pullet lays the first year is no criterion as to what she will lay the second. In January, then, I should go through the hens and remove those handling very fat at abdomen, pelvis bones, and breast. I always want hens in the moult thinned of all fat and built up for laying, but if a hen remains out of lay too long she gets excessively fat at pelvis bones, which will denote low production the next year. I turned down a hen once with a 300-odd record the first year and she had pelvis bones I did not like as regards fat, and she laid 27 eggs that season, having graded her out in February. If you will handle your low-grade hens each year you will find as they get into three and four seasons that they take a longer time to get over the moult and are very fat at pelvis bones (even as thick as one's thumb) in

January and even February, lay badly in that year, and then moult early. Early to moult and late to restart is the key to the low-grade layers. Thickly-coated pelvis bones will see such birds ruled out. Trap-nesting tells us what a hen has done in her pullet year by way of number of eggs; hand-grading looks ahead. Hence by hand-grading the hens for likely second-year production we can toe-punch the chicks from all hens promising well for longevity.

The vent will again help us in hand-grading. When a pullet sets in bone and develops her ovary, the vent increases in size, as mentioned. In like manner, when she leaves off laying and starts the moult, the vent decreases in size. Now in the pullet which spends months over the moult, the vent will go right small and dry, and yet in the best layers it will hardly have got back to its normal size before it begins to increase again, dictated by the developing ovary. If I buy a flock of hens in September and know nothing of their history I would just take note of the vent, because if one bird had a dry, small vent and an empty abdomen, it would denote that she had not been laying for months. If another had white shanks and the vent was just decreasing with abdomen half-full, I would know that she had laid well and was only just stopping. The bird to be guarded against is the pullet which moults very early and is coming into lay with large vent and filling at abdomen, which means that you are dealing not with a bird that has just stopped and is showing a decreasing vent, but one that has stopped early and is just getting back her vent-size and abdomen for re-laying. But handling and the yellow shanks would help you, as such a bird, if of a yellow-legged breed, would have deep yellow shanks.

In the second year I would again hand-grade the hens, because I am always keen on keeping and breeding from any hens which lay well for three seasons, as longevity of production should be one of our main aims. After two years of production one would select the likely third-year hens by the moult and their handling capabilities when starting their third season of production. One should always study longevity of laying, and there are many birds which you should keep on for several years in order to breed from them because they are plodders as layers and not sprinters.

Where one does not trap-nest but just hand-grades his flocks, he will buy in cockerels out of high-laying hens from a stock-breeder of repute. If he wishes to mate up his own pens and use home-bred males, he will need a little guidance as to what are the cockerel-breeding hens. As I have stated, the way to found a strain of sound producers is to mate the hens laying 180 to 220 large eggs per annum to cockerels out of hens laying 230 eggs or over. That is a general rule I adopt. This must be read with discretion, because when one starts, hens laying from 160 good-sized eggs might have to be used. One works up the numbers until one could use hens from 180 up to more than 220, providing the latter have still the good breeding frames. *Capability tells us the number of eggs likely to be laid in the nest-boxes*, but we meet with high capability on hens of Bantam size with the heart-shaped body. Supposing we have a Bantam frame, then the birds should be graded out, as they will breed textured males too small for use. Treat them as finished articles, *i.e.*, just layers to be well fed up for production. What they lack in frame and stamina make good by liberal rationing. Keep them together as a laying flock that calls for special dieting. The medium-framed hens which handle like the "rabbit" will breed good cockerels if a shade on the small side, but they are usable. We keep in mind always, however, the hen with the good frame. I like a breeding hen to score on my capacity side about 58 or more of the 70 marks, which to me means a breeding frame. The cockerel-breeding hen is the one of nice frame and yet not super in that direction, but wonderful for texture or capability. See that she lays standard eggs.

The average hen which lays 180 eggs to 220 large eggs in the twelve months will generally have a good frame and will be broody once in the season at least. These are the hens I use for general breeding purposes when grading flocks for breeders. They handle nicely without being coarse (low in capability). If we need a few hens likely to breed stock males we have in mind those doing 230 or over standard eggs in their pullet year. Here we shall be dealing with the best handling females which are continuous producers. The hand-grader will be well advised to single out a few of the hens which pass all grading tests at each season throughout the first laying year and handle well for the second after the moult, just as cockerel-breeders. Having done their 230 eggs or over for the first year they will supply us with high-textured sons, which we must have. At all times must we have the textured son out of the high-record hen, whether we are breeding layers or utility males. You will find the two types here, or rather the three, *i.e.*, among those hens which lay on and on and answer high for capability—handling like "rabbits." I ask you to select the nice-framed hen and to discard the Bantam. If you must use the mediums, do so, but keep the good-framed ones in mind as the ideal. You will always have Nature's check on frame, because when you go too large you lose the "rabbit" handling and get the "sheep" touch. It is quality that decides the handling.

On every farm, then, there should be the pen or so to produce the future stock males and the rest to breed the laying females and future strain-builders. Where you trap-nest you will place in the general pens hens of the 180 to 220 stamp and mate them to males out of hens which laid 230 upwards. Then you might keep the daughters of this mating and send the sons to table. In the pens set aside to breed high-grade cockerels for stock purposes or for sale as stud males will be hens laying 230 or over the first year and handling well for second-year production, and they will be mated to males out of hens also doing 230 or over. You need not worry about insisting on the son of a 300-egg hen, because that is where you so often get the Bantam's frame. As long as the son is out of a high-laying dam, you will be working on sound lines. The pullets bred from such matings would be cockerel-breeders and likely to be high layers. You can check them up for frame when deciding if you breed from them, because if they are Bantams you will let them go as layers and not breed from them.

Bearing this in mind, and not knowing the actual laying records of the hens we hand-grade, mating will follow the same lines. Our cockerel-breeding pens will have those hens which handle for high numbers of eggs and yet have the reasonable breeding frames. Then from the cockerels bred we will select those with the good frames which handle high for capability, discarding the Bantams and preferring good frames to mediums, although using the latter if needs be. At the same time we will have an ideal of keeping up the size of frame on both sides and in every bird. Size of frame is the salvation of stocks bred for high egg-laying. Texture or capability can look after the numbers of eggs to be produced, but we want the desired frames for stamina, size of egg, rearability, and the building up of the strain, together with maintenance of the high-laying factor.

Every bird has a use, so to speak, but we should note carefully the best ways and means of obtaining and keeping our objective. We are after all the eggs we can obtain, and Nature wages war against us because of our artificial ways. As we are working artificially we must adopt sound or common-sense breeding plans at all times. *Nature says that when you increase production in hens you will lose body size, size of egg, hatchability, and rearability.* You will also take with you in an increased form any weaknesses or deformities in the strain. That being so we must be ever on our guard and breed on sound lines. It has been the mating up of hens solely upon their laying totals and the use of males just on the number of eggs laid by their dams which have sent us along wrong roads with our utility stocks and weakened them. I aim at having every bird of either sex thoroughly overhauled before being graded in for the breeding pens, which surely is not unreasonable. Each must answer the necessary requirements of utility, but of greater importance is the fact that before being bred from they should pass the common-sense breeding standard.

I may be handling a hen which is rather fine in the bone (superfine throughout) and yet she handles so splendidly for capability that I know her to be a high producer. Such a bird with Bantam frame throws super-textured sons also bantamised in size. The latter could be mated to very large-framed hens, but how many large hens could a small male manage? How many years, too, would it take to get the necessary size up again in the majority of the offspring? Nor can you mate up a large, coarse male to such Bantam hens and get back size in the progeny. Hence I say grade out the " little gems " and, while holding on to texture, keep every bird of either sex up to standard for frames. In like manner the cart-horse hen has a use, but it would take years of mating to textured males to get our ideal. Common sense leads us in hand-grading to see our objective through the birds we mate up, and to make full use of the raw material in hand; it is as much a science of poultry work as any other branch.

Hand-grading has the advantage of enabling us to handle the pullets before they start to lay, and to say which are worth trap-nesting and which are likely to lay well or badly. We can see into the future. Again, when we take in hand stock which have already been laying, we can look back on what they have done for a time and go forward to what they are likely to do. Also, we can be quite disinterested, and will breed from the best breeders and not solely from those layers which ruin strains readily but happen to turn out a large number of eggs. The breeder who sees in eggs alone is asking for trouble!

In earlier chapters I have dealt with exceptions in handling, and advised methods of locating hens which are likely to be poor layers or not to be laying at all although giving one the idea that they are active. I would advise all who are dissatisfied with their egg-average to hand-grade the flock and locate the causes and the culprits. When flock-average is low there should be some birds graded out without a moment's delay and sent ripe to table. The number of real passengers I find in every flock is appalling, and one cannot be too strict over this grading. It is not fair to allow laying hens to carry on their backs a lot of sister-birds which are idle all the time. The sound way is to feed well those only which give a full return for their keep, accommodation, and the labour bestowed upon them. Those which are idle should be graded out for market without delay, the outlay being saved. There we have the road to high flock-average and the best profits. Flock-average is the thing that pays, and whether you keep six or six hundred in a flock I am quite sure I would find a few to be graded out as passengers. I have noticed how keen poultry-breeders are to get by other means the best laying returns, but grading out the idle birds is surely as important as breeding for increased results. Ten active hens will always bring in a better profit than ten active hens and an idle one. Most will agree over the suggestion that ten good quality birds will be a better investment than twenty inferior ones, so why not be broad-minded and reduce all flocks down to those only which are active performers? I want to see the poultry-keeper constantly grading and cutting out waste in food, management, labour, housing, etc. This he can do by constantly watching his stocks and grading out any which go wrong or cease to be profitable. Laying is such that hens will go wrong even apart from quality, and the moment they go that way they should be hand-graded out before a farthing is wasted on them. The time to part with any bird is the moment it ceases to be profitable, whether that be four months or four years.

At the Shows I find hundreds of birds with eggs, cysts, and other obstacles in abdomen, birds with abdominal dropsy, etc. The owners are keeping them for want of knowledge as to what to handle for. I hope this book will show one and all the way, and get more supporters for my system. Constantly grade the flocks and always be guided by the egg-returns. I am constantly called in to grade out birds that are doing badly only to find that there is nothing wrong with their quality but that management and feeding are hopelessly at fault. Overfat birds laying badly when put on a thinning-down ration soon begin to get back to condition and to lay as many eggs in a week as they have been doing in months. Underfed birds, too, respond to liberal rations in like manner. Handling will show if it is wrong management or low grade of stock.

At the end of the first laying year be on the lookout for the birds that show signs of being drones. All with dropped abdomens can be placed together for the second season, and be fed mainly on grain to keep the abdomens light and pliable. The egg-average of the flock can be watched and some birds, if needs be, graded out. The graded-out birds can then be trap-nested to see which are idle. I would always spend time trap-nesting the "doubtfuls," because I shall be repaid amply for my labour, unless you prefer to check the egg daily by feeling for it, as already mentioned. Birds with badly-crooked breast-bones, and especially those where the bent bone goes into the body or becomes so short as to let the abdomen down, and all with spurs, can be collected together for future testing through the eggs they actually lay. When hens of all sorts are together, the chances of checking are rendered more difficult, hence my advice to place "doubtfuls" together for testing. Many we may decide to keep for the second winter of production only, then sending them to table after fattening them. If I were due to send to table a lot of hens I would fatten some of them up, knowing that while getting fat they would lay even better until excessive internal fat had arrived, when production would slow down. It is then that I would market the birds in a ripe condition, getting an extra egg-supply and doing the fattening part at the end in one stroke.

At the end of the full year of production, *i.e.*, just before the moult, keep an eye on all birds that look seedy and mope about. Handle such "doubtfuls" at abdomen and many may be found with hard yolk-matter or shelled egg inside abdomen and like defects. Treat with suspicion, too, those hens which show pain in the eye and sad or worried facial features.

At all times, too, remember that feeding has some very important bearing on the laying or idleness of the flock. If you have the heavy cart-horse type of hens, you can, by special feeding to keep down internal fat, get more eggs. In like manner, those hens which are on the fine side can by liberal feeding be kept hard at laying, whereas any underfeeding would see them break down. I suggest very strongly that the grading or grouping of the pullets at maturity should be on sound lines which will see the several lots kept to themselves. "Weeds" will be killed as not being worth caring for or wasting expensive food upon. Only by grouping into level types can the best results be achieved by equal feeding. To-day it is common to find all sorts, sizes and types mixed, and surely the same method of feeding cannot suit them all! Group well and feed exactly as each group dictates by its type.

I am often told that a hen lays badly although she measures "a couple of handfuls" at abdomen. The first rule to be learned in hand-grading to my system is that floppy abdomens are exhausting to the birds and a hindrance to production. Always have a tightly-braced-up abdomen which is full and in control.

If you should buy a ready-stocked farm, always hand-grade severely the birds you take over, being guided by the egg-average being returned. It often pays to clear out plenty if the stock have been mismanaged, and I am never happier than when grading such purchased flocks. I have graded out more than half the birds on many occasions and still had more eggs with half the feeding-costs.

Grading applies to every class of poultry on the farm. Nothing should be kept a moment after it ceases to be profitable. Why keep dozens of males on the place when they will not be needed? You may decide to give up White Wyandottes the next season, which should suggest that the males are, from that moment, surplus and should be sold. A male bird may show "rattling" in the throat or become bronchial, at once rendering him useless for future stock. Such a male when ripe should go to table without delay. The other day I was told about a wonderful ten-year-old duck which had proved a splendid layer. Her ducklings, numbering two dozen or more, were pointed out to me, and I expressed a wish to handle the old dame. I just wished to find out how she handled for utility. Upon handling her I found that her abdomen and oviduct were full of hard-boiled yolks, and that she had not laid for a long time and would not lay again. There was nothing else for it but to kill her, although we had proof that she laid well in the spring owing to her ducklings that were running about the place. I expect the oviduct muscles had, with age, weakened and been unable to pass an egg out, which, being held up, had caused the mishap. Such a duck would die in due course when the abdomen and oviduct had been completely filled with yolks, but meanwhile there would be the loss of so many pennies per week over many months for feeding without any return. In this case, symptoms of disorder were rather apparent, because before I handled her I told my friend—Mr. W. Collins, of Bulwell, who happened to be with me—that she was wrong at abdomen. I rather suspected by observation that the weighty abdomen was interfering with the walk of the duck. In hens one may suspect the abdomen full of dropsical fluid or hard-boiled yolks if the legs are forced wide by the over-full abdomen.

A hen will often become seedy because the comb falls into the eye and inflames it. In such cases the comb should be cut off, or that part which obstructs the eye. One must ever be on the lookout for hens and ducks which ail so that the direct cause can be traced and the remedy immediately applied. It is for every poultry-keeper to get the best egg-returns, and these will go to the observant owner. My handling system will teach observation, and one has always to remember that the smaller the flock and the better the chances for observation and the noting of any that ail. That is one reason why back-yard poultry-keepers have the advantage over the poultry-farmer, in that with but a few hens kept each is under close observation, thus helping to bring out the personal side of poultry-keeping. On the other hand, with larger flocks it behoves the poultry-farmer to be very observant and always to be on the lookout.

When one is grading, he is also able to set aside certain birds to breed table progeny, if he is interested in table-chicken production. The various types in birds, whether male or female, have been fully dealt with, and if we are after table progeny we can mate up coarser birds in both sexes. To use coarse breeding hens means to sacrifice eggs; to employ Bantam hens means to sacrifice flesh. My medium hens are the *dual*. Their sons, not needed for stock or quality, will weigh up to 4 lbs. in some heavy breeds at sixteen weeks. Thus we obtain an extra pound or more of flesh per chicken over the finished article's sons. All I wish to point out here is that hand-grading will help us to locate any type we may need from time to time and for the different objectives. You would not mate a Bantam cock to big hens to produce table chickens, just as you would be unwise to let a coarse beef-type male head your pen to breed layers.

When grading a flock of birds, I pay special attention to weight, passing out those that are too heavy or too small and going for my ideal compact bird. As I take the bird in hand, I note very carefully if the hen is heavy at rear as if seeming to weigh down the whole body. Often this makes me suspect hard-boiled eggs inside the abdomen or oviduct, the weight of the yolks putting the bird out of balance when in hand. I am, too, not fond of a leathery feel about the abdomen, nor do I like a dropsical touch about it. When the pullet starts to lay, the abdomen may feel a little fat and yet nice and silky to the touch, whereas at the close of a year's laying the abdomen seems full and yet is pliable, as are the pelvis bones. And near the base of the pelvis arch the flesh is soft and pliable. If we study a hen which drops into moult in July, say, when she is as fat as butter at abdomen and pelvis bones, she will be a long time out of lay and engaged in the moult if she happens to be a low-grade performer. And when she is ready to lay again she will be very fat, not having lost the internal fat. With the better producer we will notice the opposite takes place, the bird being pretty free from excessive fat when she starts to moult, and the fat is of the easily-melted type. Consequently it soon leaves the body, and in a short time the abdomen is empty. The layers of the most eggs therefore handle fine at pelvis and abdomen when they are ready to restart production in their second or third year. The amount of time between the laying of the last egg for the season and the first egg of the next has bearing on the grading of the layer, for the lowest-quality hen will have a long rest in the moult, and the best (for number of eggs) will go right through with but a week or so's rest from eggs, laying and moulting at the same time. Handling at abdomen will therefore help us. Supposing I am handling a hen in December which has a vent quite small and dry and shanks yellow, I know it has been ages since the last egg, and with the abdomen still empty it will be weeks and weeks before the next egg arrives. On the other hand, if the shanks are still whitish and the vent is large, the bird has not long left off laying. If the shanks are yellow and the vent still large, I check up the abdomen for filling again, which shows that I am judging a hen coming into lay and not one that has just left off. When one is grading hens in the autumn and they have not been previously handled, one has to use discretion on the lines mentioned to get an idea of quality. If the hens have been ruined for laying by bad management, one has to consider whether the handling for type declares them to be good in quality even if they do not handle as advised.

Handling will be helpful when buying stock as a check against any unprincipled seller. For instance, if I buy hens in August which are guaranteed to have laid 250 eggs for the twelve months, I would begin to doubt their quality if they handled at abdomen as having left off production for many weeks. Such recorded hens would not be in moult in July. Again, if they went 8 lb. I would be sure of my suspicions, as I would if they were coarse and handled badly. With males, too, I would not admit that a cockerel was out of a 250-egg hen if he was a coarse-boned, feathery brute of the "sheep-textured" type.

I base my handling on many years of hand-grading large flocks and tracing pedigrees of the birds of both sexes, also those winning under me at the Shows. The best and silkiest Light Sussex of the year of Mr. Honey's was the son of a 238-egg hen. I have proved that many breeders have been mating their high-record hens to beef males to get back size solely because of the coarse handling of the sons, which have taken after the beef-type sire and not the "rabbit" dam. *We must have our utility males bred from high-record lines for many seasons to get the most certain results.*

It is rather difficult to tell hens from pullets, but handling will help us. In the first place, a pullet handles soft in bone not only at end of breast-bone but in pelvis bones. Her face, too, is smooth and fine in texture and eye prominent. As a second-year hen, the bones will be harder and the face will, in most birds, be somewhat puffed. But if the bird is a grand-quality one, she may have features of a young pullet (which is the quality part), although handling coarser in bone than the first year. In handling, while one tries to get the birds which last, one allows the bones of a hen to be a little coarser than the pullet's.

While our ideal male will, as an adult cock, look in features like a young cockerel, the cock's eye becomes more sunken as a general rule with age, and the face more wrinkled. The spur grows longer during the second year and takes upon itself a point, while later on in the third season the spur turns up.

Make a practice of grading as you proceed, so that good and bad birds can be specially rung during the year. At the close of the season one will then be able the better to group up those needed for the various purposes. Grade and group for preference when out of lay, *i.e.*, as regards wholesale shifting of stock, because a general move round of birds in lay will only tend to stop production. Rather remove from a pen those not wanted than the better ones. Let every bird wear a numbered metal ring, so that anything for or against her can be recorded in a note-book kept for the purpose. Coloured rings can be attached to shanks just when you sum a bird up as regards quality or when grading in general.

At all times be observant and handle mopish birds to see what is wrong with them. Note any that do not come up to the trough at feeding-time, whether hens or ducks. Something is wrong and experienced handling will help to solve the problem.

CHAPTER XXII.

HAND-GRADING OF DUCKS AND BANTAMS.

As pioneer in arranging classes for utility Bantams and championing them for a place in those Show schedules I have drawn up or advised upon, I propose to devote a little space in this volume to the "wee ones." I will, in the first place, deal with what I had in mind originally. It is not every person who is so situated that he can keep large poultry, whereas almost everyone finds himself with sufficient space for a pen of Bantams. I have known many members of poultry societies who have left the district and gone to reside where facilities for large poultry have not been available. Rather than see them lose interest in poultry I have persuaded them to have a pen of utility Bantams. At first they have thrown doubt upon the proposition, not thinking that Bantams laid eggs. But having in so many cases picked the Bantams for them, and explained the type they should aim at, they have been amazed at the baskets of eggs collected in the winter from just a small pen.

Again, those with children who need small eggs can have the ideal from Bantams, and if the children are allowed to keep Bantams they are brought up to routine and a liking for animals which will stand them in good stead as they go into the world. Educate the children to be kind to animals and I consider that it brings out the good points in them.

The egg I aimed at was the one weighing about 1½ oz., and when we bear in mind the low cost in housing and feeding as compared with large fowls, Bantams become a very paying proposition so long as they turn out full egg-baskets. At first I was severely heckled over my utility Bantam idea, and many have stated openly that there could be no such thing as a utility Bantam. I remember a "scoffer" at a Show taking me up over the Bantam section there, but I was later able to publish the laying record of the winner, which sounded more like a winter record of a 300-egg White Leghorn pullet. I have always traced the laying of my winning Bantams for my own interest, and have got some splendid letters from exhibitors. As I write, I have a letter from an exhibitor whose White Wyandotte Bantam hen won First under me at Northumberland Heath and First at Crayford, to say that the bird finished the year with over 200 eggs, all over 1½ oz.

As I have always judged utility Bantams at the Shows by handling, I will deal with that and my ideal type. Many have asked me from time to time how I judge them, and I state frankly that I score-card them just as I do the large fowls, both in males and females. I should have hit home the "Powell-Owen" score-card for Bantams and ducks years ago, but I have had to fight for it on behalf of the large fowl and so left the Bantams and ducks alone for the time being in order not to mix the issue. I remember a judge asking me how to find out the best of the utility Bantams which he had to judge, and I gave him my score-card and told him to score the lot in detail and place on totals. He did this and his winning hen was one of the best laying Black Wyandottes I have so far handled.

In utility Bantams I do not aim at a small bird which is all feather and no weight. At the same time, I avoid any that are really too large to be called Bantams. As a result, *my ideal is a Bantam which handles large and yet looks small.* I can hand you a Bantam which looks big and yet handles small to excess, and these I will not have. They handle feathery at abdomen and thighs, and have very narrow necks to back, with hardly any width at back or between legs, no abdomen and the "triangle" body already described. Some will be very fine in beak and shanks, lacking stamina.

As I take my ideal in hand, the pullet handles full at abdomen, with wide back and good neck part, wide pelvic arch and ample space between legs. Measuring well at abdomen, she has neat shanks and beak, stands wide and straight, has tight, silky feather and all the texture of the "rabbit." Head points are of fine texture, as in the large hen, and when I take the bird in hand it is plump and compact, with a proper heart-shaped body. If you will handle plenty of Bantams you will be amazed at the handling the good layers show. For instance I can give you a Bantam hen which, handling large yet looking small, may measure four fingers at abdomen, or as much as a Leghorn. Even large hens often look small because they have tight feather, and I remember a pullet winning a White Leghorn class under me at Westminster which more than one critic described as a Bantam, and yet she scored very high on my capacity side, telling me she was really big (by handling, if not appearances; the latter being always unreliable), and she laid over 270 first-grade eggs in twelve months. So it is with Bantams that have the ideal tight, silky feather. You look at them and they appear midgets, but when you handle them you find to your astonishment that they measure FOUR fingers at abdomen where you would expect TWO. Such are the Bantams you should aim at—big to handle, small to look at.

Score-card them to my card in exactly the same way as the large fowls, and keep to the capability side for the number of eggs laid. Also, have the same type of arch and breast-bone. Measurements will be lower than in hens, but after handling plenty of Bantams you will see what is the ideal. Bantams in full lay might measure three or four fingers at abdomen, two between the pelvis bones, and one from tail socket to pelvis bones. Treat the males in the same way, but they may measure two or three fingers at abdomen, one finger between pelvis bones and one above. Get the width of back well up to the wings and, except for a smaller edition, the same handling as fully described in earlier chapters will apply. In Bantams, too, you will make up your mind as to what is your ideal when in hand.

When we come to ducks, my handling system and score-card will be found to apply. I have always used it for judging utility ducks, and at every Show get splendid entries in this section. On the occasion I gave the cup for "best bird in Show" to a Buff Orpington duck she laid over 250 eggs for the year, while a noted White Runner duck which laid 315 eggs won First under me at Leicester. My students who keep ducks are taught my system of handling and mate by it, and I have just seen a student win the gold medal at Bentley Duck Test with Khakis, the ducks being graded to my system before dispatch to the competition. As I write, a fresh Test has started at Bentley, and students report that they lead in Buff Orpingtons and Khakis, the ducks being picked on my system.

I have not been keen till now to push my system for ducks and Bantams because I believe in one thing at a time, but I have now beaten down the bitter opposition to my handling system for hens that we will get ahead. The pelvis bones in the duck are rather differently placed to the hen's in that they point down, so to speak, instead of straight out. We handle in exactly the same way as hens, keeping to my card for capacity and capability. Never take up a duck by the legs, but hold it by the neck when catching it. Next, hold it in the hands at shoulders like a hen, and when measuring the abdominal capacity with the right hand have the fingers of the left hand under the breast of the duck as a support, with the body under the left arm and the duck's head held down. One will find ducks difficult to handle at first, but practice makes perfect. Hold the duck's head downwards under the left arm, with legs in the air, leaving the right hand free. When in a show-pen I usually take the duck by the neck with the left hand and gently draw it towards me. I then slip my right hand under the body and between the legs and can readily handle for the approximate space between and for type of the pelvis bones. The pelvis bones are not coated with fat in the same way as those of hens owing to different construction, the fat within, as in an overfat duck, being round the abdomen but not encircling the pelvis bones, which remain free of excessive fat. But we need to handle at pelvis for the width of arch, the texture of the bones, and their type. Often one finds the arch narrow and the duck laying small eggs, so that we should take the width in all ducks sent to the Tests in case of letting in a small-egger. In the past, duck eggs have been much over standard, but nowadays one sees in some ducks a loss in size of egg by reason of our having increased the number and lost the frame or capacity of the individual duck concerned.

We are losing size of frame in some ducks just as happens in the hens, as dealt with in detail in previous chapters. We are getting here and there "little gems" which have all the texture and yet no size—all eggs and no breeding frame. We must prevent a loss of frame by keeping to ideal types as in hens, namely, frame with texture; or capacity and capability high in the same duck. Have a wide pelvic arch in the ideal duck, with ample space between legs at all times, a wide back, also ample abdomen. Keep the pelvic arch in mind as the key to frame, and let the beak and shanks guide you as to excessively coarse or fine bone, so that you can have the medium. In our ducks which are too small we so often have very fine beak without any depth, and thin legs.

Plate 63 illustrates the Khaki Campbell drake which was the father of the ducks which won at Bentley over all breeds. He may be taken as an ideal. He stands very wide, showing great width of back, with a heart-shaped body of substance. At abdomen he is well developed, and has a very wide pelvic arch. He is tip-top for capability, having medium bill and legs, a very noble head with eye high up in skull and prominent, and neck graceful and on medium to fine lines. You can contrast the drake which is coarse with this pattern. The undesirable drake has a sunken eye; short, stout neck; long, deep and coarse bill; and excessively thick legs. The pelvis bones are thick and rounded or crooked. There is a surly expression on the face and a lack of activity, with a distinct desire not to be handled. You will generally find plenty of space between the two pelvis bones of the drake, and once you have the type you are after, check up carefully for capability. I prefer as straight pelvis bones as possible.

Friendliness in ducks and drakes is a thing to note. I always find those of either sex which are high in capability to be like the good-textured hen and cockerel—friendly to the handler. When grading ducks I find plenty with which I can do anything, and they talk in their way to me, which observation has taught me to notice. Plate 62 shows Miss Barbara Raye, of Pelling Pedigree Poultry Farm, Scaynes Hill, Haywards Heath, holding her Buff Orpington duck which has laid over 1,000 eggs in four years, made up by: first year, 260 eggs; second year, 262; third year, 243; and fourth year, 240 before she moulted. Apart from high-grade ducks being friendly, you will be well advised, if you want the best returns by way of eggs, to make them docile. Let the same person manage them, and never do anything quickly to frighten them. Never shift ducks when once in lay, and always get them used to their quarters when young. It is management with ducks which counts for so much.

Those who wish to record their ducks will be well advised to place the young ducks in their "trap" house or coops when they are four months or so old. Let them sleep in their "trap" places from that age. Miss Raye, who won the Bentley Test with Khaki Campbell ducks as mentioned, has sent me a photo of the duck-traps she uses, and they are seen in Plate 61. There is a set of sleeping boxes at the back, each duck having a separate section for use at night. There is a small wired-in run attached to each sleeping section, on the floor of which is a movable run-up board. The front of the enclosure is wire-netted, but there is a trap-door through which each duck enters by pressing against the trap-wire. Once inside, the latter allows a door to drop to prevent the duck from getting out or another from getting in. It is the same wire arrangement as found on trap-fronts for hens. Coming home at night into a large enclosure for their tea, the ducks find their way into the small sections wherein they spend the night, and the eggs are collected in the morning and credited to the respective ducks. As showing what can be done with ducks, Miss Raye lets the flocks out in the morning, and any that have not laid will return to the "traps" to deliver the day's egg instead of laying it outside. And the dams of the Bentley ducks have been taught to come to hand for scraps by number. Ducks are very clever and respond to proper management, and this pen learned their tricks in a day. When having tea in the open, Miss Raye's partner would call number four for a tit-bit and push others aside till number four came up. I have known a lady count the ducks up and, finding two missing, tell the flock to go and find the others. Off they would go and bring them home and then they were given their tea. The duck-owner will be with his ducks and maturing ducklings as often as possible and will get them docile.

Ducks need liberal feeding in the winter if they are to produce plenty of eggs, and the man who argues that they are

37

38

39

free-rangers and just want a little corn only at night will fail with them. A good deal of benefit comes to the duck-keeper who is observant. For instance, when ducks are coming into lay they become developed at abdomen. The filling of the abdomen can be noticed without handling as I practise with the hens. As you see the ducks drop at abdomen, keep up the feeding and do not shift the flock or change the feeding radically. Observe the cut-away abdomen of the drake and you will understand what ducks look like when the ovary is dormant. Do not sell ducks just as they are dropping in abdomen, and if you notice the ducks are closing in and losing their abdomens, take steps to increase the feeding to hold as many as you can and look for the cause, applying the remedy without delay.

Often drakes are reckoned as ducks, and there is much waste of good food in this direction. I have seen a man mate up a pen of ducks to another duck, thinking the latter was a drake until it laid, whereas the dropped abdomen told me it was a duck when he queried why he was getting infertiles. I have seen two drakes in a flock of a hundred ducks because at the time they were thought to be ducks and were left with the flock just to be food-wasters. I have also found, in a small flock of twelve, three drakes which were sold as "ducks" and considered to be females without even being checked. Some have no such word as OBSERVATION in their vocabulary, but observation is a glorious thing! Ducks "quack" and drakes "whistle," but I am here concerned only with pointing out the need for observing even with ducks. I have seen drakes (bought as and thought to be ducks) trapped, and their non-laying did not tempt the owner to check up the sex, just as I have found hens in September in a trap-nested flock, with abdomens full of dropsical fluid or hard yolks, which had not laid for months. On calling attention to the birds when grading with the owner, I have found the letter "M" against the number of each bird, but the fact proved that no eggs had been recorded to them for months. Lack of observation alone prompts a man to put "moulting" against every bird in a flock not laying around the moulting time. It reminds me of a few birds I often find in a trap-nested flock which never lay, and on looking for their records on the sheets I find that no such numbers exist. Solely because each bird is rung at the start and no check made as to whether or not all have started by the end of three months. I have known a pullet lost in a Laying Test because, while the record-sheet said she was there, no egg had been laid to prove it. When she was found dead on the drop-board she became officially "present."

Plate 64 is of a pen of White Runner ducks owned by Mr. C. H. C. Partridge, of Little Lambswick, Tenbury, who has won very many prizes under me with White Runners. You will notice the stylish ducks, with neat bills and fine necks, bold eye carried high in the skull, width between legs and full abdomens. Activity should be the key-note to selecting laying ducks, and there should be good width, depth and power at shoulders and chest.

In ducks, as with hens, there will be the natural tendency to lose frame as we go for numbers of eggs. The Bantam duck that is all texture but too fine should be discarded, as should the cart-horse, inactive, waddling type. Aim at the medium, with good frame and yet splendid to handle as regards capability. Ducks have wry tails, deformed backs, etc., like hens, so grade very carefully for deformities.

The duck is going to boom in the future, as I prophesied in my "Duck-keeping on Money-making Lines," although its success will depend largely upon personal management even more than with hens. Some never will keep ducks successfully. and I have always advised each person when thinking of taking up ducks to start with a pen and see how results turn out. If good, the flock can be increased. I must admit that the ladies do best with ducks, as they do with trap-nested hens. All who want eggs in plenty should remember that ducks must have no upsets!

Returning to Bantams, one notices how the "wee ones" have prominent eyes and texture in a high order, and I suppose there is some truth after all in the saying "little and good." But keep an eye on frame in the Bantam, and see that the beak and shanks are not too fine. You will remember with all stock that the width of pelvic arch holds the key to frame, the beak and shanks give a guide to amount of bone, while the width between legs provides the key to width of back and body or narrowness thereof.

It has always been a controversial matter as to utility type in Indian Runner ducks. As is well known, the exhibition breeders favour an upright type with legs well back and abdomen between legs. It is this very upright carriage which is claimed by them to be the true characteristic of the breed: It has never been my ideal for utility, because so many of such ducks do not stand for good handling. Such quite upright ducks are very slim in body and narrow across back and between legs, which helps carriage but detracts from utility, because a good body must be needed for heavy production. I have had many friendly arguments with lovers of this straight-up type, many of whom have declared them to be good layers, but when we have got down to actual ducks on their farms they have been upright but well developed in abdomen, and would not have won in the best exhibition competition. One can always find layers among any flock of hens or ducks, but rather am I interested in having a high average of production from the whole. Hence my striving for a desirable type, every hen or duck answering to same and being a profit-getter on her own as a producer.

Taking the exhibition or upright Runner, there are here two distinct types, the one which would win at any classic exhibition for show points not supported by me for utility ideal because of the very slim build, nervous temperament, and (to me) absence of abdomen and injudicious carriage of same. Then we have the second type, also bred by exhibition men and which has often won under me in utility, namely, the good upright carriage but the broader body and the well-developed abdomen, not carried between legs but at rear. This type will be active producers but would not win in the best exhibition company, although I consider ducks of such type when judging utility classes. It retains its carriage, so to speak, but loses its style, broadening out at abdomen or base in a distinctive way. The whole build of this type of Runner is more substantial than the Show pattern.

The ideal I have always had in mind is the utility Runner which has not quite the upright carriage of the Show specimen, yet it is no waddler. We must have the active free-ranger, but I am concerned with the abdomen of hens and ducks, seeing that, as so amply proved in earlier chapters, it is here where the eggs are made or lost. I like the medium beak, the bold, high-placed eye, the slender and graceful neck, because such stand for texture or capability. But I next like the well-built body and the well-developed abdomen, carried at rear to make the whole symmetrical and well balanced. Directly we depart a little from the upright we get our objective. Some winners in utility under me of this type get cards in exhibition but not premier awards, I find. I would like to see the utility part supported by the exhibition men, so that the Runner will take its rightful place among the popular breeds of ducks of to-day. It rather seems to be losing ground in utility, no doubt due to the push being made in Buff Orpington and Khaki Campbell duck circles, both breeds now having strong Clubs. The Buff Orpington Duck Club, too, has taken under its wing the utility breeders, having decided to aim at utility and beauty combined in their ideal. Coarseness is to be supplanted by quality or texture. Magpie ducks are also making great headway, a club having been formed with a utility standard.

If ducks are in show-pens they can be better judged for utility by observation than can hens and pullets. One is rather concentrating in ducks upon texture, temperament, activity, and build. Temperament and texture are so very important. Looking at a duck in a show-pen I would be attracted first by its balance. It would stand wide between the shanks, showing that the back was broad at shoulders. It would be symmetrical in build without any part being out of proportion. The carriage would be alert and graceful and not sluggish and inactive, like a cart-horse. The abdomen would be roomy for development at egg-making times, and yet it would be well carried in order not to upset the balance or symmetry of the whole. There would be depth at chest and in body to represent stamina and power for production and breeding. One soon loses this in its entirety if the duck stands narrow, seeing that with this loss will go narrow chest and body, also back. One bears in mind grace, and that soon prevents a cart-horse Buff Orpington or Pekin type of Runner from attracting you.

Texture or capability comes in as the department responsible for the number of eggs laid. The beak and shanks will be carefully observed, seeing that coarseness will be detected by very thick shanks and coarse beak just as superfineness (also undesirable) will be manifested by the very fine beak and shanks. If the beak and shanks are coarse, we will find the skull thick at back, with the eyebrow heavy and surly, the eye low in face and sunken, and the facial expression sheepish. The high-textured duck or drake will have bright and bold eye carried high, skull narrow, so that both eyes are easily seen standing out when viewed from both back and front of duck. The face will be smooth and the neck will be slender, not short and thick.

We can check up for temperament in order to get a docile or friendly specimen, and here all who keep ducks will realise that they should keep their young ducks tame and friendly, because it means so much towards full egg-baskets. Ducks are ever ready to moult (wild ducks take twelve moults per annum, I believe) if disturbed unduly. Trapping of ducks and the consequent handling help production, and when you enter a pen you will be sure to train the ducks to be accustomed to trap-boxes before dispatch to the Test. They will also be made accustomed to the foods that will be used at the competition, and at an early age, say when four months or so old, they should be placed in a small lot, *i.e.*, the selected ones, and be accustomed to handling. They must be made docile. Never send to Tests ducks in full lay, but just timed to start soon after arrival. They will be filling in abdomen when sent, and one should keep an eye on any which commence before dispatch, locating and discarding them for others.

One should ever keep an eye on ducks that are wet at rear in case of vent disease, as it spreads more quickly and generally with ducks than hens when mated. I omit measurements of vent on my carding of ducks, and you will find that measurements differ a little to those of hens. For instance, one can get four fingers or so between the pelvis bones of the duck when in full lay and three in drakes. On the other hand, you will handle plenty of ducks and drakes to get a full idea of how they handle. You will cut out the small ducks with narrow pelvic arch, etc. Handling will become the easier to those who have studied my other chapters. I like a duck which, when handled, like a hen, at the shoulders, keeps its head up and is well balanced, with good heart-shaped body, and with back wide and powerful, being nice and flat as it lengthens out.

In Aylesburys, etc., where one wishes to increase the number of eggs, one aims at the modified edition with high texture or capability and with a good frame, striving for activity and avoiding a cart-horse, sluggish type.

Personal management is ever a vital factor in the laying of ducks, and that is why the owner who moves quickly, bangs gates and buckets, fails with ducks. They need a quiet field, and do best in medium-sized flocks.

When mating, keep texture high in the drake, with ducks of nice frame and texture. Avoid saggy abdomens in ducks. Keep the right type, feed liberally, and insist on docility and quietude.

If I were handling a flock of adult drakes with an eye to selecting a tip-top specimen for pen 1 to breed layers, I should study closely the type of pelvic arch and the pelvis bones for type and quality. I know I must have high capability by way of bold eye, narrow skull, tight quality feather, but there are types to avoid. I shall get my high quality or capability in a drake that is too fine. His bill will be very finely drawn, while he will lack frame, being narrow across back and between legs, very fine in shanks and narrow in pelvic arch. I may be able to get only one finger between the pelvis bones in Khaki Campbells and Runners and, say, two fingers in Buff Orpingtons and Magpies. Such will point to narrow width of pelvic arch (if pelvis bones are straight) and a narrow frame. I want to be thinking more of two fingers in Khakis and approaching three in Buff Orpington drakes. I need that very wide flat back that betokens broad shoulders, with a good type of straight back keeping some width down to the tail. With it I shall obtain nice width between legs approaching, say, four fingers when standing firm. I am anxious to avoid the pigmy or "bantam" drake that is all texture and no frame.

I prefer a nice type of straight pelvis bones in a drake or very slightly curved, in the latter case the bones to be pliable. I do not desire razor-blade pelvis bones if I am to have all texture and no frame in exchange. On the other hand, I wish to avoid the coarse drake where the pelvis bones almost touch at the ends in rounded or cow-horn bones. I avoid the very coarse pelvis bones of this rounded type, checking up to find such a drake coarse throughout—thick at back of skull, thick and short in neck, sunken eye, low eye, coarse bill and shanks, inactive, and too heavy in hand. Pelvis bones in ducks and drakes will not be coated with fat as in hens, and one readily arrives at their texture. Go for the happy medium that is neither very thick nor very thin, *i.e.*, in both sexes. One can check up by shank and bill, weight of specimen, etc.

When ducks or drakes are getting superfine and losing capacity or frame, weight is lost, body is narrow, chest is flat and shallow from shoulders through to front, and the bone is too fine. The ideal duck and drake will handle plump and healthy as if there is something behind it to help it against any strain on the system. Spanning its body through from back to breast-bone, the body will fill the hands well and in heart-shaped fashion. It will be compact and readily settled when taken in hand at shoulders. That will be the capacity side, and you can next think of texture or capability.

Ducks run in exactly the same types and one should avoid the cart-horse and coarse pattern, also the superfine which handles light, like feather and bone, with very fine bill, legs and pelvis bones, together with very narrow skull, very narrow long neck, and lack of depth at shoulders from back to chest, *i.e.*, a cut-away appearance. In Khaki Campbell ducks one might aim at three fingers or so between the pelvis bones and four fingers in Buff Orpingtons. Abdominal capacity varies according to the size of the individual. At one time ducks and drakes generally measured well, but since increased egg-production has been an objective I am finding too many that are all capability and down in capacity. The size of egg in ducks has decreased with the breeding for numbers as Nature dictates, and in like manner has the frames of the layers. The same sound rules apply as to the hens—that we must have capability high but must not neglect capacity.

Abdominal capacity is not everything, as mentioned for hens and male birds, but one must know just when one has gone too low in general capacity, breeding frame, etc., because then we have decreased measurements, as you will follow, between pelvis bones (width of arch) and from end of breast-bone to pelvis bones. We do not want excessive abdominal capacity that may give us sagging "geese" abdomens; the latter at all times should be tightly braced up and in support.

In ducks and drakes one might place three or so fingers between end of breast-bone and the pelvis bones. Once I am satisfied with the handling at abdomen, I judge for girth of body, compactness, heart shape, bone, and general texture. It is more difficult to master the handling of ducks and drakes than ordinary poultry, but practice is essential with trapped stock. Laying records of ducks vary too much with environment and personal management, so little upsetting a flock and turning good specimens into "duds" as regards number of eggs laid.

One should keep up frame or there will, I consider, be the additional troubles of egg-binding, soft eggs, etc. When you *dwarf* producers of *large* eggs you will meet with egg-laying disorders. We find it in Bantams that are dwarfed from the larger breeds of poultry. You will find the very small ones soon go wrong in laying, no doubt, through a large egg and a small channel and passage being inconsistent with Nature's ways. A friend of mine bought a trio of Sebright Bantams and the two females never laid an egg. If ducks are bred too small in frame, I see the same troubles ahead. I meet with plenty of excessively small Bantams that are amiss as regards the functioning of their laying organs. Many do not lay at all and plenty die over the first egg. That is why I support the medium in Bantams and not the midgets, and am against those who state that the smaller Bantams are the better. The same say of the large breeds, "You cannot have them too large," without mentioning a word about quality. Never lose the latter, but, once secured, see that all utility stock have good frames and capacity.

CHAPTER XXIII.

FOR THE DEFENDANT.

I SUPPOSE no system has been more severely criticised than that of my score-card, and no one has fought the critics with greater earnestness than I have.

Critics there *must* be, but one never minds constructive criticism. Destructive criticism has, however, been the rule rather than the exception. I am supposed by the critics to be a fanatic—a destroyer of trap-nests—and a wrecker of utility stock. The reader should weigh up closely this criticism, and inquire first of all if the prosecutor is a fully interested party and has a corn that I might be stepping on. When letters appear in the Press attacking my methods, inquire who the writers are. In the majority of cases such letters are unsigned, and no notice should ever be taken of any letter in the Poultry Press on any topic unless it carries the name of the writer at the foot. A person who is not afraid of his views publicly expressed is not ashamed of his name. Many critics are exhibition men of the old school who do not move with the times and were NEVER one iota interested in EGG-PRODUCTION. Some of them think that the more popular the utility bird of my type gets, the less popular will become any exaggerated exhibition pattern. One of my keen critics among the fanciers is the very man who, by putting tremendous combs and wattles on Leghorns and Minorcas, ruined the utility and laying points of the breeds, and he is more responsible through his actions for my UTILITY type of bird being bred than anyone. He writes under a pseudonym! In a recent issue of our best poultry weekly a critic declared that I was splitting hairs when I wrote that capability and capacity were different; to him they meant the same! I have never said that *capacity* meant the *capability* to lay. A close study of this volume will explain these terms of mine and their exact meaning.

It is well to trace the history of my score-card, which is also wrapped up with utility Shows. For years prominent utility breeders have been teaching the public what points went towards making the layer. Coming from such notabilities, every publication was pleased to give space to same. In fact one publication has thought the hints so good that it has published them year after year in its yearly issue. If profitable layers had no distinguishing points from low-grade drones, why tabulate them? And why not a furore when they were published? The only crime I have committed has been to place the production characteristics into a scale of points which took years to complete.

I originated my score-card as a simple teaching card, using it as a means of tutoring my poultry-breeding students as to the class of bird to trap-nest, breed from, and enter in the Laying Tests. For years all my students who have learned my system have been well up in the Laying Tests. I took their fees as a tutor and did not wish publicity for their wins. Utility Shows then began and I am happy in knowing that I was one of the pioneers. At Tottenham, solely for the educational purpose from the view-point of the small poultry-keeper, I placed my score-card on the pen of each bird. I wished to teach the small poultry-keeper the points of a layer, the way to keep her fit, and the road to good management and a real interest in his small flock. Tottenham Show—of which society I have been President for many years—set the ball rolling and a few breeders started the fight. At first *one could not pick a good layer* from a drone, they declared. Then *good layers had no characteristics*, although the same men had been telling the public they had and explaining what exactly they were. So poor me and my score-card had to go through the mill. When a teaching card I kept my score-points to 150, namely, 70 for capacity and 70 for capability, with a casting vote of 10 for size. The Shows brought about my addition to 200 points to include 10 for health, 10 for show condition, and 30 for breed characters. Naturally, if birds were to be shown to the public, they had to come up respectable. You could not put a scaly-legged old hen straight from the coal-cellar into a show-pen for the public to come round and be attracted.

Next followed a fight as to type, the breeders declaring that *layers had no type*. Some even declared me to be a fancier of the deepest dye. Fancy admitting that a utility White Leghorn had no type! Surely something had gone wrong earlier on when the utility Leghorn was being made. But the surprising part was that even the critical breeders condescended to judge (and still do) utility classes. And they take out the birds and handle them before they give their decisions. What do they handle for—capacity or laying pointers? Again, do not these very breeders handle very closely all their pullets sent to the Tests? *Are not Laying Tests won on handling?* Do breeders not spend very much time when picking their pullets for the trials? I can assure you that they do not tell the poultry manager to send on pedigree alone without handling the birds. Watch, too, the breeders at Harper Adams

Conference when handling competing birds. What things do they handle for—lice at abdomen or defective eyes?

Consequently I accepted the fight, and with back to the wall I have won through. The critics gave utility Shows a year to live, but they have made good. I have seen Tottenham Utility Poultry Show grow from 84 entries, including carrots and rabbits, to 1,500 all-utility entries, representing the largest utility poultry Show in the world. The accomplishments of our little Tottenham Utility Poultry Society would make sufficiently good reading to warrant a small brochure for all other societies to follow. In 1917, Tottenham Hospital benefited through our Show; in 1918, £50 was handed to St. Dunstan's from our utility Show; 1919, £60 to the Prince of Wales' Hospital Extension Fund; 1920, £50 to St. Dunstan's; 1921, £71 to the Prince of Wales' Hospital Fund; and 1922, £50 odd to the Tottenham Invalid Children's Aid Society. We were the first society to hold meetings every month throughout the year, and when lecturing on "The Hen from Within" in January, 1922, as well as January, 1923, I had an audience of over 250. It is the little man I have always set out to help and educate, so that he could get the best from his birds.

I have judged more utility Shows than any other person, and placed the awards in utility at more agricultural events. My score-card was used for the first utility Show in Belgium, also in Denmark, and is extensively employed for utility classes abroad, especially in India, while it is taught in institutions across the water. It has received a thorough test both here and abroad. I have personally to my credit many records for total utility entries both at Shows and for individual classes. I have judged very often 600 and 700 birds in a day on my own, and my record is 790 for a single Show. My best "class" entries have been 201 birds in a breeding-trio class and 125 White Wyandottes in a single-pullet class. I would here put it on record that I have never written and applied for any Show that I have judged. I was appointed utility judge at the Dairy Show the first year classes were provided for our section. I judge again in 1923, making twice in four years. Twice out of three years I have been appointed utility judge at the Royal Show of England. And I have been more influential than any other in getting together utility schedules and sections at our classic and agricultural Shows. I write not in boastful fashion, but to show that I have fought for a place for the utility bird at our poultry Shows. At Bodmin Show recently, although only three weeks were set aside for organising the event, and despite the breeding season being on, I drew 600 birds for a one-day Show. The local fanciers' society were asked by a few utility enthusiasts to put on a section at their exhibition Show for utility birds, but they refused, only offering two classes. As a result, a few utility breeders met and started a utility poultry society which ran the Show without any capital and with great success.

How different it all could have been had there been co-operation. Are not all working in utility anxious for the one aim—increased demand for utility stock? Here was the glorious chance to hit the utility bird home in this country, to show our specimens up and down the country in order to attract buyers. The war had made the utility fowl, but breeders set out to lose the peace! Instead of fighting, they should have met with one voice and set about drawing up standards for utility poultry in all breeds and colours. No, they had tasted high prices for foundation stock from the multitude that rushed into poultry-farming and had an idea that such a trade and at the same high prices would always exist. During the boom I heard of a man paying £50 for a utility cockerel!

The backbone of the poultry industry is and always will be the small poultry-keeper, and he cannot pay £5 for a pullet just for the eggs she will lay. He might as well buy his new-laid eggs. High prices for pullets just for laying would see him buying cheap first-crosses. He might pay pounds for a bird that can lay and, in addition, win a prize for him at his local Show. I wanted quite sincerely every utility breeder to go all out for that demand and new channel which I could see was ahead of us. Up and down the country I anticipated thousands of utility poultry societies like Tottenham sprouting up, and one of the first things any poultry society decides upon after formation is to run an annual Show and to include in the year's programme young stock and monthly shows, also egg shows. What were they to show? I have always made up my mind that no one will ever stop a Britisher from exhibiting something, whether pet mice or rats, or even a "baby" at the local fête. Were the members of the societies (which I could see coming) to interest themselves in showing exhibition poultry with exaggerated points, or to concentrate on utility poultry? If they went for the former, then utility would go back to its old place. That is one reason why in all honesty of purpose I went right out for the utility Shows and the utility-Show bird. The old price meant anything from 5s. to 10s. 6d. for a matured pullet, 5s. for a sitting of eggs, and 10s. 6d. for a dozen day-olds from the backyarder buying utility stock for eggs alone, yet he parted in those days with his pounds for exhibition birds for show. To-day, thanks to my energies, add beauty to the layer as in my ideal and the backyarder will pay his pounds to utility breeders for a far more useful article—a bird that can lay and win. It has always been my contention that, apart from a very few purely ornamental breeds, every bird kept in this country should lay sufficient eggs to pay for its keep, housing, labour, etc., and then yield a profit. I have in my judging and from the start aimed for a pullet that would lay plenty of eggs and large ones at that, would be typical of its breed, and would make a good breeding hen after attending to production. Utility with beauty has been my ideal.

Many breeders declare that you cannot have breed-type and eggs, but mistake me not, as I am not fighting for fancy or exhibition type where it interferes through exaggerated points with high egg-production. *When utility Shows started, there had been but one aim by utility breeders, namely, more and more eggs.* No thought of breed-type. Crossing had been rampant in most utility breeds, and one had but to watch the Black Leghorns in the Laying Tests towards the close of the trials to see them turning with the moult into Anconas. Tests had to be won and stamina maintained, and crossing was a custom. Laying Tests have not helped one iota, because the birds have never had to fight through any breed-type test. Only this year for the first time has a pen been returned from a laying competition for being untypical of the breed. Why did not breeders play their cards and decide at once on standards after the war, and see that the birds in the Tests were not accepted unless they were reasonable specimens of the breed and variety concerned? Why were specialist utility breed clubs not inaugurated at once to show the strength of the utility bird? We could have had a Utility White Wyandotte Club that would have stood alone for membership. We lost organisation and co-operation through certain breeders putting the gloves on and fighting against progress.

The craze has been for eggs and eggs alone, without regard to exceeding the speed-limit. I have just been to Denmark, where roup is unknown. Is it because they have not yet gone after the sprinter or record individual layer? Are we free from roup in this country, when immediately after nearly every Laying Test starts roup is reported to have broken out? Some birds are sent but a short journey and have roup a few days after arrival. Has there been sufficient sound breeding in utility?

Let us work on different lines, so that in several years our stocks will have improved a hundredfold. Our activities and work for better utility poultry can be rewarded in a few years. *Already the class of utility bird seen at our current Shows is miles ahead of what first appeared a few years back.* No one can deny that. I have been attacked by breeders because utility

Shows teach the public to ask for high-grade stock. Surely if a breeder sells stock the public have a right to ask for graded articles. Of ten cockerels of equal pedigree, surely one is in some way better for grading than another. If I buy a cockerel for £5 5s. all I want is a graded article. At the moment it seems to be a question solely of what the dam laid; if 220 eggs in her pullet year, then her sons cost £3 3s., and so on, up to £20 for the son of the 310-egg dam. *It is all quantity before quality.* It is severe grading that is wanted with the sons and daughters of high producers, and that grading should decide the price paid for the individual bird. Why should there be a rule that every son must be deemed up to sample because his mother laid over 250 eggs? Are we not pushing forward to the public too much of the finished article and not enough of the raw material? When a man starts poultry-farming, does he not want to start with raw material and gradually build up a healthy strain? Should it not be a case of making and building, also maintaining? Buyers who have the finished articles complain that out of each sitting they get but one or two chicks, whereas at one-tenth the price they get from the local smallholder twelve chicks from twelve eggs from his hardy stock. Trace one's thoughts back to the thousands of beginners in poultry-farming who made heavy losses at the start through buying weakly finished articles. Imagine the valuable capital they lost by getting outbreaks of roup. Don't blame their ignorance of poultry matters, blame the stock as well.

I am often accused of attacking the trap-nests. No one has worked harder than myself for trap-nesting, even recommending it strongly to backyarders in my books written expressly for them. I have, however, pointed out to critics the shortcomings of trap-nesting, as so often practised, in defence of my handling system. Trap-nesting is supposed by many to be a cure-all and a positive. It has consequently become quite automatic, so that the trap-nester believes that the bird laying the most eggs in twelve months (the pullet year, above all) is his best layer and breeder. If one just keeps hens commercially for eggs alone (in the minority in this country), one has no right to suggest what type of bird he should keep. He can cross Leghorns with Wyandottes, Rocks with Rhode Island Reds, Black Leghorns with Anconas, etc. His Light Sussex need not have striped hackles, and can feel proud with feathered shanks. But stockbreeders upon whom falls the onus of keeping breeds pure and hardy surely have a clear duty to buyers and to the industry.

To be carried out properly *trap-nesting should be employed for all birds that are to be bred from and also for longevity.*

Every egg should be weighed and tested for hatchability and rearability of the chick, and there should be thorough matings for testing. We all know that testing is difficult because fowls are so short-lived, and by the time matings have been tested the stock are rather aged for service. There should be severe grading and a firm determination to grade out of the breeding pens all birds that will not be useful in strain-building. How many work on these lines? Is it not the general idea that the more eggs a hen lays and the more will her progeny produce, making out the layer of the most eggs to be the most valuable bird on the farm? Does not every breeder's catalogue start off with the highest producers in Extra Special Pen 1?

Trap-nesting tells us, however, what a pullet has laid in the nest-box. Handling also has its shortcomings, but it allows full play for the breeder's powers of observation. He selects to a type of bird that will be useful, and has no temptation to slip in a hen that has laid 280 "Bantam" eggs as he would have were he, as in trap-nesting, acquainted with her egg-yield. It is human nature to be greedy. Those who criticise hand-grading have an idea that *handling is a failure if it fails to place in exact order twenty birds according to the exact number of eggs the trap-nest has recorded to each bird's credit for the pullet twelve months.* Even the number of eggs laid in the trap-nest for twelve months is no key to every bird's grading. She may have gone wrong for a spell, losing so many eggs from her total. The owner may have been lucky in getting one bird into the broody-coop before she had lost her abdominal contents, but unfortunate with another. It will just mean 30 to 50 eggs difference between the two. *Again, a pullet may carry an egg in her shelling department through fatty internals for a month and lose 30 to 50 eggs therefrom.* What, too, of a pullet which has all the good fortune and lays to her full total of 180 eggs but, being a low-grade bird, lays 89 eggs the second year? And of the pullet which has bad luck and, producing but 179, is turned down for breeding, only to yield 220 the second year on her quality?

Every stockbreeder will, of necessity, trap-nest his stock, but I am asking him to add to trap-nesting the valuable asset of hand-grading. He will hand-grade his pullets put up for trap-nesting, rejecting all that are too small to be good breeding hens after their laying, and any which are so obviously untrue to type or character. He will waste no time or labour on deformed birds, and it is useless for the critics to say that this is not at present done. If I cared to mention names, *I could trot out a long list of breeders in whose poultry houses I have noted deformed birds all being trapped.* Why not find out by handling that they are deformed before time, money, and labour are wasted on them?

When the full year's record is complete, hand-grading, which looks ahead, will give the fully experienced an idea as to future possibilities in good time to remove from the breeding pens any hens not likely to be good for longevity of laying. And *when choice has to be made between two good hens and two good males, why not support that one with the better breed characters?* I do not mean pure exhibition type but true characteristics. We all know that a Light Sussex with a pure white hackle (and I have had in the early days plenty of this stamp shown under me) ceases to be typical, just as a White Wyandotte male with white lobes ceases to be typical. There must, after all, be some difference between Black Leghorns, Black Minorcas, Black Bresse and Noir du Berry. Utility must come first, and no one can deny that in all my judging I have placed utility first. I have passed through some very tricky times in working white-hackled Sussex up to something typical, if not classic in stamp, for hackle. A visit to any important utility Show to-day will show the great improvement in utility stock over that exhibited at the Shows years ago, proving the influence of my judging methods. The birds sent to Laying Tests have also improved as a like result.

The broad-minded poultry-keeper will see the good that hand-grading offers; the narrow-minded one will be still determined not to. But *what of the vast number of poultry-keepers who have neither time nor facilities to trap-nest their birds?* I am up and down the country constantly, and see therefore the varied conditions under which poultry-keepers work. The great majority cannot trap-nest, and I ask if they are to be denied a good method of selection by handling. Surely some method is better than none at all! I have often been amazed at the way some keen poultry-keepers carry on under such disadvantages. A man may keep poultry and only see them during the week-ends, the wife going up to the farm just to collect the eggs. Many work in factories and in the mines, and it is dark in the winter when they go to work and also when they return, but they keep a large head of poultry. Trap-nesting is impossible, just as it is with the backyarder in the majority of cases. *In Tottenham, hand-grading has become quite a fine art, and I would like to back three members of that society against any three from another at picking out the layers.*

Execution is very important, and too many think that they can master handling methods in two minutes. The best way to learn hand-grading is to trap-nest some birds of all grades and in that way to understand exactly what capability means. If you cannot trap-nest, you can visit a utility Show and obtain some lessons from the judge, or join a society and secure help from any good hand-grader. That is one benefit exhibitors have in utility Shows in that they can have educational lessons in selection. At all Shows where I judge I make a point of remaining almost to the close to help any interested exhibitor. But one must bear in mind that considerable practice is needed to get the proportion of a good layer. Above all, too, make up your mind that you are going to master selection. You need not concern yourself with what critics say against me or my methods, and need not take it for granted what I say, because you have some birds at home and can test my system on them. You will be the best judge after all and can well act as the reliable and disinterested jury between plaintiff, Mr. Critic, and the defendant, P.-O.

I am anxious that all who keep poultry will make use of those parts of my handling system which apply to them and afford benefits. *It astonishes me to meet so few who know even how to hold a bird properly.* When you get well on your way with handling you will begin to use the powers of observation, and that is another important aim in my teachings. When a prominent poultry-keeper sends me a liver of a hen that died, and tells me that it is studded with wheat which must have passed through the gizzard—thinking I would be so interested—and the wheat kernels were tubercles, I am in a better position than most to know whether or not there is room for handling in this country. When I have shown under me a humpty-backed hen which has been washed and rubbed down with towels without the deformity being noticed, birds blind in one eye, and others with deformed shanks, I think there is a strong claim for handling. I will not mention those birds with a few eggs in the abdomen or with the latter full of dropsical fluid and ready to burst.

On my suggestion, many who show under me in utility and are successful have entered the Laying Tests this year and are doing well. Already I see that in the first Test finished (winter competition) utility exhibitors have led the way. I refer to the Bewdley Laying Test, the final report for which (just published) reads:—

WHITE LEGHORNS.

1st—C. H. C. Partridge
2nd—F. W. Barley.
3rd—F. Heydon.

WHITE WYANDOTTES.

1st and medal—F. Heydon.
2nd—W. B. Lloyd.
3rd—C. H. C. Partridge.

A.O.V. HEAVY.

1st and cup—T. Osborne.
2nd—Mrs. E. Farmer.
3rd—H. D. Barley.

A.O.V. LIGHT.

1st—W. Burke.
2nd—H. G. Badger.
3rd—R. D. Thompson.

40

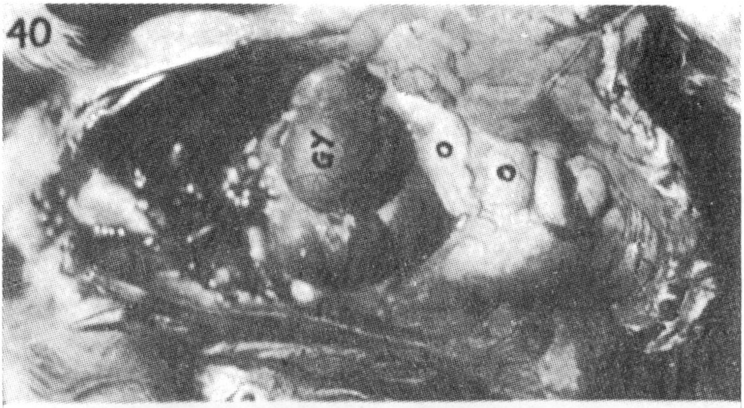

41

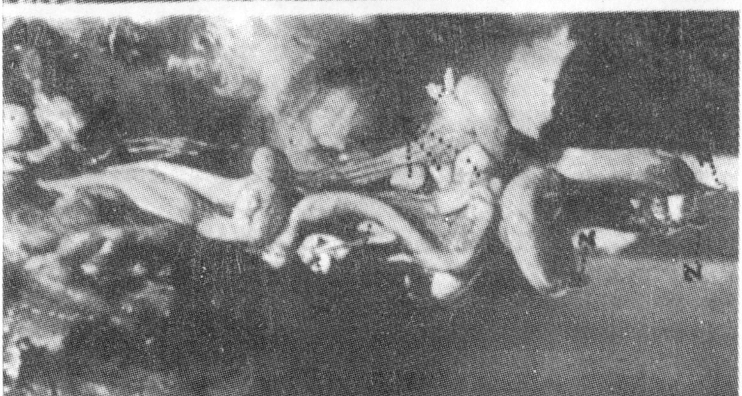

Taking White Leghorns, all three places are filled by those who show utility birds. Mr. Partridge is a very heavy supporter of utility classes under me, and I remember at Worcester Show Mr. F. Heydon told me he had shown all kinds of breeds under me and every bird had secured a ticket. He also told me incidentally that Mrs. Heydon was going to send me her bills for oil, as he was so often out handling his birds at night. It was left to Mr. Heydon to win in White Wyandottes and for Mr. Partridge to obtain third prize with a pullet which I placed in a class of 87 pullets at Yeovil Show. She moulted in the second month, laying only a few eggs that month and, of course, sending pullets to a Test when they have been to a Show is rather risky in face of a likely moult. Many send my winning birds to Tests, in fact Mr. Wace tells me that my Yeovil winning Columbian Wyandotte pullet of 1921 was his best layer at Harper Adams and never broody. She also had a rest at the Test and did not lay till December. Birds to win Tests should be especially selected at the last moment. Mr. F. W. Barley is also a winning exhibitor at utility Shows, and secured second in White Leghorns.

Mr. H. D. Barley, who is also a successful exhibitor in utility, secured third in A.O.V. Heavy Breed, and Mr. R. D. Thompson's pullet—third in A.O.V. Light Breed—I picked myself, Mr. Thompson having shown Brown Leghorns under me most successfully in 1922, his first Gloucester Brown Leghorn under me still laying well in her third season.

I ask all my friends who support handling to enter the Laying Tests so that critics can see that our birds can do the laying and yet answer to breed and beauty. I would be glad to have notice of such entries before the Tests commence, as I shall of all pens selected on my system. So far I have not made use of such performances, but nothing must be left undone to offer evidence of proof of the help hand-grading can be to poultry-keepers of every description. The critics have often stated that score-carded birds mated together would only breed "mongrels." Mr. Partridge mated the first Tottenham 1921 cockerel (under me) to the first Westminster 1921 pullet (laid 273 eggs), also my score-carding, and bred the pullet which won Bewdley Winter Test. Other pullets bred from these two won firsts and cups at Leicester, Worcester, New Barnet, Olympia, etc., in 1922.

For many years I have selected pullets for Laying Tests, and, having obtained my fees, it has not been my wish to give publicity to the wins secured. I have left the breeders concerned to attend to the publicity side on their own behalf, being quite satisfied with my part of the contract. It has often been said against me that my card and system are based on theory, but I have always understood that once a thing works out in practice it ceases to be theory. For instance, I have judged the same birds over and over again and have still been successful enough to obtain record entries, Show after Show, the same exhibitors sending me their birds. I was elected the club judge for the Austral Orpington Club with a view to the usual one club Show, but, as things panned out, there were eleven club Shows during 1922 and I took them all. What is more, every bird at every Show had to be score-carded by me, even though at most Shows I had hundreds of other utility exhibits to get through. I heard of no complaint as regards scoring one bird high at one Show and low at the next. Certainly I was complimented often on scoring birds within a point each time and out of 200 maximum. *And at the close of the year after the Olympia judging, those at the general meeting voted again for my score-card to be used at their* 1923 *club Shows, and when selecting their judges graded out all nominations if they were not score-carders to my card.* This chapter is not a free advertisement for myself, but all who know what attacks I have had to bear will realise that this volume is my own platform and from it I can have my say without having the editor return the letter to the sender because the facts supporting my handling were too strong. It is easy to return a letter as being a free advertisement! It is easy to blue-pencil matter as irrelevant! It is easier still to add "This correspondence is now closed!"

I have carefully gone through all Test reports and secured the data of all pens I personally selected in 1921 for the 1921-2 Laying Tests at home and abroad. It must be remembered that I am not so well placed with my selections as a breeder picking his own birds at the last possible moment. My critics often say that they never see my name in the Tests, but if they do not they see pens in all of them which I have picked and others which have been selected on my system. After I have taught a breeder selection to my card, he picks his own. I have Shows to judge and am pretty busy in other directions when selection time comes round, and often I pick months ahead, nominating reserve birds in case any should lay before date of dispatch. I know nothing of pedigrees and am not in a position to watch the birds or tend them, or to make a final rejection or replacement. I have to select far ahead, timing the pullets to lay when the given Test starts. Any fair-minded person will agree that I work under difficulties. Let us see what my selections for 1921-2 competitions did by way of laying, all picked just by handling without regard to pedigrees whatsoever:—

My first selection was a pen of Rhode Island Red pullets for Mr. G. F. Bell, of Mindrum, Northumberland, for Bentley Test, 1921-2. These five pullets laid at Bentley in the N.U.P.S. official eleven-month Test as follows, despite a very slow commencement:—

No. 1—149 eggs.	No. 4—160 eggs.
No. 2—162 ,,	No. 5—178 ,,
No. 3—223 ,,	

Here, then, we have an average of 174 *eggs per bird for the forty-four weeks* during which the competition ran, and over 86 *eggs per bird for six winter months,* despite a slow start. What is more, the pullet with the 223 eggs to her credit was the *second best out of the* 200 *Rhode Island Red pullets* competing. Four out of the five pullets were officially given as in full lay at close of Test. Award secured was FIFTH POSITION and FIRST-CLASS CERTIFICATE (forty pens of Reds competing), with but two eggs below the third pen, eight behind the second or silver-medal winning pen, and thirty-eight behind the leading or gold-medal pen.

The full year's trap-nested record (twelve months) of these pullets was:—178, 181, 250, 196, and 191, or an average of 200 *per bird per annum.*

I also picked for Mr. Bell a pen of White Leghorns for Bentley, 1921-2, but when they were sent off they caught cold *en route* and were returned. Despite the journey to and fro, the traps were put on them and they laid 231, 190, 190, 180, and 150, but such were not official and will be ignored. But having selected the pen, I include their home record in order to cover the complete list of my selections for the year.

Next, I selected the following pullets for Mrs. Ward-Jackson, of Street Court, Kingsland, Herefordshire.

Australorps received first attention, and it is well to remember that the Australs had just been introduced to this country and none had had experience in their selection. Therefore I applied just my ordinary handling system to them. Again, the parent stock had only arrived in this country from Australia in April, were mated on arrival without any rest, and were pullets that were bred from. Consequently, when I was picking, I had to deal with May-hatched pullets in a heavy breed. It was no fault of the importers, seeing that the parent stock had been ordered in ample time to arrive months before the breeding season, but dispatch was delayed from the other side. When importing stock from Australia, what with the seasons being just the opposite and the journey a long one, imported Australs have not had the best of good fortune. That is why they have been laying small eggs, but now that there are in the country plenty of well-matured hens which have become acclimatised one need not worry about the small egg.

I had in the first place to select four pullets for the Indian Test starting in November, 1921, and ending on January 31st, 1922—a three months' winter trial. If you can picture me selecting from May and June hatched pullets in a heavy breed like the Black Orpington, in order to be ready to commence production on November 1st and to be picked in good time for dispatch to India, you will have an idea of some difficulties which I have often to work under. At any rate, these four pullets averaged sixteen eggs per month for the three winter months and finished FIFTH AGAINST ALL BREEDS, beating all the popular breeds. Had the total eggs counted for the four, it would have been just one egg behind the best four pullets. But it was a single-bird trial.

Award: FIFTH POSITION FOR HIGHEST INDIVIDUAL TOTAL (four single pullets averaged *forty-eight eggs per bird for the three winter months*).

Next I selected eight Australorp pullets from the same May-hatched youngsters bred the same way and full sisters for the *Daily Mail* Laying Competition at Bentley, 1921-3, which runs for two years. You can imagine my difficulty with May-hatched pullets over the small-egg factor, but there were no March or early pullets for me to select from. What one has not got one cannot send, and my selections have to follow the birds placed before me.

At the time of selection many well-known local breeders were present, and I did not like the proposition of selection, but concentrated on pullets likely to thicken as they became older, hoping thus to lessen the number of second-grade eggs. I promised responsibility for satisfactory laying, but issued a warning as regards size of egg. I also had to include one or two pullets in the eight which would not begin to lay till the Test was many months old, and hoped they would get through the selection committee sandwiched in with the other more forward ones. I think one of this brand did not commence to lay till the fourth month and another only laid a dozen for the first three months.

With the first twelve months over I can well be pleased with my selection. The eight pullets stand 105TH OUT OF 200 PENS of all breeds, having produced 1,616 *eggs for the twelve months, or an average of* 202. Unfortunately, I have had to carry 196 second-grade eggs which do not count, otherwise I would have been about twenty-sixth for position. But, as I anticipated, the pullets have thickened out (Croads, Orpingtons, and the like show such a tendency with age, as I knew) and when other breeds in the summer went down for egg-size this pen of

Australs was the only one as officially reported to go up. During the eleventh month the pen laid 126 eggs, of which but 4 were second grade, and during the twelfth month (October), 89, of which but 4 were second grade. Therefore the eight birds may have a chance for the second year of the competition.

For the twelve months the eight birds laid as follows:—

No. 1—201 eggs.
No. 2—214 ,,
No. 3—152 ,, (did not lay till fourth month).
No. 4—247 ,,
No. 5—208 ,,
No. 6—167 eggs (only laid twelve eggs first three months).
No. 7—193 ,,
No. 8—231 ,,

An average of 202 eggs per annum as against 208 for White Wyandottes, 197 for Light Sussex, 194 for White Leghorns, 145 for Anconas, 170 for Black Leghorns, 188 for Rhode Island Reds, and 217 for Buff Orpingtons, *or a six-months' winter average of 97½ eggs per bird, despite two being inactive.*

When you see a critic running down the Australorp for small eggs, just think of the drawbacks that the breed has had to undergo in fetching them over with the long journey and the opposite seasons. Allow time and you will find that there is sterling merit in the breed, although critics shout it down, not being behind the scenes. Learn the lesson, too, from late hatching, which cuts down the capacity side of the pullet and tends to produce small eggs even though the pullet may have been hatched from a 2½-oz. hen's egg. The majority of small eggs in Tests can be put down to the lateness in the season when they commence. Think carefully over this problem on the following lines. The competition you have in mind commences about mid-November, and you are to send pullets that have not laid prior to dispatch. I hope you will not have any March-hatched Wyandottes or April-hatched Leghorns that have not commenced production by mid-November. All your best will have started production long since. Should you have any early pullets which have not begun, they will be "duds" and not competition winners. It compels one to send late-hatched birds, and there is the tendency to small eggs. One must therefore (being robbed of sending your best early birds because they are laying away merrily) hatch in plenty and not force the birds along. The greater the number to select from at the last moment and the better. There *must* be a continual hatching programme maintained to leave birds of the right type and development if one has many entries. It is often my bad luck to find a breeder with plenty of pullets hatched too early or too late, but none of the intermediate kind, which is in itself sufficient to rob him of a chance of success. Some take the bull by the horns and send the best pullets whether in lay or not, but, while a few birds lay on, others take two or three months' rest. Some rely upon early pullets which have laid in July and August and take a moult, but such present a lottery, and may not start re-laying till December or January.

One cannot say I am out for destructive criticism when I point out the lottery side of laying competitions. Constructive criticism suggests an earlier beginning for all trials, with all birds to be in before they are ready to lay. It is December before the pullets get going, and even January (second month) in some trials, and where, I ask, is the winter record for the birds as from October to March? Again, when the birds come back, being originally late-hatched every breeder is encouraged to breed from late-hatched stock year after year, which does not help stamina. The birds, too, have to be selected and sent off when the owner is fighting against colds at home among the late-hatched stock of pullets, so that no wonder Tests have trouble from colds and roup directly they begin and soon after the arrival of the Test birds. I am constantly on farms at this time and think I should be in a position to know what the conditions are generally. There would not be so much trouble from colds and roup on home farms but for these late-hatched pullets intended for Tests. In some counties, however, one can hatch earlier for late Tests because the pullets do not mature so quickly as in the South, the latter thus having the disadvantage. What handicap did the southern breeders suffer after hatching late for a certain November Test which was brought forward at the last moment to start early in October?

The best advice I can give to all breeders is to grade very severely the late broods and never hesitate to kill on sight any "bad doers." Next, avoid changes in housing for the flocks, but rather take out from the main flock the selected pullets for the Tests and place them in small lots in small houses, making the final selection at the last moment. Let the main flock then be liberally fed so that they remain graded and plump ready to fight any colds, etc. And do not shift this flock but let them settle down for winter production in ripe condition.

Often when picking pullets I have found birds with roup and colds in the very flock from which must come my selected birds. In handling all the pullets, to pick the best out I may hit upon a pullet that handles all feather and bone, and when opening the "mouth" find it full of diphtheritic growths unbeknown to owner. The selected pullets have to go from this flock, but I can quite understand why Test managers have a warm time through outbreaks of colds and roup. When I take such birds from a flock I always insist on treatment before dispatch and isolation, but such does not help the egg-laying of the chosen pullets.

I also selected for Mrs. Ward-Jackson eight White Wyandotte pullets for the *Daily Mail* Two-year Competition, 1921-3. *With* 859 *eggs for the eight pullets for the six winter months, we have an average of* 107 *per bird. For the twelve months the total was* 1,723 *eggs for the eight pullets and five still in lay (November), or an average of* 220 *per bird for the year.* During the first month of the second year the pen is in the twenties for position, but ended the 110th out of 200 pens.

Their first-year individual totals were:—

No. 1—221 eggs.	No. 5—195 eggs.
No. 2—255 ,,	No. 6—238 ,,
No. 3—236 ,,	No. 7—237 ,,
No. 4—234 ,,	No. 8—96 ,, (went wrong).

Of course, I had my usual "lottery" touch, No. 8 producing but 96 eggs divided up into 2 of first grade and 94 second grade. Don't smile when I point out that these 94 eggs were no larger than a blackbird's, containing albumen only and no yolk. But one takes it all in good part to come up smiling the next year. Of course the bird is wrong somewhere, and may have an internal watery cyst in oviduct or at ovary to irritate her, but in the eyes of some critics it would be just the thing to condemn my hand-grading system.

Five White Wyandotte pullets were selected at the same time for Mrs. Ward-Jackson for the Midland Laying Test, 1921-2. They finished twenty-ninth out of sixty pens with individual records of 148, 99, 94, 65, and 114. All readers will be aware of the considerable number of deaths at this Test from roup, etc., and without a doubt, as at Tottenham, remedies tried affected the laying organs of the birds, considering the immense number which only laid two or three eggs month after month. It furnishes us with some data as regards production, namely, that whatever be the grading of a pullet, conditions make or mar her production.

Tottenham single-pullet Test gave us also some valuable data. Being the first attempt, we decided to reject for colds only and not immaturity, etc. Some birds arrived with roup and colds, and were returned. Colds, however, started within a week of the birds arriving and trouble began. In the first place, far too many pullets sent in were late-hatched, even from breeders, and some mere chickens. I gave one till March to lay its first egg, and the bird began to lay in April—six months after the competition began. A large number were not even set in bone, and could have been sent back as immatured. Many were sent in lay, and within a month all were at sea as regards condition. Those that had been sent in lay moulted and started those just coming into production; the young and backward pullets began to take colds, and drastic treatment had to be carried out, such as dipping, etc., as generally recommended, in order to prevent any outbreak of roup. We soon discovered that, after taking three months to get an ungraded flock back to condition, we had prevented roup but, on the other hand, had taken too much out of the birds—production thus suffering severely. I think the highest egg-total for the twelve months at Tottenham was 197, and at Midland, 205, both records being taken by birds from my own students and picked by my system, and both White Wyandottes. When the Poultry Institute gets going, the first experiment should be to see if vaccination will cure roup without affecting egg-laying of the birds. I am sure that, *while I can prevent and hold roup by the old methods, I can only do so by interfering severely with production and ruining the birds.* If to prevent roup we have to lose the production of the birds treated, so that many doctored on old lines drastically lay under fifty eggs for the year, then we want newer remedies, and I am sure our road lies in the direction of serum.

Tottenham Test should, however, teach all readers that, unless pullets are graded into degrees of development when laying time approaches, it will take months to get condition general and a graded flock. Moral: To get the best out of feeding, etc., support graded flocks and grade drastically before the winter weather comes in to attack all weakly and backward pullets. The larger the flock and the greater the need for graded stock in that house and flock, because one can condition six pullets in a house sooner than six hundred. In a small flock, too, every bird can be under observation at feeding and other times. In a large flock, some ailing birds may go unnoticed until they have spread trouble. Moral: Always remain a short while when feeding a large flock, and always be on the lookout for mopish birds. At Tottenham Test pullets moulted right through and even moulted again, whereas had all birds been in small pens and houses of six instead of in one large house, many lots would have been got fit to carry on with the laying. It is, as I have already stated, easier to condition small flocks and to get more eggs from them than from large flocks. When you are working to large units be very careful to concentrate on grading and conditioning as laying time approaches. In the Midland Test of 1921-2, also at Harper Adams trials some years back, the outbreak of roup started in and clung to the large-flock houses. It is only reasonable to contend that grading should be very strong under such conditions, for if a few weedy birds are allowed in the flock will suffer, the weaklings soon falling a prey to ailments and spreading them. I have always said that pullets in lay and taking a moult can start other pullets into a moult before they commence production. I was the first to draw attention to this partial moulting of pullets and the critics then declared that I was wrong. If I mix ten early-hatched pullets in lay with ten later ones out of lay and the former should take a partial moult, the latter can readily follow. Then as one is getting some of the birds fit again they are sent off into another moult. Some pullets just moult at neck, whereas many take a complete moult, especially the highly-strung ones, which show fright in eye. Do not select such pullets for a Test, and be sure that all pullets sent are up and level in neck-hackle, because when there is a neck moult the hackle will be straggled and loose or parted, and there may be new feathers coming underneath. Note this, because when a flock is not laying well you might need to check up for a partial moult as the likely cause.

It is because one man's pen may arrive in lay or moulting that it is not fair to mix at a Laying Test another exhibitor's birds with them, because the only fair test can come when each man's birds are placed on their own. Again, the conditions should be the same as regards housing, and at the Midland Test of 1921-2 some pens were in small houses and yet competed against those in large-flock houses. As mentioned, trouble from colds, etc., readily starts with large flocks and takes longer to blot out than small pens. This year's (the second) Tottenham Test has been very successful, the new rules allowing for all pullets sent in lay to be rejected on arrival. Hence we had better graded flocks and no trouble. The Midland Test has also been a greater success, and good results have been reported for treatment of roup cases by serum.

Three Australorp pullets for the single-pen section at Harper Adams, 1921-2, did not complete my list for Mrs. Ward-Jackson, but they laid, respectively: 188 (started second month), 159, and 177.

The final selection was very interesting, seeing that I picked them out after dark by handling aided by a storm-lamp. Mrs. Ward-Jackson had forgotten that she had entered a pen of Golden Buttercups at Harper Adams, 1921-2, and, discovering this at the last moment when I was due to leave, mentioned it, but said she would not bother to send the pen. But I went down to their roost and picked out my selected and rung them. Imagine my amusement when told by letter afterwards that I had included her best exhibition pullet which had been earmarked for exhibition at the Dairy Show. As there was time I advised her to send the pullet to the Dairy Show and, on its return, straight to Harper Adams. So the pullet went to the London classic, secured second prize, and went with the other pullets to the Test. They were pure exhibition stock and competed in the A.O.V. section.

Again I hit upon an unlucky corner and lost four of the seven pullets. This A.O.V. light-breed section was a sad one, seeing that *out of the fifty-six pullets competing, no less than twenty died*. Lucky indeed were the three who finished, each with six birds in the pen, and they stood first, second, and third in the finals. The lottery side of the Laying Test is again seen here. The tenth monthly report credits my pen with 626 eggs, whereas the twelfth and final report gives 597. This is easy to understand when you know the key. At Harper Adams, when a bird died, with her departed all eggs she had laid. And a pen of Black Leghorns, I am told in the Press, after standing second till the eleventh month, lost a pullet twenty-four hours before the closing second. (I suppose on the stroke of time an assistant visits each house to make sure none has died.) My 597 eggs therefore is the total for the only three pullets left standing after the final round. One laid 210 eggs for the twelve months, another 188 (started second month), and the third 197 with two eggs unrecorded. Not so bad for exhibition-bred Golden Sicilian Buttercups (bred for beautiful barring and up to classic standard).

We captured an award here in a SECOND-CLASS CERTIFICATE for exceeding the 200 eggs.

My next trip was to Mrs. N. Wilkinson to select some White Bresse. One pen had to go to the N.U.P.S. Test at Bentley, 1921-2, and the five pullets began well and continued for four months, when up came the little "D" to show that a pullet had died. So we jogged along with but four pullets and yet finished twelfth out of twenty-five pens in A.O.V. against Anconas, Black Leghorns, etc. The laying records for the eleven months (44 weeks of Test) were:—

No. 1—203 eggs.
No. 2—144 ,,
No. 3—160 ,,
No. 4—51 eggs (died).
No. 5—159 ,,

Three were officially reported in lay at the close, and the average for the four would be $166\frac{1}{2}$ *eggs per bird for eleven months*. Had the pullet lived and laid on at just the pen-average rate there would have been 830 eggs at least instead of 717, and the pen would have secured fifth place and a certificate. But Tests do not run that way!

Two or three pullets were also picked for Tottenham single-pullet competition. One finished *seventh for light breeds*, only beaten by White Leghorns, and *beating all other non-sitting breeds*. She stood twenty-third out of 105 against all breeds, heavy and light. Another finished *eleventh for light breeds* and thirty-fifth out of the 105 pullets competing as single entries These two pullets laid 102 and 80 eggs as against the winning pullet's 155 eggs, and a third laid but 50 odd eggs. In order to be quite fair, I mention all my selections but have explained the circumstances which befel both Tottenham and Birmingham Laying Tests. While all birds had a hard time, one accepts the positions made by the birds and also their laying, as all had the same severe conditions to fight. But when reading the actual number of eggs laid by the pullets I selected for these two Tests, any broad-minded person will take all things into account. To me, it shows clearly enough that environment and conditions have a vital say in what a pullet will lay, no matter what her grading may be. And that when you have the trap-nested record of a pullet you merely have details of the number of eggs she actually laid in the nest-box and not what is her key-standard—proof positive that all who trap-nest should use discretion before using or discarding any recorded hen, and that handling will be a valuable aid to trap-nesting; in fact the combination is the ideal procedure.

My final selection was for Mr. Leslie Williams and here I had to pick out a large number of pullets, but luckily had plenty to select from. But I had to keep in mind the objects of this very progressive breeder, namely, size of egg and breed-type.

In the first place, there were five White Wyandotte pullets for Burnley (Northern U.P.S.) Laying Competition, 1921-2. All pullets were selected on September 27th and 28th. *After leading for many months and by a goodly number of eggs, including the sixth winter month*, two birds died about the ninth month. The final twelfth month, however, saw them only down to EIGHTH POSITION with but three pullets left to carry on for the last few months. Despite a slow start, the total yield was 831 eggs, made up as follows:—

No. 1—172 eggs.
No. 2—211 ,,
No. 3—125 ,, (died).
No. 4—194 eggs.
No. 5—129 ,, (died).

The average for the five would be 166 *eggs each*, counting the two that died, while the three *averaged* 192 *eggs each for forty-eight weeks*. What is the odd part is that while certificates were awarded down to the seventh pen (fifty-six pens of White Wyandottes competing), honours stopped short of my pen, although they had laid 831 eggs with two birds short for the last three months or so, whereas the seventh successful pen scored 847 eggs. Any broad-minded reader will see the force of my contention that there should be no first, second, third, and "also ran," but proper gradings into two or three classes according to the all-round performance of the individual pen.

To lead the whole section for SIX WINTER MONTHS makes the pen to many people a good one; finishing down among the "also rans" at the twelfth month proves, I suppose, to many that the pen and strain are low-grade. That is the lottery side of Tests which should be altered if competitions are to serve their useful purpose. There should be broad-minded official awards by way of class certificates or gradings at the end of the competition, any bad luck not condemning a breeder and his strain but being taken into account. To-day you win or you do not; if you win you have a good strain, because the public do not buy Test reports but just see the names of the winners printed as first, second, and third. If you lose you have a low-grade strain. Poultry papers with so many Tests to report have no space for the "also rans," and with more competitions springing up everywhere reports will be still further curtailed.

The score-cards of these five pullets may be of interest. Selected on September 27th and 28th, with the competition commencing on November 1st, the birds would be dispatched about mid-October. No. 1 laid three eggs first month and No. 4 eight eggs, the others beginning in the second period—November 29th to December 23rd. They would be picked by me two to three months before laying, and five or six weeks before the Test commenced. One can follow the scores by going straight through my score-card.

No. 1.—Laid 172 eggs. Capacity: 10, 1½, 8, 8, 4½, 8, 8; total, 47½. Capability: 10, 3, 15, 4, 4, 8½, 9, 8; total, 61½. This bird when handled would be two-and-a-half fingers at abdomen and one finger between pelvis, and, increasing up to four at abdomen and three between pelvis, would score around 57½ for capacity when in full lay, so long as she did not begin to lay too soon. She was excellent in capability but rather too fine, which would act against her laying if capacity ended short through early laying. That is what happened to her, I should say.

No. 2.—Laid 211 eggs. Capacity: 12, 1½, 8, 8½, 5, 10, 8; total, 52½. Capability: 10, 3, 15, 4, 4, 8, 8, 7; total, 59. This is the stamp of bird I like and aim for. Measuring three fingers at abdomen with one between pelvis bones she would increase probably to five fingers at abdomen and three between pelvis, getting well into the sixties for capacity, while capability is high.

No. 3.—Laid 125 eggs (eight months). Capacity: 12, 1½, 8, 10, 5, 8, 8; total, 52½. Capability: 10, 3, 12, 4, 5, 7½, 8, 7½; total, 57. Useful all-round sort, with three fingers at abdomen and one at pelvis. Not such a good handler as No. 2 for capability

No. 4.—Laid 194 eggs. Capacity: 8, 1, 7½, 8, 5, 8, 8; total, 45½. Capability: 7, 3, 15, 4, 4, 8½, 8½, 8; total, 58. A backward bird; two fingers at abdomen and hardly one when handled between pelvis; would probably get up to 58 or 60 for capacity; nice texture.

No. 5.—Laid 129 eggs (eight months). Capacity: 12, 1½, 8, 8½, 4½, 10, 8½; total, 52½. Capability: 8, 3, 12, 4, 4, 8, 8, 8; total, 55. A similar bird to No. 2. Three fingers at abdomen and one between pelvis bones, getting capacity up to the sixties in full lay, and with capability high for a plump bird (well-fleshed pelvis bones) which would increase in points with production.

Four of these five pullets were pen bred and only one individually bred from special pen 2.

Eight White Wyandottes for the *Daily Mail* Test, 1921-2, were next to be dealt with. *Holding fifteenth position at the close of the six winter months, one pullet died in the seventh month* and away "west" went the pen. Seven good pullets cannot lay as well as eight good ones! The final placing of the pen at the end of the twelfth month, despite the loss of one bird, was *fifty-fourth out of the* 200 *pens.* Here are the individual records of the eight pullets for the twelve months:—

No. 1—169 eggs.	No. 5—190 eggs.
No. 2—267 ,,	No. 6—244 ,,
No. 3—126 ,,	No. 7—227 ,,
No. 4—236 ,,	No. 8—109 ,, (died).

Here we have a total of 1,581 eggs for the twelve months for the eight birds, or *an average per bird of* 197 *eggs.* Taking the production of the seven pullets that lived, we have an average of 210 *eggs per bird.* Had the pullet not died we would have landed in the "money" in the twenties, but——!

All the eight pullets were pen bred and none individually bred, although I was picking, as is my custom, without knowledge of pedigrees. Naturally, all the birds were well bred from good stock hens mated to the best cockerels. But I contend that if a pen wins they will bring home their pedigrees. I prefer to leave things to my practised hands and the handling of the birds, and when selecting I have all pullets through my hand whether bred from pen 1 or pen 10.

Five pullets for the National Test at Bentley, 1921-2, were also to be picked. These White Wyandotte pullets laid 866 eggs for the eleven months (44 weeks), or *an average of* 173 *eggs for the eleven months* per bird. They finished SIXTEENTH POSITION OUT OF SIXTY-FIVE PENS of White Wyandottes.

All five were again pen bred and the individual records were for eleven months of Test:—

No. 1—140 eggs.
No. 2—195 ,,
No. 3—176 ,,
No. 4—162 eggs.
No. 5—192 ,,

A pen of White Wyandotte pullets were included for Harper Adams, 1921-2, also a single pullet for the single-pen section. It was three months before the six pullets got going, which ruined their chances, but they finished *forty-fourth out of sixty-three pens* of the variety. The six pullets laid 1,051 eggs in the twelve months, or an *average of* 175 *eggs per bird*, despite the bad start. The individual records were:—

No. 1—166 eggs.
No. 2—194 ,,
No. 3—177 ,,
No. 4—162 eggs.
No. 5—160 ,,
No. 6—189 ,,

No. 1 scored on September 27th as follows, the Test starting on November 1st:—Capacity: 8, 1, 10, 10, 5, 10, $8\frac{1}{2}$; total, $52\frac{1}{2}$. Capability: 8, 3, 12, 4, 4, 7 7, 7; total, 52.

No. 2.—Capacity: 8, $1\frac{1}{2}$, $6\frac{1}{2}$, $7\frac{1}{2}$, $7\frac{1}{2}$, 8, $7\frac{1}{2}$; total, 46. Capability: 10, 3, 12, 3, 5, $8\frac{1}{2}$, 8, 8; total, $57\frac{1}{2}$.

No. 6.—Capacity: 10, $1\frac{1}{2}$, 8, 8, $4\frac{1}{2}$, 8, $7\frac{1}{2}$; total, 47. Capability: 9, 3, 10, 4, 4, 8, 8, 8; total, 54.

All pullets were again pen bred.

If one were selecting in September one would roughly hand-grade all the pullets and set apart the selected ones. Then the owner would go through the pullets at the last moment and pick those which were just up to the ideal for development, etc. An individual owner could do this. I have to select my birds, pass on to the next place, and fit in a few Shows as well.

The White Wyandotte pullet for the single-pen section rather interests me. It shows how awards operate at a Test like Harper Adams, where values are concerned and not the number of eggs. Just another case to support the lottery side of laying competitions. This pullet failed to start till the second month, when she went on to yield 210 *eggs for the eleven months* or periods, only the first egg being second-grade. She finished only THIRTEENTH OUT OF TWENTY-SIX ENTRIES for the variety. Eggs are counted on market values just as they are laid and if all started production at the same time or if the counting commenced when all had begun to lay it would even things up a bit. But 210 eggs in eleven periods is not bad production. She lost the value of twenty eggs for November—the first month of Test—when she failed to get going. The final report now reads:—

Position.	Eggs Laid.	First Grade.	Second Grade.
11th.	201	168	33
12th.	200	198	2
13th.	210	209	1

The average reader would probably find it hard to reason out why my bird which holds thirteenth position does not beat those above which have laid fewer eggs. When I take 20 eggs as the key, I find that

20 eggs laid in November (first month) were valued at 7s. 5d.
20 eggs laid in May to June were valued at 2s. 6d.
20 eggs laid in September to October (last month) were valued at 4s. 7d.

Therefore this pullet lost 7s. 5d. or more the first month, and she struggled through the year trying to pick that up at 2s. 6d., etc., a dozen, and while making up a lot of ground she found the handicap too much. Luckily, Harper Adams have a grading prize for birds laying over 200 eggs, so that this pullet secured a SECOND-CLASS CERTIFICATE. I think a first-class certificate is awarded to every pullet laying 250 eggs or over, and a second-class certificate for 200 to 249. Therefore had this pullet laid twenty the first month she would have still secured her second-class honour and in consequence lost nothing in that direction.

I will just deal with another like case. Mrs. W. Beale entered a Rhode Island Red pullet selected by my system in the single-pen section at Harper Adams, 1921-2. She laid 142 *eggs in the six winter months*, when she stood *third for her section out of twenty-seven entries.* These eggs were valued at £1 14s. 9d., and going wrong internally the pullet lived to finish the Test but never laid another egg. And *without a single egg to her credit for the last six months of the trial she only went down to seventeenth position.*

If we take this lesson seriously we will picture what it means to a pen of six pullets, each bird of which gets a flying 7/5 start. Where is the pen that gets a bad start and settles down from a partial moult about the third month? I quote this case to impress upon all readers who enter Tests placed on values that an early start and a good finish are imperative. Harper Adams, I am happy in fairness to report, has a secondary system of awards so that any pullet in a pen can get a first or second class certificate according to whether it lays 200 to 249 or 250 and over. But five such certificates with six pullets would not compensate a man who lost first position among the prize-winners through the lottery of when the pullets begin to lay.

The pen in the Midland Test finished fifty-first, producing numbers as follows:—118, 33, 109, 33, 115. I often smile over the Midland selections with their two, three, and four eggs per month.

My final selection for Mr. Leslie Williams was a pen of White Wyandotte pullets for the North American Test, 1921-2. I think seven pullets were picked out by me to include two reserves. I find I have no details of their production, except that at the end of the winter and for many months they stood second place. One bird died there, I believe.

When reading the totals of eggs already given, one must allow for the different periods during which the Tests ran, some being for forty-eight weeks, and Bentley (N.U.P.S.) forty-four weeks. Also that the total is the number recorded during the official run of the Test. If a pullet begins production in the third month, that is not allowed for, but such birds on arrival home would be trap-nested for the completion of the twelve months from first egg. Therefore *many of my birds would show increased egg-numbers for the full fifty-two weeks from date of first egg.*

Summed up in a nutshell, then, from the view-point of Laying Tests, all my selections only drew the modest two certificates. I leave the individual reader to picture whether the selections were good, bad, or indifferent. *The fifty odd pullets selected by me and sent to the main Laying Tests at Bentley, Burnley, and Harper Adams without any allowances for when they commenced individually to lay, and despite the N.U.P.S. Competition being for* 44 *weeks only and all except the "Daily Mail"* 48 *weeks, averaged over* 190 *eggs each officially at the Trials.*

JUDGE: You have heard what the defendant says, gentlemen of the jury, and it is for you to decide. The defendant declares in reply to the critic-plaintiff that he has neither time nor labour nor £ s. d. to trap-nest his poultry. He advocates the use of the trap-nest for all who can adopt it, but with discriminate use and a knowledge of handling and of the hen from within. Defendant says that if he were better situated he would never hesitate to combine trap-nesting and hand-grading as being the perfect whole. Having studied his selections I put the following questions to you. If you answer them in the affirmative then the defendant wins his case and the plaintiff pays costs.

Gentlemen of the jury, these questions for your consideration:

1.—*Bearing in mind that the defendant is so placed that he cannot trap-nest, has he succeeded in selecting good birds by hand-grading?*

2.—*Would he on his selections by handling have bred from good hens had he earmarked his choice for the breeding pens?*

Then there is a supplementary question. The defendant contends that by hand-grading pullets an experienced handler can save expense, labour, time, and housing, because he is

43

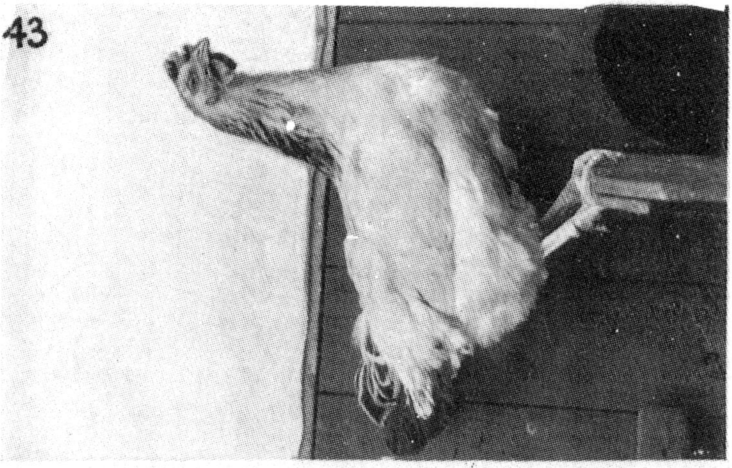

44

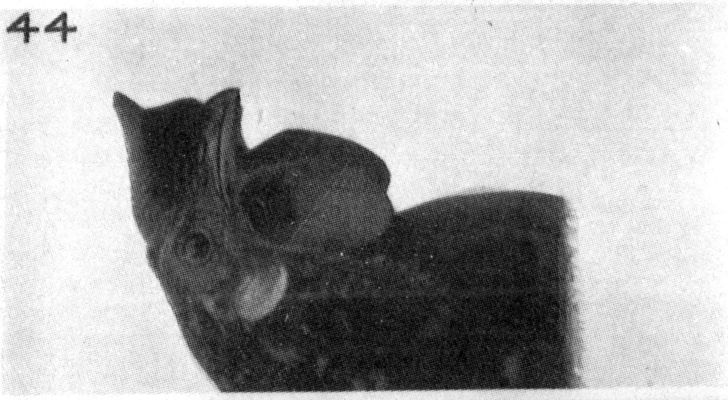

45

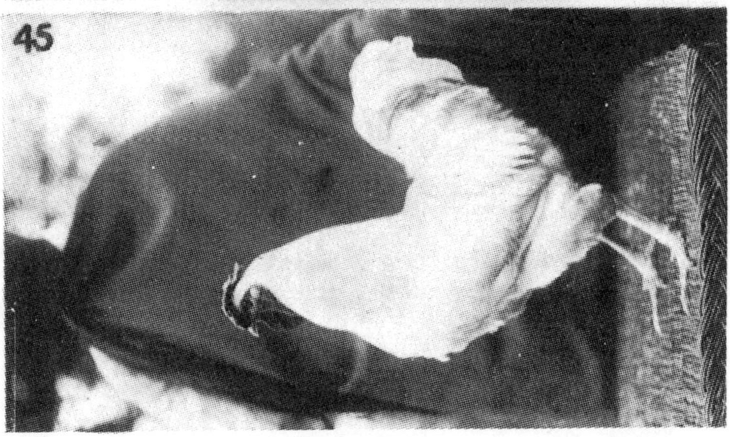

able then to trap-nest only the trappable birds. The plaintiff, Mr. Critic, has sued the defendant for such a libellous statement. Will you answer in the affirmative or negative the following question:—

3.—*Would the defendant by putting all his selected birds into a trap-nesting house have saved any time or labour as against trap-nesting indiscriminately* 600 *laying pullets? Would he have achieved good results therefrom and discovered that by trap-nesting his choice he had got together some good breeders and not wasted his energies?*

JURY: We find for the . . . . . . (That is for you, dear reader, to fill in.—Author.)

CHAPTER XXIV.

WITNESSES FOR THE DEFENCE.

HAVING dealt with my own selections it will not be out of place to publish instances of where my system of hand-grading has been of great help to others as regards selection for the Laying Tests.

CUP AND GOLD MEDAL:—Miss Barbara Raye secured both gold medal and cup for highest scoring pen of ducks at Bentley Duck Test, 1921-2. Her five Khaki Campbell ducks during the eleven months of the competition (forty-four weeks) laid 1,188 eggs. Their individual totals were:—

No. 1—228 eggs.
No. 2—251 ,,
No. 3—281 ,,
No. 4—272 eggs.
No. 5—155 ,,

The average for the eleven months works out at 237 per duck, while the two secured the highest two totals for Khakis, fifty-three pens each of five ducks competing. I publish one of Miss Raye's letters to me:—

PELLING PEDIGREE POULTRY FARM,
SCAYNES HILL, HAYWARDS HEATH.

DEAR MR. POWELL-OWEN,

You will remember the pen of Khakis picked out on your system and sent to Bentley. They are going along nicely and will, I hope, be among the leaders at the end, but I expect you will be watching their laying month by month as you do the pens sent by all your other students.

I intend trap-nesting even more ducks this year and have picked out some gems from my large 1921 flock of youngsters similar to those sent to Bentley. I have now, I think, thanks to your kindly help, mastered the selection of likely layers in both ducks and hens, and I save a lot of labour by adopting your handling system and trap-nesting only those ducks and pullets worth recording and which will be good future breeders after I have trapped them. I carefully check the laying of those I select by trap-nesting, and the records prove me to be very near the mark.

Some day I hope to be able to pick out the best laying ducks by observation as they run in the enclosures, as you did so amazingly correctly when you were last on the farm.

Yours sincerely,
BARBARA RAYE.

Miss Raye specialises in both Khaki Campbell and Buff Orpington ducks, and I am glad to see, as I write, that she leads at Bentley in Buff Orpington ducks in the current 1922-3 competition.

ANOTHER GOLD MEDAL.—Mr. F. H. Baker, of 7, Silver Street, Broughton, Kettering, picked, on my system, his Rhode Island Red pullet with which he won the GOLD MEDAL and "BEST HEAVY BREED" in the National Poultry Council's Laying Test,

1921-2. She laid 237 *eggs in the twelve months* with a total of 79 for the twelve winter weeks, leading her section easily by 333 points. In one of his letters, Mr. Baker adds:—

> I send you photo of *Red Empress*, my pullet selected on your system. As regards handling, I used to be a critic but to-day I am proud to have become a disciple of W. P.-O.

This competition was held at Blackpool. Winter record for breed.

GOLD MEDAL AND FIRST-CLASS CERTIFICATE.—Another breeder who selected her birds in 1921 on my system, namely, Mrs. W. Beale, of Manor House Farm, Claybrooke, near Lutterworth, secured at Harper Adams in the Test just concluded 1921-2 GOLD MEDAL and FIRST-CLASS CERTIFICATE for winning the Rhode Island Red section against twenty-eight pens. Total of eggs laid by the six pullets was 1,225, or an average of over 200. *Four of the pullets also secured a second-class certificate each for exceeding* 200 *eggs, a fifth missing it by one egg, laying* 199.

Mrs. Beale also, with a Red pullet picked at the same time by my system, led Harper Adams single-pen Test, 1921-2, for the winter, the pullet having laid 142 *eggs in the six winter months*, when she went wrong internally.

Her winning six pullets at Harper Adams laid, during the twelve months:—

No. 1—221 eggs.
No. 2—225 ,,
No. 3—208 ,,
No. 4—213 eggs.
No. 5—199 ,,
No. 6—156 ,,

Her pen at Bentley (N.U.P.S.) 1921-2 finished twelfth with a total yield for the eleven months (forty-four weeks) of 810 eggs, or an average of 162 *eggs per bird for the eleven months*. Their individual totals were:—

No. 1—153 eggs.
No. 2—171 ,,
No. 3—150 ,,
No. 4—178 eggs.
No. 5—158 ,,

This pen took three months to get started, which must be allowed for in the totals. All five pullets were reported officially as being in lay at the close of the competition.

Writing to me on August 22nd, 1921, Mrs. Beale says:—

> I shall have to pick my own birds for the laying competitions this year. Shall be most grateful if you will help me all you can.

Writing to me on November 6th, 1922, Mrs. Beale says:—

> I am pleased to tell you I have won first at Harper Adams with a total of 1,225 eggs. I am well pleased, as it is the first time I have selected my own birds, and I picked them on your methods.

SILVER MEDAL AND DISHES.—In the Tottenham Laying Test, Grainger's Poultry Farm, Farringdon, Alton, Hants, secured the following awards: Silver medal for second highest total in breeder-members' section of 105 entries; silver dish for best Rhode Island Red; silver dish for best White Wyandotte;

and silver special for highest total during the winter months. Herewith a letter from them:—

FARRINGDON, ALTON.

DEAR MR. POWELL-OWEN,

I have been studying very closely your system of handling, and I feel pretty confident now in picking out the layers. I check up my selections by the trap-nest records for the birds, and it is surprising how few mistakes I make.

I am sending a few pullets to the Tottenham Laying Test; it will be my first attempt, but, armed with your system of selection, I feel quite confident I shall pick birds that will not disgrace me.

Many thanks for giving me your final instructions *re* picking the pullets to lay to time.

Yours sincerely,
FRED J. GRAINGER.

A £19 HEN.—Captain G. C. Heseltine, Woodside, Ropley, Hants, who is so well known in Buff Orpington duck circles, writes:—

My White Wyandotte pullet, which won the Cheshire-section Test, laid 224 first-grade eggs and 1 second-grade egg at Test, and 241 for full year, with weight of egg 2½ oz. She was picked on your system, though of course I took her ancestry into consideration. She was ex a 225 2⅝-oz.-egg hen by a cock ex a 252 2½-oz.-egg, with sire's dam 254 2½-oz.-egg. She fetched £19 at the sale.

The competing pullets in the Lancs. and Cheshire competition were sold for the National Poultry Council Fund, competition being open to all.

A WINTER RECORD.—It has not been left before to the Light Sussex to lead all breeds for number of eggs in the winter months. One of my students has already set up this record for Reds at Bentley, but at the Northern Trials, 1921-2, Mr. M. W. Slade's pen of Light Sussex laid more eggs than any other breed, namely, 482 for five pullets for the six winter months, or an average of 96 *eggs per bird for six periods of four weeks*. Unfortunately, two of the birds died later in the broody-coop, I believe.

At Bentley (N.U.P.S.) Test, 1921-2, his pen of Light Sussex laid 488 for the same six winter months for the five pullets. This pen finished fourth in the Light Sussex section of twenty entries and secured a FIRST-CLASS CERTIFICATE, producing in the eleven months of the competition 794 eggs, or an average of about 160 per bird. The individual totals were:—

No. 1—173 eggs.
No. 2—162 ,,
No. 3—182 ,,
No. 4—92 eggs.
No. 5—185 ,,

STOKENCHURCH, OXON.,
May 1st, 1922.

DEAR POWELL-OWEN,

The small-egg problem wants investigation; it is most interesting and yet difficult. It is very curious that at Bentley neither of my pens are laying small eggs, whereas at Northern they are. Yet all picked on exactly same methods.

The way we went to work was this. To get five pullets Bowker (the manager) would first of all pick out ten to twenty of about the right age which he judged by handling to be likely. Then he and I went through the selected birds, handling them and judging them on your system. We generally managed to pick three or four as certainties the first time through; for the last place or two we had to go over the possibles again and again, using your system solely. Having picked five we then studied their breeding, examining rings and toe-punches, and found that in every case except one the chosen birds came from my best pens. In that case, I think that as she was only a last choice we rejected her for the pullet which came next her in the handling test and on her parentage was likely to be the better layer.

The net result is that all my birds this year except the one were picked on your system, and the exception was picked mainly on same lines. That the birds so selected have done as they have is to my mind pretty conclusive proof that your system of judging gives a pretty correct estimate of the relative egg-laying powers of the birds tested by it.

Yours very sincerely,

M. W. SLADE.

I had almost forgotten that Mr. Slade entered a pen of eight Light Sussex pullets in the *Daily Mail* Two-year Test, 1921-3. They stand 112th out of 200 pens, having laid 1,526 eggs, with five of the eight in lay at the close, or an average of 190 *eggs per bird for the eight pullets.*

TWO CUPS.—In the Tottenham Laying Test, 1921-2, Mrs. F. G. Martin, of Pedigree Poultry Farm, Scorton, Yorks., who is also a successful exhibitor at utility Shows, won against all sections, securing the silver cup for best bird laying greatest number of first-grade eggs, cup ditto in breeder-members' section, and silver dish for best White Leghorn. This breeder also picked her three White Leghorn pullets on my system. One of the pullets secured the above wins, a second finished seventh out of 105 (fourth for White Leghorns), and the other twelfth for White Leghorns. Dr. Martin wrote:—

PEDIGREE POULTRY FARM,
SCORTON, YORKS.

DEAR POWELL-OWEN,

I am sending three White Leghorns to Tottenham Test and enclose their score-cards. I make No. 15 the best. The birds are all sisters bred from our hen No. 8, a bird of lovable type and bone, whose record from January to end of June was as follows: 22, 18, 23, 21, 16, 9. In July she started to moult, but laid seventeen and was deep in the moult in August. She is now (September, 1921) practically newly decked and will be in lay about end of month. Her record is not brilliant, but the fact that she was mated to the 307-egg cockerel which you got me from your student, Miss N. Clayton, and which you liked so much on handling, may result in these daughters being better than their dam.

Yours sincerely,

F. G. MARTIN.

This was the first entry in any Test. At Bentley at the same time Mrs. Martin's pen was balloted out.

SIXTH AT BURNLEY.—Messrs. Pratt and Park, of 6, Eden Terrace, Sunderland, secured SIXTH AND FIRST-CLASS CERTIFICATE at Burnley (Northern U.P.S. Test), 1921-2, against all breeds in the small breeders' section, as many as fifty pens competing. The five White Leghorn pullets laid, despite a slow start, 906 eggs, or an average of 181 *eggs per bird* for the forty-eight weeks of Test. Individual totals were:—

No. 1—185 eggs.
No. 2—189 ,,
No. 3—213 ,,
No. 4—160 eggs.
No. 5—159 ,,

Mr. Sydney M. Park is a prominent member of the Sunderland U.P.S., and often acts as my clerical steward when I am judging Newcastle and Sunderland way.

SUNDERLAND.

DEAR MR. POWELL-OWEN,

Discussion being prevalent on the merits of handling for selection, etc., it may interest you to know that my pen of five pullets which finished sixth at Burnley were handled on your system before being dispatched (*verb. sap.*).

Had I the knowledge then which I have to-day, thanks to your tuition, I am confident I could have selected birds which would have finished higher.

Kind regards,

SYDNEY M. PARK.

SIX PULLETS: ALL OVER 200 EGGS.—It is a distinguished performance indeed at Harper Adams College to secure a pen certificate on performance and also a certificate for every bird of the six in the pen. There is one thing to praise in the Harper Adams awards in that they allot more certificates for high performances than any other competition. Credit where it is due. To allot but half a dozen awards with fifty or sixty pens competing is mean, especially in the popular breeds, where one has to go all out to win. Naturally, competition is keener in White Leghorns and White Wyandottes than in, say, Orpingtons or A.O.V.

The six White Leghorn pullets sent to Harper Adams Test, 1921-2, by Messrs. Welford & Sons, of Ashley Poultry Farm, Clifton, Oxford, finished the year's official Test with 1,318 eggs for the six pullets, or an average of 219 *eggs per bird*. This performance secured for these noted breeders a second-class certificate. And as every pullet exceeded 200 eggs, each bird received a second-class certificate, making *six certificates with the six pullets*. The individual totals were:—

No. 1—208 eggs.
No. 2—213 ,,
No. 3—238 ,,
No. 4—222 eggs.
No. 5—222 ,,
No. 6—214 ,,

Scored out of lay, their score-cards were:—

	Capacity.	Total.	Capability.	Total.
No. 1	9, 3, 7, 8, 5, 8, 6	46	9, 3, 10, 8, 9, 9, 7	55
No. 2	12, 4, 8, 8, 5, 8, 7	52	9, 3, 12, 8, 7, 7, 7	53
No. 3	8, 2, 6, 7, 5, 8, 6	42	7, 3, 10, 8, 8, 7, 6	49
No. 4	10, 2½, 8, 8, 5, 8, 6	47½	8, 3, 10, 7, 8, 8, 6	50
No. 5	10, 3, 7, 7, 5, 6, 6	44	8, 3, 10, 8, 8, 7, 6	50
No. 6	9, 2, 6, 8, 5, 6, 6	42	7, 3, 12, 7, 7, 8, 8	52

This pen finished FOURTEENTH out of SIXTY-SEVEN PENS of White Leghorns, with SECOND-CLASS CERTIFICATE.

Six White Wyandottes entered at the same Test, 1921-2, laid 1,062 eggs, or an *average of 177 eggs per bird*. One pullet laid 240 *eggs*, and secured a *second-class certificate* for producing over the 200 eggs.

Of the pen of Rhode Island Reds entered two died (one sends seven pullets, the seventh acting as reserve) and the five left laid 951 eggs, or an *average of 190 eggs per bird*, one securing a *second-class certificate* with 217 eggs.

A White Wyandotte pullet entered in the single-pen section beat all in its section with 136 eggs for the six winter months, finishing the year with 228 and securing a *second-class certificate*. Her score-card, taken out of lay but on the point of starting, reads:—Capacity: 16, 5, 8, 8, 4, 8, 10, total 59. Capability: 10, 3, 12, 7, 7, 9, 8, total 56. This would be a nice score to aim at when about to start laying for any heavy-breed pullet.

The five White Leghorn pullets sent to the National Test at Bentley, 1921-2, took three months to get going, yet put up totals during the competition of 182, 152, 149, 181, 158, or 826, making an average of 165 *per bird* for the forty-four weeks.

At the Midland Test in White Leghorns (professional section), THIRD PRIZE was secured out of EIGHTY PENS of White Leghorns competing in the section. In White Wyandottes FIFTEENTH POSITION was taken at the close out of fifty-seven pens competing. In Rhode Island Reds, too, FIFTEENTH PLACE was secured out of thirty-two pens.

At Burnley, their pen of five White Leghorns finished TWENTY-THIRD out of EIGHTY-SIX PENS of the variety, laying 803 eggs, but here one pullet died, reducing the chances of the pen. Their laying for the forty-eight weeks was:—

No. 1—215 eggs.
No. 2—164 ,,
No. 3—190 ,,
No. 4—203 eggs.
No. 5—31 ,, (died).

They produced 803 eggs, despite not getting going till the third month, or an average of 160 *eggs per bird*. Take away the total of the pullet that died and the other four *averaged* 193 *eggs*.

A pen of Reds at Burnley completed the selections of Messrs. Welford & Sons, but here again a bird died. The five pullets laid 143, 62 (died), 170, 145, and 197, again getting going only in the third month.

I think I am right in stating that at Burnley some new houses were installed for their 1921-2 Test, and as the houses were not ready in time the birds in some cases had to be shifted again.

ASHLEY HOUSE POULTRY FARM,
CLIFTON, OXFORD,
September 12th, 1921.

DEAR MR. POWELL-OWEN,
Many thanks for your copy of new edition of "Poultry-keeping on Money-making Lines." Am looking forward to a few winter evenings with it.
We have entered all the four main Tests again and hope luck will come our way this time.

Yours faithfully,
W. J. WELFORD.

October 16th, 1921.

Many thanks for your kindness in sending me your notes of score-carding. I will certainly give you particulars of the pullets we send out. I am still a little doubtful of this means of selection.

W. J. WELFORD.

March 4th, 1922.

So sorry not to have replied to your letter of January, but really have been nearly run off my feet. The enclosed score-cards of pullets we sent to the Tests have been got out for you a long while.

W. J. WELFORD.

SIXTH AT BURNLEY.—Miss Nancy Clayton, of Hadlow Wood Poultry Farm, Willaston, near Birkenhead, is another student, and she secured with White Leghorn pullets picked on my system SIXTH PRIZE and FIRST-CLASS CERTIFICATE. Her pen of five pullets finished *sixth out of eighty-six pens* in this variety, laying 910 eggs, or an average of 187 *eggs per bird* for the twelve months (forty-eight weeks). Individual totals:—

No. 1—175 eggs. No. 4—196 eggs.
No. 2—207 ,, No. 5—191 ,,
No. 3—141 ,,

At Harper Adams Test, 1921-2, her White Leghorn pullet entered in the single-pen section finished FOURTH with 240 *eggs* for the forty-eight weeks, securing also SECOND-CLASS CERTIFICATE. Her score-card may be of interest, and I append it in detail.

DAM NO. 15.—Laid 251 eggs in pullet year (79 winter), 186 second year, and 145 third year; weight when selected, 3 lb.; hatched late April and due to lay in November.

CAPACITY.		CAPABILITY.	
12	(3 fingers tight).	9	
2½	(2 fingers tight).	3	
7	(1½ fingers).	10	
9	(4½ ins.).	4	
4	(4 ins.).	4	
8	(4 fingers).	10	Beautiful head
6	(length 3½ ins., depth 3½ ins., width 2½ ins.).	10	and outstanding
		8	red eye.
48½		58	

These figures may not right away be readable to those not up in my card, but one should study my score-card elsewhere in the volume, and endeavour to read the two in conjunction. This pullet measured three fingers almost at abdomen from end of breast-bone to pelvis, with two fingers between pelvis. Being out of lay by, say, several weeks, she would increase a finger in each measurement, measuring when laying started four fingers at least at abdomen and three between pelvis bones. Not being in lay by several weeks the abdomen would be half full, and inches are given as a key. Length of abdomen would be from thigh to end of abdomen, depth from tail socket to end of breast-bone, and width from side to side of abdomen in flesh, *i.e.*, just spanned. This bird would pick up 4 (at least) more marks at abdomen, 2½ between pelvis, and 2 (at least) at abdomen, making, in lay, another 8½ marks to be added, which would mean 57 for capacity minimum. She has very high texture, especially as she is built up for laying. As you will see, she handles a little fleshed at abdomen as well as pelvis bones, which is exactly what she should do. As a rule, when fleshed up ready for production, a pullet is often lower for capability than this, and then with laying shows an increase just as fat or flesh leaves the abdomen and pelvis. In pullets we have the variations in abdomen and pelvis according to whether they are in or out of lay. It is for the handler to get the proportion of the other points by care. When a pullet is ready for laying she must not be very fine at pelvis and empty of all fat at abdomen, or even finely fleshed at breast-bone, which are really signs denoting heavy laying has taken place. She should be plump, as explained fully elsewhere. When ready to lay, too, there might be about one finger space between wattles, with the latter close together after months of production. You must avoid razor-blade pelvis bones, which denote so often super texture and lack of stamina and capacity.

In the small-flock section at Harper Adams, 1921-2, Miss Clayton's pen of seven White Leghorns was not fortunate, as three died. The remaining four laid 746 eggs, or an average of 186 *eggs per bird* for the forty-eight weeks. Individual totals were: 185, 190, 175, 194.

Eight White Leghorn pullets were sent to the *Daily Mail* Two-year Competition, 1921-3, but after being seventeenth for the fourth winter month with only seven laying, one bird died and the pen finished 103rd out of the 200 pens, producing 1,591 eggs, or an average of *nearly* 200 *eggs per pullet*. As the one that died laid but 7 eggs, it would mean 1,584 eggs for the seven pullets that lived for the full twelve months (fifty-two weeks), or an average of 226 *eggs per bird* per annum. The individual egg-totals were:—

No. 1—271 eggs.	No. 5—193 eggs.
No. 2—268 ,,	No. 6—236 ,,
No. 3—180 ,,	No. 7—257 ,,
No. 4—170 ,,	No. 8—7 ,, (died).

When one passenger that dies has to be carried in a Laying Test, the pen's chances are wiped out at one blow. That is the lottery part I have so frequently referred to. One is satisfied in that from the pen there will be, as above, some splendid birds with official records which can be used at home when the two years are up, but in these days of "first or nowhere," with the buying public so educated that way by the system of awards in our Tests, the compensation is very small. I publish the score-cards of these pullets:—

	Capacity.	Capability.
No. 1.		
	12 (3 fingers easy).	9
	3½ (2½ fingers).	3
	10	12
	10 (5 ins.).	5
	4½ (4½ ins., 2 fingers tight).	5
	8 (4 fingers).	10 (red eye).
	8 (5 ins., 5 ins., 3 ins.).	10
		8
	56	62
No. 2.		
	12 (3 fingers).	9
	2½ (2 fingers).	3
	9 (2 fingers tight).	12
	10 (5 ins.).	5
	3½ (3½ ins.).	5
	8 (4 fingers).	10 (red eye).
	8 (5 ins., 5 ins., 3½ ins.).	10
		8
	53	62
No. 3.		
	8 (2 fingers).	9
	2 (1½ fingers).	3
	7 (1½ fingers).	12
	10 (5 ins.).	4
	5 (4 ins., 2 fingers tight).	4
	8 (4 fingers).	10 (red eye).
	6 (3½ ins., 3½ ins., 3 ins.).	10
		8
	46	60
No. 4.		
	12 (3 fingers).	9
	2½ (2 fingers).	3
	7 (1½ fingers).	12
	10 (5½ ins., 2 fingers tight).	5
	4 (4 ins.).	5
	8 (4 fingers).	10 (red eye).
	8 (4½ ins., 5 ins., 3 ins.).	10
		8
	51½	62
No. 5.		
	12 (3 fingers).	9
	2½ (2 fingers).	3
	10 (2 fingers).	12
	10 (5½ ins., 2 fingers tight).	5
	4 (4 ins.).	5
	8 (4 fingers).	10 (red eye).
	8 (5 ins., 5 ins., 3 ins.).	10
		7 excess.
	54½	61

	Capacity.	Capability.
No. 6.		
	12 (3 fingers tight).	9
	2½ (2 fingers).	3
	7 (1½ fingers).	12
	10 (5½ ins.).	4
	4½ (4½ ins.).	4
	8 (4 fingers).	10 (red eye).
	7 (4 ins., 4 ins., 3 ins.).	10
		8
	51	60
No. 7.		
	12 (3 fingers tight).	9
	2½ (2 fingers tight).	3
	6 (1½ fingers tight).	12
	10 (5½ ins.).	4
	5 (5 ins.).	4
	8 (4 fingers).	10 (red eye).
	7 (4 ins., 4 ins., 3 ins.).	10
		10
	50½	62
No. 8.		
	12 (3 fingers tight).	9
	2½ (2 fingers).	3
	7 (1½ fingers).	12
	10 (5½ ins.).	4
	4 (4 ins.).	4
	8 (4 fingers).	8
	7½ (4½ ins., 4 ins., 3 ins.).	10
		8
	51	58

Where for width of back "2 tight" two-fingered neck appears, it means that the wide part of shoulders is well up to wing socket, which I aim at in all birds, as I have already explained and illustrated. The idea in this pen was to keep my red eyes and to pick pullets around 3½ lb. each at this development stage. No. 3 would not be so forward as the others.

At the Midland Test, 1921-2, Miss Clayton's pen of White Leghorns was 5th for six winter months out of 80 pens (finishing 16th), and 10th for White Wyandottes (59 pens), finishing 11th for the twelve months. At Tottenham, her White Leghorns were 3rd (5th out of 105, all breeds) and 6th. At Bentley, after being 11th out of 101 pens in the National Test for the five winter months with 380 eggs for the five pullets, or an average of 76 per bird for the five months, one pullet died. The four pullets completing the Test laid 693 eggs, or an average of 173 eggs per bird for the eleven months.

For the fourth, fifth, and sixth winter months Miss Clayton's White Leghorns in the six Laying Tests entered stood 5th (68 pens), 7th, 6th (105 entries), 5th (80 pens), 11th (101 pens), 17th (200 pens), 16th (86 pens), and in White Wyandottes,

10th (59 pens). Fatty degeneration of the liver was the cause of death of the *Daily Mail* bird.

HADLOW WOOD, WILLASTON,
NEAR BIRKENHEAD,
November 2nd, 1921.

DEAR MR. POWELL-OWEN,

At last I can reply to your last letter, having been so rushed getting the birds off to the Laying Tests. I enclose you score-cards and notes, and am lucky in not being balloted out of any Tests entered. Last year, with my first entry in any Test, I won third at Midland, as you know, with my first execution of your system. Since your visit here in August and further tuition from you, I feel I have mastered your score-card a bit better, although there is a lot to learn still. But I never could have picked all these pullets without the score-card.

Am sending you a budget including egg-record cards for the month, also for the twelve months, and perhaps you will study them as last year and make out my breeding pens for 1922. Pullets bred from the pens you mated up last year and laying wonderfully well, and of course the Test pullets are sister-birds, so we will be able to see what they do for laying away from home.

(MISS) NANCY CLAYTON.

"SCORE-CARD MARVEL."—I had intended keeping solely to current-year performances, but feel that I must give space to the wonderful performance of "Score-card Marvel," to which this volume is dedicated. I have many records put up by my score-card, but this is the one I am proud of. Mr. H. W. Honey, of Pedigree Poultry Farm, Alton, Hants, was personally tutored by myself into my score-card system, and in 1920 sent an entry in White Wyandottes to the Midland Laying Test. The pullets were picked by my score-card and won FIRST and GOLD MEDAL, leading all sections and breeds, 1920-1. One of the pullets laid 315 STANDARD EGGS IN ELEVEN-AND-A-HALF MONTHS, and readers may be interested to see her egg-chart. By this performance she set up a WORLD'S RECORD for White Wyandottes (official) and a BRITISH RECORD for all breeds. I append her score-card sent to me at the time the bird and her sisters were dispatched to the Test:—

CAPACITY.		CAPABILITY.	
Breast to pelvis	18	Pelvis bone	7½
Between pelvis	5	Straight	3
Pelvis to tail	7	Flesh	12
Width of back	8	Vent—size	5
Length of back	5	,, —fineness	5
Width between legs	10	Head	6
Abdomen	6	Bone	8
		Feather	8
	59		54½

PEDIGREE POULTRY FARM,
ALTON, HANTS,
November 10th, 1920.

DEAR MR. POWELL-OWEN,

You will find score-cards enclosed. I score-carded the pullets available and sent the best. Scored on October 1st and allowed a little as they were not in lay, but cut them down as much as possible. Think you would have given them a bit more. I went strongly after your minimum of fifty capability and sent the best scorers.

H. W. HONEY.

"SCORE-CARD MARVEL." **H. W. HONEY'S MIDLAND TEST WORLD'S RECORD BREAKER.**

W. POWELL-OWEN'S COPYRIGHT TRAP-NEST & PEDIGREE CARD.

REMARKS— Month.................. Age of Birds..................

Pen No.................. No. of Birds..................

Breed.................. Strain..................

I = Ill. A = Additions. RR = Returns. R = Removals. S = Sold. L = Broken or soft egg. B = Broody. M = Moulting. D = Dead.

REMARKS—

Date Hatched	1st egg, date & weight.	Hen No. or colour.	1	2	3	4	5	6	7	8	9	10	11	12	13	14	15	16	17	18	19	20	21	22	23	24	25	26	27	28	29	30	31	Total.	Br'ght For'd.	Grand Total.	WINTER RECORD Oct. to Dec.	Jan. to Mar.	Oct. to Mar.
		Oct. ..	.	.	.	.	.	.	.	.	.	.	.	.	.	/	/	.	.	/	/	.	.	/	/	.	/	./	.	/	/	/	.	11			69	83	152
		Nov. ..	/	/	/	/	/	/	/	/	/	/	/	/	/	/	/	/	/	/	.	/	/	/	/	/	/	/	/	/	/	/	X	29					
		Dec. ..	.	/	/	/	/	/	/	/	.	/	/	/	/	/	/	/	/	/	/	/	/	/	/	/	/	/	/	/	/	/	/	29					
		Jan. ..	/	/	.	/	/	/	/	/	/	/	/	/	/	/	/	/	.	/	/	/	/	/	/	/	/	/	/	/	/	/	/	29					
		Feb. ..	/	/	/	/	/	/	/	/	/	/	.	/	/	/	/	/	.	/	/	/	/	/	/	/	/	/	/	/	X	X	X	26					
		March ..	/	/	/	/	/	/	/	/	/	/	.	.	/	/	/	/	/	/	/	/	/	/	/	/	/	/	/	.	/	/	/	28					
		April ..	/	/	/	/	/	/	/	/	/	/	/	/	/	/	/	/	/	/	.	/	/	/	/	/	/	/	/	/	/	.	X	28					
		May ..	/	/	/	/	/	/	.	/	/	/	/	/	/	.	.	.	/	/	/	/	/	/	/	/	/	/	/	/	/	/	/	27					
		June ..	/	/	/	/	/	.	/	/	/	/	/	/	/	/	/	/	/	/	/	/	/	/	/	.	/	/	/	/	/	.	X	27					
		July ..	/	/	/	/	/	/	.	/	/	/	/	/	/	/	/	/	/	/	/	/	/	/	/	/	.	/	/	/	/	/	/	29					
		August	/	/	/	/	/	/	/	/	/	/	/	/	/	/	/	.	/	/	/	/	/	/	/	/	/	/	/	/	/	/	/	30					
		Sept. ..	/	/	/	/	/	/	/	.	.	/	/	/	/	/	/	/	/	/	/	/	/	/	.	.	.	/	/	–	–	–	–	22					
REMARKS—		Unrecorded Eggs																																					
		TOTAL																																315					

REMARKS—

Obtainable from **W. POWELL-OWEN, 47a, High St., Hampstead.**	Total 1st week	Total 2nd week	Total 3rd week	Total 4th week	Full Monthly
	Av'age " "	Av'age " "	Av'age " "	Av'age " "	Average ..

This score-card and those of other birds were submitted, at the time the Tests started, to the editor of a well-known poultry paper, and this bird's record followed up throughout. An interesting item is the fact that her dam was not a good layer by trap-nest record and was relegated out of the special pens. But being mated to a tip-top cockerel he carried the laying power on to the daughters. Ignoring what the dam had laid, Mr. Honey found that this pullet handled best of all his pullets and sent her to the Midland. I had always told him to rely upon his hands.

There always will be a controversy over this item, and every person is entitled to his own honest opinion based on his experiences. Some place pedigree first, and many who use my system do so, too. I have seen such wonderful layers bred from low-record hens, without a doubt taking their laying qualities from the male mated to them, as did "Score-card Marvel"; what the dam laid does not influence me one iota. I am thus able to concentrate on getting some body with the selected bird. My capability side of score-card will attend to the quality or breeding of the bird concerned and her egg-laying, and if we get this high let us have with it as high capacity as possible. The bird from the best-bred dam is so often small and down in capacity, with a tendency to small eggs. To get the high texture I do not mind cockerels being a little down in capacity, but all interested in utility should keep up the frame of the females. Only in recent years have the broad-minded breeders seen the force of my argument that you cannot have utility birds too high in capacity if you hold capability high. But with so many breeders not yet mastering this aim, selections will be varied.

I consider that there must be a limit for egg-number when it comes to adding reproduction to production. That is why I favour coming back every time to hens laying 180 to 220 large eggs per annum mated to males ex dams laying 220 and over. Naturally when you have been breeding high layers for many years, and the heavy laying has been worked into the strain, you can go a little higher than 220 for the breeding hen. But who can say how much higher? When you begin to mate up the high fliers that are small you have roup, colds, deaths, bad hatchability and rearability, small eggs, etc. There must be some limit, and yet my matings will breed pullets that lay very high totals.

In the 1921-2 *Daily Mail* Test I see that Mr. Honey has found another in "Score-card Gem," which laid officially 290 EGGS IN THE TWELVE MONTHS, of which only 5 were second-grade. At Harper Adams Test his pen of White Leghorns finished TWELFTH, securing a SECOND-CLASS CERTIFICATE, with

1,267 eggs from the six pullets in the forty-eight weeks, or an average of 211 *per bird*. *Four of the six pullets secured a second-class certificate* for laying over 200 eggs at the Test. Individual records:—

No. 1—204 eggs.	No. 4—170 eggs.
No. 2—224 ,,	No. 5—241 ,,
No. 3—188 ,,	No. 6—232 ,,

In all cases where these individual totals do not tally with the pen figures the odd eggs will be booked as "unrecorded" in the official reports. No. 4 did not lay in the first month, and it is often such loss that makes some of the pens odd in numbers.

ANOTHER THREE CERTIFICATES.—Mr. W. Roebuck Staniforth, of Todwick Manor, near Sheffield, also selected his birds on my score-card system. At Harper Adams his pen of White Wyandottes laid 1,184 eggs, or an average of 197 *per bird* for the forty-eight weeks, *three pullets securing second-class certificates* for over 200 eggs. Individual egg-totals were:—

No. 1—222 eggs.	No. 4—173 eggs.
No. 2—168 ,,	No. 5—201 ,,
No. 3—184 ,,	No. 6—232 ,,

Some were slow at getting off the mark the first month.

A White Wyandotte pullet was also entered in the single-pen section, laid 116 eggs in the six winter months and finished up the forty-eight weeks with a total of 224 eggs, securing FIFTH PLACE and a SECOND-CLASS CERTIFICATE. She scored 59½ for capacity and 58½ for capability when in lay, as she had produced four eggs in five days before dispatch. Luckily, this sending in lay did not upset her for long, as she managed to get settled and laid ten eggs the first month. She probably rested a week or so at Test before the competition commenced, and took several weeks off at the beginning of November (first month). But I record the fact of being in lay when dispatched. I prefer pullets not to have laid, although one or two eggs would be different to, say, thirty or so. It is just the risk I do not like to take, as I think we have enough hurdles to overcome as it is, in other directions.

As an experiment, Mr. Staniforth sent five pullets to Bentley picked out as "little gems"—all capability and little capacity. They were picked to test two types of birds, this breeder being anxious to master my system and types. Five "little sprinters" were selected from the early layers starting in June and July, and one had laid 96 eggs from June to September (four months), another 82 in like period, and so on. After laying like rapid firing they were in the moult by October and were sent to the Test. It took them three or four months to get started, and the total number of eggs laid was only 694, of which about half were second-grade. We need more breeders of the kidney of Mr. Staniforth, who will do anything to get to the bottom of things. I had condemned this "little gem" type, as I have always done, and we glean valuable data from this trial test. To send pullets to a Test which have laid early and are in the moult is dangerous, fearing a late start in production. Wherever it is done, each bird should be on the point of laying when dispatched, and clear of the partial moult and loose neck-hackle feathering.

TODWICK MANOR,
NEAR SHEFFIELD,
November 1st, 1921.

DEAR MR. POWELL-OWEN,

I send you the score-cards for Test pullets. The Harper Adams birds run about four to four-and-a-half pounds, and I consider the single-pen pullet one of my best, which is why I sent her, although she had laid four eggs in five days before dispatch. All grade over fifty for capability.

I note your remarks *re* type. There are several heavier types which you like, and these will go into the breeding pens. Am sending to Bentley "little sprinters" which have put up high totals since June, but they were not over moult when sent.

Personally, I like your card, as it is easy to score a bird, especially for an amateur at handling. Hope to get to the Palace Show to see you.

Yours sincerely,
W. ROEBUCK STANIFORTH.

Expect you have been watching the Bentley pullets. These were those "little sprinters" which moulted out. I have some of their sisters which soon exceeded the 200 eggs from July, but I realise now they are useless as breeders. Small eggs, etc., as you said.

W. ROEBUCK STANIFORTH.

SECOND AT MIDLAND.—As already mentioned, a pen from Messrs. Welford & Sons secured third position in White Leghorns at the 1921-2 Midland Test. Messrs. Blaksley and Blyth, with score-carded pullets, obtained SECOND POSITION in the same section of EIGHTY PENS. The remarkable part is that this pen of four pullets was housed in the large-flock house and had to fight against the others with four pullets in a small house to themselves. It is the only pen housed in the large unit that is within "miles" of any leading section-pen. In very few pens at this Test did every bird exceed the 100 eggs, but this was also achieved by these pullets. And they lost first place by 2¼d!

PEDIGREE POULTRY FARM,
WEST MOORS, DORSET.
March 6th, 1922.

DEAR MR. POWELL-OWEN,

You may perhaps be interested to know the results of our use of your score-card method of grading fowls. We may say that for some time we have used your method on our own stock, and in every case the trap-nest has proved the great value of your system.

Of course, we know that the trap-nest will tell us what a bird is doing during the time she is doing it, but this does not enable us to get rid of a poor layer before she has eaten more than she has earned.

We consider the greatest value of an efficient system of grading (and we have proved your own to be all that) is that one can learn beforehand which are the likely profit-makers.

Last autumn I selected birds for other people for the Tests, and in most cases they are doing well. Also our own White Leghorns are competing at the Midland and doing well, and they were picked by your method quite independent of their pedigree.

46

47

48

These pullets were leading at the end of December in spite of being under the disadvantage of being housed in the large-flock section. We exhibited many birds under you last year, and every bird which you placed high has proved her right to her award by the trap-nest.

We send these few lines to you because we think that after the many foolish remarks made by people in the correspondence columns of the various poultry papers you ought to know that the value of your system is properly appreciated by unbiassed poultry-farmers who cannot afford to keep "culls."

Yours faithfully,
Per pro Blaksley & Blyth,
A. BLYTH.

YEOVIL WINNER LEADS.—Mr. L. H. Wace, the popular secretary of the Columbian Wyandotte Club, included in his pullets sent to the Tests, 1921-2, a Columbian pullet which had secured first at Yeovil utility Show, 1921 (September), and so placed by me by score-card. He also sent a pullet first at another Show under another judge. I am glad to learn that my winning pullet, although not starting to lay till December 10th, laid 171 eggs at Harper Adams in under ten months and never broody.

KINGSLAND LODGE,
BEAMINSTER,
October 29th, 1922.

DEAR POWELL-OWEN,

Have been analysing my Harper Adams pen; nothing to shout about, but the best bird was 1st utility, Yeovil, under you, and she did not start laying at Harper Adams till December 10th, and to August 7th laid 145 eggs and never broody. I wish I had made a decent start, but was last for first three months and then 2nd, 2nd, 1st, 1st, 1st, 1st, 1st, other monthly positions for monthly totals, but hopelessly behind for making up the loss of winter eggs.

The above bird has been in lay from July to September, so should not have been sent to Harper Adams. I sent a single pullet to —— Test that was 1st —— under —— and she was last but two in her section, so your score-carded bird at Yeovil worked out a good bit better.

I return your notes on final selection with many thanks.

Yours sincerely,
L. H. WACE.

I am glad to have this letter from the secretary of the Columbian Wyandotte Club as I am a recently-appointed club judge, and as the Club has guaranteed so many classes for the variety in 1922 for me to judge by handling. This pullet was also first exhibition at Driffield, showing that in the Columbian Wyandottes we have a breed which stands out among the few to-day for both eggs and beauty on the one type, and that the Club ideal. The Harper Adams record also shows how impossible it is to catch up winter loss in eggs at high values. With no eggs for the first month my first Yeovil pullet was the first to commence the second month, and the totals were:—118, 125, 158, 171, 157, and 140. From the fourth month till the twelfth the monthly yield of this pen was first or second for numbers in the section, but throughout the pen remained in the same position, viz., sixth out of eight. The odd part is that it had laid but two eggs less than the third pen but is 30s. behind in values.

CERTIFICATE FOR WINTER EGGS.—Another first-class certificate was obtained in the Wilts. Laying Test for winter output by Mr. W. G. Arkell, of Cricklade. Picked on my system these four Rhode Island Red pullets finished up FOURTH with an average of 160 *eggs for forty-four weeks.*

MARSTON MAISEY POULTRY FARM,
CRICKLADE, WILTS.,
July 1st, 1922.

DEAR MR. POWELL-OWEN,
We secured a first-class certificate at Wilts. Laying Trials for winter months, our Rhode Island Red pullets laying 519 eggs in thirty-six weeks for the four birds. Our White Leghorns at Bentley, wired for at last moment, were young and only laid five eggs first month, but after starting 95th are 48th at seventh month.

All these birds were graded on your methods, only we did not know half as much then as we do now, thanks to your further help.

Yours truly,
W. G. ARKELL.

August 12th.

Our Reds finished up 4th in Wilts. Test with average of 160 eggs for forty-four weeks, only four behind third pen. We have won, since last November, 35 first prizes in utility and 10 specials, including Wilts. Diploma, best male.

Have just averaged out the flocks from November 1st to June 30th, and White Leghorns and Light Sussex are over 140 eggs per bird and others over 130, and all laying hard. This over hundreds of birds, all ages. To-day we had 360 eggs, our highest being 500 for a day from 600 laying stock, half pullets and half hens, run in flocks of about 110.

W. G. ARKELL.

SEVENTH BEST WYANDOTTE.—Apart from having the best White Wyandotte in the Midland Test, 1921-2, I had the 7th best in Captain Leslie Smith's pullet, which laid 170 eggs under severe conditions, his pen of four pullets finishing EIGHTEENTH of FIFTY-SEVEN PENS. This pullet was 7th out of 230 odd. Again picked on my system.

DALSWINTON, MAWGAN, ST. COLUMB,
CORNWALL,
August 26th, 1921.

DEAR MR. POWELL-OWEN,
We have sent the money for our entry for five pullets for Midland Test. In your letter you speak of them being accepted by ballot; does that mean that we shall not know until last moment that the birds are to compete?

I wonder if you would start your guidance as to choice of birds now. Supposing you were a novice like ourselves and you had forty-eight White Wyandotte pullets feeding in front of you and all more or less difficult to catch, how would you start about choosing the best five? What are you to look for?

(CAPT.) LESLIE SMITH.

BEST OF 1,240 BIRDS.—At the Midland Test, 1921-2, another White Wyandotte pullet headed her section as well as beating the best in every section, in fact beating the 1,240 odd pullets competing. Under the severe conditions already mentioned she laid 205 eggs as against the highest individual score for other sections, being White Leghorns 198; Reds 182; Light Sussex 144, etc. Thus two years running I have had the best individual pullet at the Midland Test, the former being " Score-card Marvel " with 315 eggs. Miss Vestey, as a first-year student and with first entry, apart from picking out on my system the best out of 1,240 pullets of all breeds, finished in the section TENTH OUT OF FIFTY-SEVEN PENS.

PARK HOTEL, PRESTON,
July 14th, 1921.

DEAR SIR,
Miss Vestey wrote to you about your Postal Club some weeks ago. I am going to manage her farm, and we think it better I should join your Club. I have had about the same experience as Miss Vestey, *i.e.*, nine months' training.
We would be glad if you would tell us when the Laying Test you refer to starts and what we have to do to enter birds.

Yours truly,
(MISS) D. MACARTNEY.

NEVILLE LODGE, NINEHAMS ROAD,
UPPER CATERHAM, SURREY,
September 30th, 1921.

Would you kindly let us have your score-card and fullest details *re* selection as soon as possible.

D. MACARTNEY.

A pen of White Leghorn pullets sent to Bentley (N.U.P.S. Test) laid 813 eggs in the eleven months (forty-four weeks), or an average of 162 *per bird* for the five pullets, of which four were in lay at the close by official report. Individual totals for eleven months were:—

No. 1—186 eggs.
No. 2—163 ,,
No. 3—110 ,,
No. 4—222 eggs.
No. 5—128 ,,

BEST AMATEUR'S BIRD.—In addition, another student, Miss D. P. Foster, of Trevillis, Liskeard, was *second in the amateur White Wyandotte section* at the Midland Test, 1921-2, and supplied *the best pullet in the whole of the amateur section*, against all breeds, some 230 odd pullets competing. Housed in the large-flock house, this pullet laid 161 eggs as against best Red, 93; best White Leghorn, 157; Light Sussex, 98, etc. This was another score-card pullet, so that I had best in both sections and best bird in the whole Test.

FOURTH IN REDS.—Mr. R. E. Eiloart, Little Zeal, South Brent, Devon, secured fourth in Rhode Island Reds open section at Midland Test, 1921-2 (thirty-three pens competing), and the fourth best pullet in section.

LITTLE ZEAL,
SOUTH BRENT, DEVON,
September 15th, 1921.

DEAR MR. POWELL-OWEN,
Have been score-carding some Test pullets. I hope you will enlighten me again on a few points.
Will see you at Tottenham Show.

Yours truly,
R. E. EILOART.

THIRD AT WYE.—White Wyandottes score-carded to my system also secured third place in the Wye College Winter Test for Mr. and Mrs. Robert Jacobs, Cartref Poultry Farm, Walderslade, Chatham.

CARTREF POULTRY FARM,
WALDERSLADE, CHATHAM,
October 29th, 1921.

DEAR MR. POWELL-OWEN,

I have been selecting birds for Tests, and if care and patience are going to do any good we should win something. I enclose score-cards.

The Bentley birds had to be rushed off owing to telegram. I was balloted out and money returned, but was then offered a pen of Sussex if I could send them off at once and allow the ringing to be done the other end.

Yours sincerely,
R. JACOBS.

February 17th, 1922.

We have ended third at Wye College Test; fourth at Harper Adams with Light Sussex, but some birds have died. Cannot understand it, as we have not had any deaths (touch wood!) on our farm among the pullets.

R. JACOBS.

At Harper Adams, Mr. Jacobs' pen of Light Sussex started off well with fourth position, but he lost five pullets out of the seven during the Test. Despite this, the pen finished eighteenth out of twenty-five pens, the only two living on laying 164 and 210, a SECOND-CLASS CERTIFICATE being obtained with the latter. Out of twelve pens of Light Sussex at Harper Adams in this section no less than FIFTY ODD PULLETS DIED OUT OF EIGHTY-FOUR BIRDS competing, which would be news to all who happened just to see the first, second, and third winning pens announced in the Press and not to subscribe to the official monthly reports.

The Light Sussex pullet entered in the single-pen Test laid only four eggs the first month but finished the twelve months with 197 eggs, missing a certificate by three eggs.

At Tottenham, ninth position out of 105 (third for White Wyandottes) was secured; together with ninth for Light Sussex and thirtieth for White Wyandottes (fifty-seven pens) at Midland competition. It is interesting to see how the pen of Light Sussex fared which were rushed off by telegram to Bentley. Only the owner and Test officials would know such was the case, because no mention is made of such pens in the official reports, but in fairness to pens a note should be given in final report. It shows how dispatch and selection have a say in winning laying trials, and not birds picked at random and on pedigree alone. Anyone reading the final report and seeing only 634 eggs recorded against these five pullets, or an average of 127 eggs per bird, for the eleven months would say they were "dud" birds. But I have to trace their laying for the whole period, and it took the pen more than three months to get going, all due to hurried dispatch and the owner not knowing that he was to send birds, therefore not holding any in reserve *and not in lay*, which is a common occurrence. Often Tests advertised to start on a certain day are brought forward several weeks at the last moment, or even put back a like period.

Pullets selected for a certain date are thus severely handicapped, except where the fortunate owner is a big breeder and has a host of other pullets available.

My critics would seem to have us believe that they do not select their birds but just send on pedigree alone, the junior poultry lad finally sending off the birds! Being behind the scenes I know different. If Tests were so run that the exhibitor had to send in a dozen hatching eggs, the resultant chicks when matured at the trial ground to compete, so that the best average yield of the grown-ups won, we would see some interesting results, just as we would if it were compulsory for a novice to pick at random every exhibitor's pen of birds. The big men who used to be always in the money rather take second place to-day, and, in many cases, to the newer men who are studying hand-selection. Competition is getting so keen that wins in the future will go to those who make handling a science.

FOURTEENTH AT BENTLEY.—In White Wyandottes at Bentley in N.U.P.S. Test, 1921-2, Mrs. M. C. Arnold finished fourteenth out of sixty-five pens, the individual totals for the eleven months being:—

No. 1—217 eggs.
No. 2—128 ,,
No. 3—186 ,,
No. 4—171 eggs.
No. 5—226 ,,

Total number of eggs was 929, or an average of 186 *eggs per bird* for the eleven months of Test. Thus with only two entries in the section I finished fourteenth and sixteenth (latter pullets I personally picked). And one has to remember that to be near the top in White Wyandottes or White Leghorns needs some accomplishing in view of the high quality and the red-hot competition. Her pen also supplied the *second-best bird out of* 325 *Wyandottes* for individual total.

In White Leghorns at Bentley, Mrs. Arnold finished TWENTY-FIRST OUT OF 101 PENS, totalling 868 eggs, or an average of 173 *eggs per bird* for eleven months and four in lay at the close—one died. Totals individually were:—

No. 1—162 eggs.
No. 2—199 ,,
No. 3—163 ,,
No. 4—153 eggs (died).
No. 5—190 ,,

THE SHIELING,
HALE, CHESHIRE,
September 22nd, 1921.

DEAR MR. POWELL-OWEN,

My White Wyandottes have won second, so your birds take first and second. I go up to second for size of egg. My Wyandottes finished fifteenth at Bentley and, as you know, ducks were eleventh, and I won the special for greatest number of first-grade eggs in whole S.P.B.A. Trials.

Yours sincerely,
(MRS.) M. C. ARNOLD.

August 24th, 1921.

Send me some more trap-record cards. I always use your cards and pick out all my birds by your system of finger-measurements.

M. C. ARNOLD.

FIFTH IN REDS.—Apart from being fourth in Reds at Midland, 1921-2, Test, I secured fifth place as well, Mr. F. J. Marston, of the Biddenden Poultry Farm, Biddenden, Kent, being so placed with pullets picked on my card. Mr. Marston was also eleventh out of 105 (all breeds) and fifth in light breeds at Tottenham, 1921-2, competition.

With the results published in this chapter I have kept to those students mainly whom I have taught (or was teaching at the time) my system of handling and score-carding. I have made no exception even where they have been picking pullets for the first time. I am able to watch a man improve each year in his selections. Naturally one has gradually to master my methods, which is not done in a day or even a season. If my system has helped them it may be of help to you.

From Mr. Marston's 1923 catalogue:—

Trap-nests are used throughout, and all birds are graded on the score-card devised by Mr. W. Powell-Owen, only those birds which score a sufficient number of points at end of their first laying season are used in the breeding pens.

BIDDENDEN,
April 6th, 1923.

DEAR MR. P.-O.,

The following results were obtained from pens of birds which after being trap-nested during their first WINTER season were graded out on your system in the following August and placed in pens. During their first year the birds were running in a flock of about 250 pullets.

The first two pens were made up more on the strength of the individual performance of the birds during their first winter season, although no bird was allowed to qualify unless they stood the test of your system of grading. I mean the first two pens of birds were not so closely graded out owing to their good performance in their pullet year.

Breed—RHODE ISLAND RED S.C. *Hatched*—FEBRUARY AND MARCH, 1921.

Number of Birds.		Oct., 1921, to Jan., 1922.		Oct., 1922, to Jan., 1923.
8	...	526 eggs.	...	267 eggs.
7	...	307 ,,	...	369 ,,
8	...	294 ,,	...	404 ,,
8	...	197 ,,	...	417 ,,
8	...	167 ,,	...	200 ,,
		Total—1,491 eggs.		Total—1,657 eggs.
		Average per bird—38·2 eggs.		Average per bird—42·5 eggs.

I think from the above figures, which I guarantee are absolutely correct, that your system of grading IN CONJUNCTION WITH THE TRAP-NEST is IDEAL.

FRANCIS J. MARSTON.

P.S.—One of my White Wyandotte pullets has laid 143 eggs (nearly all 2 oz.) from October to March. A client reports 1,763 eggs from 15 of my Red pullets for six winter months, average 117½. *My egg-production from December 1st to March 31st, 1923, was 26,074, against 6,436 in 1920, with very few more birds but an increase of nearly 20,000 in four months.* CAN PUT MOST OF IT DOWN TO TRAPPING AND GRADING. The man who buys my eggs says that he does not know of anyone who gets eggs from second-year hens during the winter months like I do. I use less than 10 per cent. fishmeal.

There have been scores of competitors in Tests with pens high up whom I have helped in selection to my system. Some have acted as stewards at Shows where I have judged, and not a few I have scored "key" birds for, while some became interested through winning under me at their first endeavour at exhibiting. I have not written them or anyone for their results or letters of appreciation. In the past I have never endeavoured to gain publicity for my selections, but feeling that we can only beat the critics by actual performances of the birds picked on my system, I hope all who have so selected their pens, or will do so in future, will be kind enough to let me have full details, also score-cards of the birds. I am anxious to have, just as the Tests commence, details of any pens entered so that I can have a register of same. Our critics never allow us any rope, and that is why I want details before the Tests start in preference to afterwards. Nor do I wish it to be said that results were sent to me after the performance through an endeavour to get a free advertisement for the birds. I do not work that way, as my friends all would know.

SEVENTH OUT OF 101 PENS.—The Revd. Charles Birkett—who has so many record wins in the Laying Tests—uses my system when picking for the Tests, balancing the handling by pedigree.

At Bentley, in the N.U.P.S. Test, 1921-2, his pen of White Leghorns finished SEVENTH OUT OF THE 101 PENS in the section, and secured SECOND-CLASS CERTIFICATE. The five pullets laid 937 eggs in the eleven months, or *an average of 187 eggs per bird for the eleven months*. Totals were:—

No. 1—203 eggs.
No. 2—189 „
No. 3—194 „
No. 4—167 eggs.
No. 5—184 „

STAVELEY RECTORY,
NEAR HARROGATE.

DEAR MR. POWELL-OWEN,

I invariably choose my Test birds to your method of grading. I don't see that one can do otherwise. Naturally I balance the handling by trap-nest records of the birds' ancestors.

Yours sincerely,
(REVD.) CHARLES BIRKETT.

Several seasons back I visited the Revd. Birkett's farm and score-carded most of his breeding-stock in detail to test the value of handling in conjunction with trap-nest records. I hand-graded the hens in February and marked those which I should earmark for toe-punching their chicks in the coming season. It was a test for longevity of laying, and when I had scored all the hens we compared their trap-nest records of the previous season. At the close of the year he sent me their full records after the grading, and many hens I graded out laid badly for the year, although having high records the season before.

Round the Shows I am constantly reminded by winning exhibitors that the bird is the "Revd. Birkett's strain." Once when he sold out his Black Leghorns the new purchaser entered some under me and won with them, paying compliment to the quality of stock sold.

His pen of White Wyandottes in the *Daily Mail* Test at Bentley, 1921-3, finished the twelve months FIFTY-NINTH out of the 200 PENS with a total of 1,785 eggs for the eight pullets, or *an average of 223 eggs per bird for the twelve months*. One pullet laid 120 second-grade eggs, or the pen would have been in the "money," but Tests do not run that way. I always pity a competing pen of five pullets if the fifth is bullied by the other four or happens to be a shy feeder. Records were:—

No. 1—255 eggs.	No. 5—233 eggs.
No. 2—187 ,,	No. 6—238 ,,
No. 3—268 ,,	No. 7—199 ,,
No. 4—189 ,,	No. 8—206 ,,

BRONZE MEDAL AT HARPER ADAMS.—Mr. J. S. Parkin, so well known in Rhode Island Red circles, is manager of the Ouse Manor Poultry Farm, Sharnbrook, Beds., for Field-Marshal Sir Arthur Barrett. His pen of Reds picked on my system won bronze medal and first-class certificate in the exhibition section at Harper Adams Trials, 1921-2.

OUSE MANOR POULTRY FARM,
SHARNBROOK.

. . . . The pullets were not hatched till first week in May owing to our coming here so late so were over a month too young; hence they were put back at start. They were picked entirely by your system. Their parents had never been trap-nested; in fact none of my strain had since 1913 till 1922.

Yours faithfully,
J. S. PARKIN.

As is well known, Mr. Parkin avoids extremes and concentrates with all his breeds on the dual pattern with preservation of type, reasonable size and colour, and heavy laying. He wins in exhibition and utility extensively. For the 1922-3 Tests he selected his many pens by different methods—by (1) records and (2) my handling system. I am naturally watching the various returns, and *so far all pens picked on my system are leading easily*. That is the best test as applied by each breeder; go by comparative results. Extract from letter:—

SHARNBROOK,
April 12th, 1923.

. . . . Everything at present points to your handling system as the best and final test birds should be put to before being sent to Tests.

Yours faithfully,
J. S. PARKIN.

In both Leghorns and Reds all birds were bred alike, so that it is just a test against individual selection by my handling system against relying solely on pedigree. Taking the fourth winter month the results are as follows:—Picked on my system—*Harper Adams*, 1st in Reds (exhibition section); *Northern*, 5th with White Leghorns in small breeders' section; *Middlesex*, 4th for White Leghorns and 9th for Reds. Sent solely on records without selection by handling—*Bentley*, Reds and White Leghorns low down; *Harper Adams*, White Leghorns also not doing well. Another interesting test was to send to Harper Adams in the exhibition section only pullets that had won first prizes at utility Shows and as reserve one that was at the Dairy Show in exhibition. They were finally hand-graded to my system, and the pullets sent are leading at the end of the fourth month. In this section, too, are Mr. Wace's Columbian Wyandottes, which stand third to the above, and were also picked by my methods.

CHAPTER XXV.

THE "POWELL-OWEN" SCORE-CARD AS AN INTERNATIONAL CARD.

IN this country my score-card system is employed at the majority of Shows. My score-card is the national one used by the National Utility Poultry Society and its branches. When a committee was formed to study score-cards in general and to adopt one, I submitted, by request, my own, and it was accepted after consideration of all cards and systems. In like manner, the Poultry Club accepted my score-card for their Shows when their utility committee met to approve of a utility score-card. Others can, of course, be approved by the Poultry Club upon submission with details of the Show, judge, etc. My card thus has the acceptance of the leading utility society and the premier exhibition club of this country.

Over the water my score-card has been used very frequently, not through any "pushing" by myself, but simply after test. Naturally I defend my card when attacked by interested critics, but no one can say that I have unfairly "pushed" it, seeing that I have never applied to any society asking to judge their Show. The following two letters will be of interest and are the only two I have had from the writer:—

UNIVERSITY OF BRITISH COLUMBIA
(DEPARTMENT OF POULTRY HUSBANDRY),
VANCOUVER, CANADA,
March 23rd, 1921.

DEAR SIR,

I have been advised by the editor of —— that you might be able to supply me with a copy of your score-card for utility poultry judging.

I am trying to work out a score-card for my senior students in Poultry Husbandry, and would be grateful for such assistance as you can render.

Yours very truly,
E. A. LLOYD,
Associate Professor.

I merely responded by sending details of my system so that any good therein might be employed by my correspondent. The following letter I am proud of because, instead of being an epistle from an interested critic and like many over here declaring that it was a complicated card they could not follow, it is the reply of a poultry man who is willing to get down to the bottom of things.

VANCOUVER,
July 3rd, 1922.

DEAR SIR,

I received your letter with regard to your score-card system, also proof of article you wrote for *Feathered World* Year Book explaining your system in detail. We are much obliged to you for all the trouble that you have taken for a stranger in a distant land.

Although we are so far away, many of us keep in very close touch with all that is transpiring in English poultry affairs, watching each new development and studying particularly all systems of utility judging and problems in poultry management.

We are going to give your score-card system a special trial with our students in the third year in connection with their judging work. We hope to be able to make a comparison of this system with others that are advocated. So far as I have investigated, the "Powell-Owen" system is the clearest and simplest so far explained by any private individual. Connected as we are with a public institution our object is to arrive at the truth and to use all the truth and facts that we can assemble in judging our layers.

Yours very truly,
E. A. LLOYD.

My score-card was used for the first utility Show in Belgium, the request for card and details being again sent to me.

LIÉGE,
December 17th, 1920.

DEAR SIR,

I have read with great interest one of your articles. As a section for utility birds has been admitted at the Exhibition of 1921 at the "Union Avicole," I would like to know your score-card and your way of judging the birds of this type.

Thanking you in advance.

Yours sincerely,
PH. QUESTIENNE.

Details were sent to my correspondent and I received in due course the following letter:—

LIÉGE.

DEAR SIR,

I thank you very much for your esteemed letter and enclosures.

I have carefully examined your score-card on the hens of my own pen selected by trap-nest and on those of some of my friends, and I have found that it gives good results.

The enclosures that you send I hope you will not mind if I publish the French translation in "L'Union Avicole."

"L'Union Avicole" has, on my proposal, admitted your score-card for the judging of hens at the practical section of Show. I have waited to reply desirous to let you know this resolution, and send you a copy of the printed translation of your card which will be utilised.

Yours very truly,
PH. QUESTIENNE
Hon. Sec., Provincial Poultry Federation.

It was of interest to note from the score-cards received that White Leghorns scored the highest, many over 160 ex the 200 and the best 181.

UNION AVICOLE DE LIEGE

CAPACITÉ.	Nombre de doigts.	Points admis.	Points accordés.
Du bout du bréchet	1	4	
aux os pelviens	2	8	
	3	12	
	4	16	
	5	20	
Entre les os pelviens	2	2½	
	3	5	
Entre les os pelviens	1	5	
et l'os de la queue	2	10	
Largeur du moyenne		3	
dos bonne		6	
très bonne		10	
Longueur moyenne		2½	
du dos bonne		5	
Largeur moyenne		3	
entre les bonne		6	
jambes très bonne		10	
Long., prof. moyenne		3	
et larg. de bonne		6	
l'abdomen très bonne		10	
TOTAL		70	

APTITUDE.	Points admis.	Points accordés.
Os pelviens très épais	3	
épais	6	
minces	9	
très minces	12	
S'ils sont droits, ajouter	3	
Viande èpaisse	5	
moyenne	10	
fine	15	
Cloaque proportion	5	
finesse	5	
Tête moyenne	3	
bonne	6	
très bonne	10	
Os gros	3	
moyen	6	
très bon	10	
Plumes en excès	3	
moyenne	6	
idéal	10	
TOTAL	70	

N.B.—Le présent tableau ne peut être reproduit sans autorisation.

ÉCHELLE DE POINTS D'UTILITÉ

Nom de l'exposition..

Race.. Sexe................

Date de l'exposition..

Classe n°........... Parquet n°........... Bague..............

Pureté de race, etc.	Points admis.	Points obtenus.
Condit. d'exposition	10	
Santé	10	
Proportions	10	
Pureté de race	30	
TOTAL	60	

Totaux Généraux.	Maximum.	Points obtenus.
Capacité	70	
Aptitude	70	
Pureté de race, etc.	60	
TOTAL	200	

Prix gagné........................ Signature du Juge,

Echelle de points établie par M. Powell-Owen, du N.U.P.S.

The following letter reached me from a reader of my weekly "UTILITY DOINGS" in the *Feathered World*:—

ARUVANKADU, NILGIRIS,
S. INDIA,
November 3rd, 1922.

DEAR SIR,

I read your articles every week in the *Feathered World* and enjoy your utility "Problems" and always try my hand at answering them. I am very interested in your system. I was home during 1920-1 and watched you at work very carefully, especially at Westminster, and I was also at Tottenham. I am a Waltham Abbey man and I wanted to have a little chat with you, but you were very busy so I did not intrude. Anyhow, I have studied your methods and read every article on the subject that I can lay my hands on and practise on my stock.

I have utility Leghorns and Australorps, so have a good chance. And it may interest you to know that I started the Shows in India on the English method, and I think I am the first man in India to use your methods when judging utility classes.

The South India Poultry Association Shows (utility) have been judged by your method—1921 and 1922—and I have written to the editor of our All-India Poultry Club advocating its use at our next All-India Show.

I have always been keen on poultry, and I am a member of Waltham Abbey club. My brother writing me this week says he has got third with Black Leghorns at Tottenham, so you see it is in the blood. I read my *Feathered World* from cover to cover and then back again.

I am very fond of exhibition birds, and know a few exhibition men, and have won the Governor's cup for best bird in show (S.I.P.A.) every time I have competed for it.

The next time I come home I hope to have the privilege of a little chat with you. Wishing you every success.

Yours sincerely,
L. W. PRIOR.

In 1921 and 1922, then, my score-card system was used at the South India Poultry Association's Show for judging utility birds, unbeknown to me till November, 1922.

My trip to Denmark to judge their first utility poultry section started with a series of letters from Architect S. Bird, of Copenhagen, one of which I publish herewith:—

KONGENS LYNGBY,
COPENHAGEN, DENMARK,
August 25th, 1922.

DEAR SIR,

I do not very much care to remember how long ago it is since I got a kind letter from you answering all my various questions.

I got the *Feathered World* Year Book and read your article on hand-grading; in fact have re-read it several times. Great as it is it always leaves me with a feeling that there is more to come. I only wish to find a way to get hold of that "more."

Evidently the best way would be to come and see you, but that is out of the question as I am in the middle of building my poultry farm. I am planning for 1,200 layers, and through the year selecting from these 200 breeders to supply sittings, chicks and for my own stock. And right here in the selection of these superiors is where I am in need of all the knowledge and information I can get from you.

I wish we had something of interest wherewith to bait an invitation for you to spend a vacation over here. Well, as I have said, I am after your knowledge, and the question is "How can I get it?"

Trap-nesting 1,400 individuals is naturally out of the question, as that must be reserved for my breeders, and therefore I must learn to grade my pullets as far as possible first, and I hope you will be willing to help me and can suggest a way.

Yours very sincerely,
S. BIRD.

With "P.-O." there is always a way of helping, and I suggested a trip to England, when I would be able to give my correspondent plenty of demonstrations and he could steward me at one of my utility Shows. I hit upon Sunderland Show, 1922, as being a classic likely to help, together with Tottenham and the Dairy Shows. And Mr. Bird came all the way over to see me.

Several months after, and I was on my way to Denmark to judge their first utility section, put on especially for me at their classic exhibition at Odense, 1923. It was a fanciers' event with over 1,000 entries, and similar to any county Show of the Poultry Club. My first item of interest was to stand and admire their White and Brown Leghorns, also Black Minorcas, all of prime utility type to please me absolutely and yet shown by fanciers. Each variety had about 100 specimens on view. I had almost started to judge them when I was told that I was in the exhibition section of the Show. They seemed to favour the American type in White Leghorns and bought heavily from U.S.A., also Barred Rocks. Their Rhode Island Reds were a picture, and while they have no liking for any chocolate birds—disqualifying them all on sight—they go for the real rich red plus utility type. Coarseness is refused, and thus we have as their ideal even among fanciers my utility-beauty ideal. I was the guest of the Danish Rhode Island Red Club, which has over 600 members, and visiting many farms of members, also Mr. Bird's establishment, I saw some splendid lots of coloured utility Reds. I was struck, too, with the types of Reds of Messrs. Lohmann Hansen and C. F. Grafte. I made many friends, including Messrs. Eyvind Elm (President of the Danish Rhode Island Red Club), J. P. Hansen (President of Copenhagen Branch), Petersen Bjergaard (one of the oldest breeders of poultry there), G. Holding (President of the Esbjerg Red Branch), K. Sundbr, J. Jeansen, K. H. Knudren, L. Lauritzen, Aa. Hansen, J. P. Anderson, J. Traberg, etc. I only saw one buff-coloured Rhode Island.

We visited many farms, and I was much struck with the progressive methods of Mr. Bird, who had installed a mammoth incubator and brooder stoves from England. He specialises in Rhode Island Reds, but has also some Khaki Campbell ducks and Roman geese, the latter laying up to over 100 eggs per annum. In fact I saw records for geese up to 130, showing that even geese can be bred to an active utility pattern if the owner so wishes. And why not? Why should one keep the parent stock for fifty eggs or under per goose and have no eggs for most of the year? Over here so many seem well satisfied with thirty to fifty eggs per annum, and I was pleased to see the Danish records, as I have always said it was possible to get geese to be very prolific. I was very interested in some of Mr. Bird's grading experiments since his return from the Dairy Show visit, where I initiated him into the A to Z of handling to my system. The Danish Red Club—of which he is a prominent official of the Copenhagen Branch—started an official breeding station, and twelve Red pullets selected by Mr. Bird started off the first month (December) with 285 eggs. Again, I visited a farm where he had graded out forty pullets with deformed backs out of several hundred reared, tracing the defect to deformed males. All the forty were just fit for killing—so deformed were they. So he had started grading very quickly.

And grading is needed in Denmark just as it is in England. I found millions of old hens were kept even to four and five years old, and learned that egg-production does not pay but even means a heavy loss on exported fresh eggs to England of some thousands of pounds, the profits being made by the exporters on preserved or pickled eggs sent to this country in their millions when our hens are moulting and the pullets have not started to lay.

Their Shows interested me a great deal, and I wish I had had a few of my critics with me. In the first place, every Show although run by and for fanciers (this was the first event with utility catered for) is a score-carded event. Every bird, duck, and goose has to be score-carded by the judge, who finds on every pen a case containing a score-card. The latter he fills in, the officials then posting copies to exhibitors. An entry fee is charged, but there is no prize-money. The judge grades his birds into classes, every bird scoring, say, 85 to 100 of the 100 maximum points secures a first-class grading, with a second-class award for from 75 to 85 marks. Judging is the day before the Show opens and commences at 8 a.m., with a judges' dinner in the evening, when printed catalogues are available. There is thus utility and education about the Danish Shows, and seeing that the Poultry Club has 14,000 members (of which organisation Count Ahlefeld Laurvig is President), that entries are large and interest keen one can regard score-carded Shows as successful there. By law a vet has to handle every bird for health on its arrival at the Show. Judges score-card at the rate of ten birds an hour.

My best utility Rhode Island Red scored 159 out of 200 points. I found Mr. Grill (a lover of exhibition Black Orpingtons) in charge of the White Leghorn station and there improving the breed on utility-type lines. Mr. Rasmussen was attending to the utility points of the Black Minorca, while Mr. Hofman-Bang is doing sterling work for the industry as director of the Government feeding experiment station. Here, Barred Rocks of the American type were kept and trap-nested, and a Laying Test organised. He has worked the average up to 150 eggs per annum, and I picked by handling twenty-eight hens out for the breeding pens, some three years old and others two years old. I was not shown the trap-nest figures for the birds I selected, but was informed that, out of the twenty-eight I had picked, no less than twenty-three had good trap-nest records. I found handling most intricate because, although it was February, the birds were only just starting to lay, and hatching is very late even up to June for heavy breeds. Again, I had no 250 or 280 egg layers to pick out. But I had included in my lot a hen with over 300 in two years, and another with a pullet record of 210. The birds had been kept for experimental feeding, and as they were in lots of twenty I had to pick in tens in order not to upset the planning and the experiments, as most of the birds could not be shifted.

The director set me one test. A pen of birds had not laid an egg since arrival, *i.e.*, for the first two months, and I was asked to locate it from the Test pens. I located it successfully, but it was not easy, as the birds were just coming into lay. However, I was able after naming the pen to say eggs would soon be in the nest-boxes.

I did not trace any sample of Sussex ground oats or fish-meal in Denmark, so that developments on English lines will be needed. America is the source of so many breeds and strains sent into Denmark because their exhibition poultry are nearer the utility ideal. Perhaps as our utility-beauty ideal becomes perfect the utility breeders may have another good market in Denmark for foundation stock, etc. I saw some good Anconas but hardly a Black Leghorn, although some nice Buff Leghorns were shown. Brown Leghorns were great, and the White Leghorns (American type) very stylish and aristocratic. I gave many lectures, even one at 9 a.m. to a crowded house. At the Land College in Copenhagen I had an audience of 500, and included in the series, " by request," the " Hen from Within," buying a hen in the market for five shillings, killing it in a taxi, and proceeding to the lecture hall. You can picture me trying in English to make the vendor understand that I would give her an extra shilling if she would let me make my own choice. I think she thought I wanted the lot for the shilling, but I handled several as if I were going to buy the lot and then took just what I wanted, namely, one coming into lay as found by handling.

Presented with a cup by the Danish Rhode Island Red Club inscribed " In appreciation and thankfulness to Utility from the Danish Rhode Island Red Club, February, 1922," and with flowers at the station on my return, together with a real welcome from all, I may have sown the seed for future visits to our Shows with an interchange of ideas.

My score-card used at Odense Show may be of interest to my friends:—

49

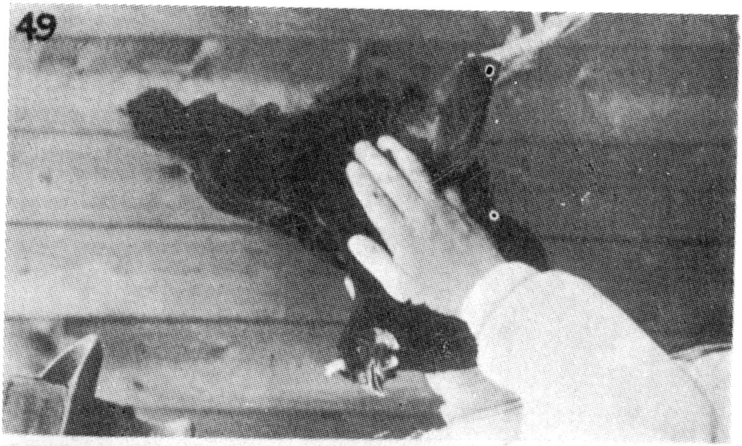

50

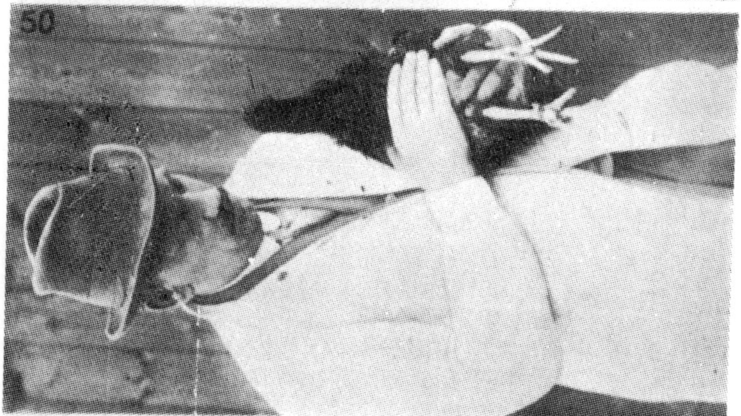

51

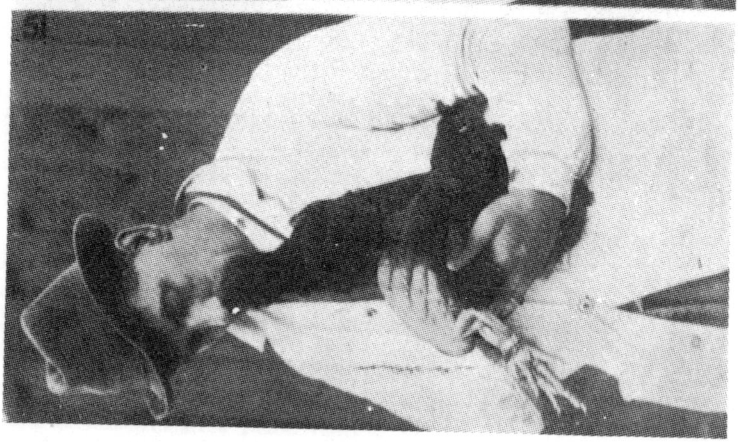

POWELL-OWEN
BEDØMMELSESKORT
GODKENDT AF THE NATIONAL UTILITY POULTRY SOCIETY OG THE POULTRY CLUB OF ENGLAND

KØN	NR.	SKUE RACE DATO

KAPACITET evne til at danne væv		Points		KVALITET evne til at lægge æg		Points	
Enden af Brystben til Skamben	1	4		Skamben	særlig tykke	3	
	2	8			tykke	6	
	3	12			tynde	9	
	4	16			særlig tynde	12	
	5	20			hvis lige +	3	
Mellem Skamben	2	2½		Kød og Hud	grov	5	
	3	5			middel	10	
Mellem Skamben	1	5			fin	15	
og Halerod	2	10		Kloakaabningens Størrelse		5	
Bredde af Ryg	middel	3			Finhed	5	
	god	6		Hoved	middel	3	
	særlig god	10			god	6	
Længde af Ryg	middel	2½			særlig god	10	
	god	5		Ben og Horn	middel	3	
Afstand m/ Ben	middel	3			god	6	
	god	6			særlig god	10	
	særlig god	10		Fjer	middel	3	
Længde, Dybde, og	middel	3			god	6	
Sidde af Gump	god	6			særlig god	10	
	særlig god	10					
Mulige Maximum........		70		Mulige Maximum........		70	

RACEEJENDOMMELIGHED Etc.			SAMLET ANTAL POINTS		
Udstillingskondition	10		KAPACITET	70	
Sundhed	10		KVALITET	70	
Størrelse (ideel Nytte)	10		RACEEJENDOMMELIGHED	60	
Raceejendommelighed	30				
Mulige Maximum........	60		Mulige Maximum........	200	

SIGNERET

N. C. ROM, KBHVN.

R. I. R. KLUBBEN · F...

As I pen these lines, appreciation comes from a New Zealand cattleman who is in this country buying large stock.

NEWARK, NOTTS.,
January 26th, 1923.

The utility section in this country is getting very strong. I think our thanks are due to you for the able way in which you handle and encourage this section. Best wishes for your future welfare.

To-day in New Zealand the utility section are coming in at a gallop, and proving stayers when they get in. If they don't appreciate you to-day, well—that day will come and before long.

I have made up my mind to see you ere I go back to New Zealand, as I am most anxious to get an insight into your methods.

Yours faithfully,
GEORGE N. BELL
(A N.Z. cattleman).

CHAPTER XXVI.

ON THE FARMS AND AT THE SHOWS.

I APOLOGISE to readers for devoting so much space to results of hand-grading, and only do so in its defence—so bitter have always been the destructive criticisms against it. Until this began I had never sought to mention any performances, leaving it to any who wished to test my system of handling and then to prove or disprove its advantages for themselves. That is the best test for any system, seeing that you can extend to it a broad-minded trial without bias. And you can utilise whatever parts are of service to you and discard the rest. When I begin to mention results I am accused of prophesying numbers of eggs.

At the Shows, in addition to egg-laying powers, I study a bird's value as a breeding hen after she has laid the eggs. Again, I have to see that the bird is well shown, and finally that it has high breed characters. While I insist on high laying powers, never neglect the fact that my score-card is my standard. If, knowing what my card is, you care to show under me a pullet as fat as butter and scaly-legged, or even covered with lice, then she will have to be severely penalised under health and show condition. The fact that societies with which I am associated hold all records for utility entries, that I personally hold many class records and Show records—having judged more birds in a day than any other judge—will show that exhibitors do not mind my score-card as their accepted standard. Again, I have judged more utility Shows than any other, and think that such in itself should convince the destructive critics. But their biassed and interested views might carry weight with many, and that is why I have devoted all this space to the same, not robbing the book but putting on extra pages at my own expense.

Many have very crude ideas of what handling is. I have known a man condemn it when he has been so inexperienced as to try to place five fingers between the pelvis bones instead of from breast-bone to pelvis. Again, I have known a man score a pelvis bone by fingers, *i.e.*, three for " very thick finger " and twelve if " very fine finger." This volume may, however, put beginners right as to what my handling system is.

Judging at Shows, however, is rather different to grading on a farm or in your backyard, because in ordinary grading one likes all good birds and detests all low-grade ones. At the Shows I try honestly and sincerely and without favours to put the best all-round bird first, as found by a minute handling for every item on my card; and it needs very great care and experience. Again, one has to get through hundreds of birds in a short time, as many as 790 having been my lot at a single Show and in one day. I plod along at the rate of thirty scored birds per hour, or fifty per hour at ordinary handling without scoring in detail. But every bird has a fair chance, because I take out and handle every bird shown under me. In being consistent to my winning birds I have been accused of just being lucky, but is it not rather a question of showing that theory ceases when actual practice finds the same birds winning so often under me? And they win under other tried judges who use my system. To realise what a Show is, one should attend as a steward or helper. Often after I have judged a class I am told that several birds have just arrived or have been overlooked or penned in wrong classes. I always go back and give them a run for the entry fees. Then one has to put a few cocks back into the cockerel classes, and vice versa, and often to make out one's own judging-books, seeing that so many Show officials in utility are new recruits and only want time and experience. But the difficulties are there when grading has to be applied at Shows. All such things will be removed if patience is granted and with extra judges to help, so that each judge has a given number of birds to handle. Time alone is needed. Again, one has to bear in mind that, at first, the birds of high grading stood out, but now improvement has become so marked that in a class of 87 to 125 pullets we have a Show in itself when it comes to placing the winners. Score-carding can at least take the credit of improving the birds shown at utility events. My score-card totals clearly prove that.

The exhibition of utility pullets in full lay must naturally tend to upset them for a time, especially the flighty ones. I always advise you to show the pullet which lays "in the wash-tub" or near the fire when being prepared—the docile, lovable bird which lays *en route* to the Show, daily at the Show, and also on her return back in the basket. A lot has to do with temperament, and all broad-minded readers will agree that showing may temporarily affect the laying of the birds. Some exhibitors show a bird to death and then expect it to lay 250 eggs and to keep fit. Take my advice and be fair to the bird, yourself, and the judge by letting each pullet have but, say, three outings per annum. Get together when maturity is being reached a pen of pullets of likely quality for exhibition, send out the fittest every time, and avoid over-showing. I have seen pullets exhibited at a Show for nearly a week returned home for another wash and off again to another Show for another four to six days. When pullets are left at home to lay they are not subjected to washing and frequent moving about in all weathers, so please allow for the different conditions as regards the egg-laying of the birds. Some send birds to a Show in open crates and draughty receptacles, and no wonder some winning birds lay badly afterwards. Again, I have seen the floors of show-pens swamped with water and wet litter, also frozen, which are unfortunate drawbacks. Realise they are there, however. A chill on the winner's liver and naturally her egg-laying on the return may be lowered.

Some think that a pullet lays up to grading or pedigree and that nothing can alter that. Again, many think that pullets sent to Tests lay up to their breeding. You never see a man advertise the fact that one of his best pedigree pullets laid but eighty-seven eggs in an official competition, yet such exist *ad lib.* Treat yourself to a few monthly Test reports. Environment and management with a big slice of luck are closely associated with egg-totals per annum. In the White Leghorn section at Harper Adams College Test, twenty-five pullets have not laid an egg for the first three winter months, and nearly ninety in the one on section have laid under ten eggs for that period. Do you think I call them all "duds"? They will each be sixty eggs short at end of Test. Some may be "never-layers."

One has to visit utility Shows to see how the birds arrive. I have seen them brought in sacks, orange boxes, and even under the arm with legs tied to prevent them from flying away. I have seen an exhibitor struggling home from the Show with half a dozen pullets crammed tightly in a basket just to sweat and chill. Can there be good laying under such circumstances? Be fair on the birds if not on the judge, and use proper receptacles and do not overshow.

My critics have unearthed one case. In the first place, a Speckled Sussex won dozens of prizes under all utility judges (I would be included, I suppose) and laid about eighteen eggs. The critic told me over twenty, and then another critic he told published the case as a Light Sussex that laid even fewer eggs.

If a pullet lays but a few eggs, *science* calls her a *dud*. It has been left to me to disclose the actions of the "never-layers," and those birds which "hold" their eggs. When I was lecturing at Ware, Mr. McEwen referred to a friend's White Leghorn pullet that had won in utility under another judge and had not been a good layer. At a later visit he told me that his friend had received a complaint from a client who brought back a " rotten egg " sold as new-laid. All birds were trap-nested, and the number on the egg showed the layer to be the utility winner referred to. As no egg had been laid by her for two months, this egg must have been held up in the shelling department of the oviduct for that period. This pullet visited the trap-nest to lay the " rotten " product, but as a rule such eggs slip out in the litter and are eaten. *My record period for an egg to be so held up is five months*, in a hen owned by Miss Barbara Raye. Finally the egg was dropped and the bird laid normally again in due course. Any pullet would put up a low yearly trap-nest record if she held up an egg for five of the twelve months. Such mishaps are constantly happening, which is one reason why there can be no definite egg-factor, and maybe why some low-record hens scientifically termed " duds " breed tip-top laying progeny. One may, of course, argue scientifically that they show a reversion back to a low-grade ancestor, but we want more definite facts than that.

It was in an endeavour to test trap-nest records as merely indicating *the actual number of eggs laid in the nests* that I hit across the " never-layer." I collected all birds with low trap-nest records which really handled like laying type, my object being to ascertain why they did not put up higher totals. My view of a " dud " is one which lays, say, sixty eggs or so because she takes twelve months to mature, goes broody several times in the season, and takes months to moult—all capacity and no capability. One of my early surprises was a pullet which had never laid an egg in her pullet year, but visited the nests regularly. Keeping her over the moult she surprised me by starting to lay normally. A post-mortem revealed nothing wrong inside. Then I had a White Leghorn hen (see Plates 22 and 23) which, hatched in April, 1919, started to lay in October, and produced 118 eggs to May, 1920. Except for a stray egg in August she did not lay again for twelve months, although visiting the trap-nests regularly. I decided to thin her of all fat, and in June, 1921, she laid a shelled egg. Killing her I found an ovarian cyst inside but still was fogged. With my more complete knowledge that was a stray egg, and she must have been laying internally. But I was on the track of the " never-layers " which visit the nest-boxes regularly but never drop a shelled egg in the nest. They handle as if in full lay, with large moist vent, full abdomen, three fingers between pelvis bones, etc. One must now cross out the term " handles in lay " and say " handles for functioning ovary!"

Such birds have a fully-functioning ovary and oviduct, but *lay their eggs within.* In some cases the yolks fully ripen and rupture out into the body, there to be absorbed into the system. In others the cases are filled with yolk and the latter is then absorbed back. One of the most interesting cases I had of a " never-layer " was a pullet belonging to Mr. Frost, of the Bristol Branch of the N.U.P.S., which had not laid an egg for six winter months, but had visited the traps regularly. She handled in full lay, and the post-mortem revealed a full ovary with several broken cases, showing that the yolks had dropped into the body. The amazing find, however, was that the bird had no oviduct. The hen I demonstrated with at Tottenham lecture in January, 1923, had been sucking her yolks back, as had the hen at the London and District lecture. Such " never-layers " when dropping their yolks absorb them into the system as fluid, the yolk not becoming " hard-boiled " as with other types of hen. Some go broody, which explains why a hen sits without actually laying an egg (*i.e.*, shelled). Some start laying again after the moult or a rest period. Hence, as discovered by myself, a hen can lay no eggs in the first year and plenty the second, or vice versa; or she can lay normally the first year, inwardly the second, and normally again the third. And, of course, some live two years and more without producing a single *shelled egg*, or contributing at all to the nest-box returns.

Far from attacking trap-nesting, I wish it were possible and usual to trap-nest every bird. I certainly shall always point out the mistaken policy of just trap-nesting thirty pullets to find out the layers of the highest numbers and then labelling them *the best reproducers*. What of the other six hundred pullets on the farm which never have a production test at all? What of the low-grade pullets rejected for breeding and put into the laying sheds, despite some being " never-layers " perhaps? If trap-nesting were employed *mainly* to locate the passengers of the farm or backyard run, and *incidentally* for the high producers to be known, that would be my ideal. Then the hand and the eye will decide with the known records which to breed from as sound reproducers.

Where trap-nesting is not used the handling test should be applied. Surely the first method to improve the stock and flock-average is by grading out the passengers! " Never-layers " deposit their absorbed yolks, *i.e.*, *fat*, on to the abdomen, so will handle fat at pelvis bones and abdomen, with gizzard difficult to feel. Some grow spurs, others call the hens in masculine fashion, while not a few develop increased comb and wattles like the male, and even crow. Parts of the abdomen too often become after a time bare of feathers and covered with chalky matter, which may be the albumen from the oviduct, irritating fluid, or the lime from the shelling department of oviduct. As each yolk drops into the body the shelling department may send out lime which settles at abdomen. Such birds, too, look heavy eyebrowed and worried.

The Utility Poultry Societies that have come into being in their thousands are all working on progressive lines, holding monthly lectures, egg, appliance, and stock shows, etc., and place education and instruction before all things. Their utility Shows are a valuable means of instruction, and whether the critics call the birds shown utility-show specimens or anything else, no one can deny that our utility events are progressive and have high ideals. If one section of the poultry public want just layers they will decide and create the necessary demand. Again, if another wants a layer that stands out for beauty and shapely form then there will be breeders to supply the new demand. If utility is always placed first it may so happen that the pullet that can (1) win at the local Show, (2) be admired for breed characters and beauty, and (3) still lay as well as the Laying Test competition winning bird, may beat the latter for popularity. I stand for one type for all the purposes named, in short the *beautiful* layer! The little man, from his purchased sitting of eggs all from carefully graded pedigree stock, has every chance of getting several winning utility pullets therefrom at an economic price.

Some of the big breeders are against utility Shows, but why do they not make full use of them as another channel for publicity? Let them enter well-mated trios for exhibition or display so that the small poultry-keepers can see the class of stock they offer and breed. Many progressive breeders *are* making use of their opportunities at the utility Shows, and even exhibition events like the Dairy, Olympia, Palace, etc. The utility breeder cannot have too many channels of demand. The utility bird can only hold its own by being always in the limelight wherever present or would-be poultry-keepers attend. To ignore progress is to take a back seat!

I am all out for the small poultry-keeper; he needs all the help and education we can give him. Our Utility Poultry Societies, meetings, lectures, conferences, utility Shows, etc., spread such education. The utility Show makes the small producer keen in his work. He enjoys the friendly combats in the show-pen against his fellow club members. He learns the value of well-bred utility stock. Greater observation comes his way, and the stock receive better housing, management, and methods of conditioning. By strengthening his management he is able to complete the good work started by the utility breeder; he does not by bad management turn well-bred layers into low-grade producers. The breeders then get the credit for stock supplied, not the improved management. I am among

58
59
60

the backyarders constantly, being connected with so many Utility Poultry Societies, and I should know their needs. To-day, as heretofore, they represent the backbone of the poultry industry, so nurse them well so that the utility bird claims their favour and whole-hearted support. We have the majority behind the *beautiful layer to-day*, thanks in a measure to my pioneer work. It is for all utility breeders to see that they do not withdraw their support from us and go for the *non-beautiful non-layer*. If extremists among the breeders persist in fighting against progress instead of co-operating we shall lose that support. Such are my personal views which dominate my programme.

I had always in mind the high ideal of our utility Shows teaching small poultry-keepers how to condition their stock and how to discern laying from non-laying type. Hence my score-card, primarily for teaching purposes. I wanted to bring out strongly the personal side of poultry-keeping, which has the last say in egg results. Utility Shows have accomplished very much in this direction. You can prove that only by visiting such events constantly, and by noting the improved quality and condition of society members' stock year by year. I have sought to strengthen observation among the small poultry-keepers by teaching them simply the internal mechanism of the laying hen. That my lectures on this subject do good work is emphasised by the large attendances I obtain. It opens one's eyes when as at Battersea lecture the hen demonstrated upon had *two gizzards;* the hen at Barnet supplied by a member had tuberculosis; while a third hen possessed two breast-bones. The Battersea hen had its normal gizzard, with outlet and inlet, but joined on by a tube was a second gizzard full of putrid food and without an outlet. To rid herself of this second organ (felt in the abdomen like an egg) the hen had tied up her intestines severely. A moment's handling would have revealed the obstacle in abdomen even if one had taken it for an egg. The bird had not laid for a long time, and could have been tested for production when suspected.

Score-carding is no easy task, but the ideal would be to have every bird score-carded in detail so that the owners would possess "key" birds. One can only score fifty per hour, which is the only "fly in the ointment," as it would mean very many judges each taking, say, 150 birds for a 1,000-bird event. In Denmark they score *all* exhibition birds, geese, ducks, etc., at *all* Shows at the rate of ten birds per hour, as they run their Shows solely for education. They have a host of judges, and keep strictly to the Specialist Club score-cards or hundred-point standards. They find it practical and award grading certificates according to marks obtained—no prize-money is given, but entry fees are charged, yet entries are very high. A day is set aside for judging, the judges plodding along all day.

In this country exhibitors have been educated to quick judging—all cards up by lunch whether 200 or 800 exhibits per judge, even with rabbits, Bantams, and pigeons thrown in. The keen utility man will find ways and means of attending the utility Shows and learning his lessons in handling. He may exhibit a few birds from time to time and seek the judge's views thereon. That is the educational side I study closely. In the summer of 1922 Messrs. C. L. and J. A. Thompson, of Bersham, Wrexham, were recommended by a local breeder to write me for details of my score-card system. I gave all the details I could, and suggested they should exhibit under me, score-card all birds sent, attend the Shows, and I would then help them all I could and compare my score-cards with theirs. At Shows close at hand where I was judging this plan was carried out, and I was able on one occasion to pay them a quick visit. The results you may judge from extracts as under, embodied in an unsolicited letter:—

BERSHAM, WREXHAM,
April 17th, 1923.

. . . . *Re* handling. You know we did our selections for the Tests by your handling system. We made six entries, and although some made a poor start, all, with the exception of a White Wyandotte at the Middlesex Test that is dead, are now doing well. Our Black Leghorns are first over all at Midland; our Sussex 2nd at *Feathered World* Trials, and making a good shot for 1st; our Sussex are 4th at Bentley; our Sussex at Middlesex 1st for November, 2nd in December, then spoilt herself by going broody; our Wyandottes at Bolsover are 7th (the two birds are only 23 points from top hen). These started badly, and suppose they both moulted.

The above is not bad considering WE KNEW ABSOLUTELY NOTHING OF HANDLING BEFORE LAST JULY [1922—Author].

Yours sincerely,
J. A. THOMPSON.

Messrs. Thompson were trap-nesting all their birds, and so were able to follow my methods with tested stock. It is only one of hundreds of like cases where utility Shows have greatly helped by way of education. Taking *the four winter months* at Midland, the "winter" report credits Messrs. Thompson's pen of Black Leghorns with first in the A.O.V. section, and heading every pen of every breed in the Test—*another breed winter record* for my handling system. The Test had been running 143 days, and the report credits Mr. and Mrs. R. Jacobs with first in Light Sussex section, and with the highest individual record over the thousands of birds of all breeds with 101 eggs in the 143 days for one of their Light Sussex—another breed record for my handling. Second to this pen in Light Sussex is Mr. Honey's pen, and third to Messrs. Thompson's pen in *Feathered World* Test are Mr. Lethbridge's Reds, both picked on my system, and I see Messrs. Thompson won the monthly ribbon for the section for greatest weight of eggs. But these are 1922-3 performances, which I had not intended originally to deal with in this edition.

As my friends know, I have always asked for laying records of birds shown under me, and have judged some Shows many years in succession, which would allow me to hear any detrimental remarks against birds winning under me at previous Shows, and if my judging had been bad I should have lost and not increased my entries. Handling and utility judging are in their infancy, and we are doing our best as judges, believing in handling and trying to master all the difficult problems that arise. Criticism is easy from those who never try their hand at judging publicly, but time will perfect any of our shortcomings.

And in my visits up and down the country none can say that I have not openly been tested on breeders' farms. In fact it has been my first aim when visiting a farm to have a few tests of handling in order to obtain converts. But when I turn down a 300-egg hen which lays 27 eggs only the year following, and such a bird is a noted competition winner—I refrain from mentioning names—I can be a gentleman in such matters. I have always refrained from publishing names of such breeders.

171 EGGS IN SEVEN MONTHS.—The 1st Trowbridge Light Sussex pullet, October, 1921 (beating the pullet that was 1st Tottenham and 1st Ilford under me), and which later won 1st Welsh National, 1922, under me for Mr. W. G. Arkell, laid 171 eggs from January 1st to July 31st, 1922.

TESTS AND SHOWS.—Mr. Chas. H. C. Partridge, of Little Lambswick, Tenbury Wells, writes:—

January 21st, 1922.

The little Westminster White Leghorn pullet of 1921 still continues to lay well, and I am putting her eggs down as they come along. I bred from her last year, and her daughter, sired by the 1st Tottenham cock, 1921 (Breeder-members' class), has just won the Bewdley Laying Test (winter-month Test) with 52 eggs for the three months. The little Westminster bird has earned her corn, for she laid, as I told you, 273 eggs in her pullet year, and from some of these eggs the Leicester, Worcester, New Barnet, and Olympia birds were hatched.

I finished third at Bewdley for White Wyandottes with the pullet placed by you at Yeovil in a class of eighty-seven. She laid 21 eggs first month, only 4 the second (neck moult), and 25 eggs the third, of which 44 were first grade.

I have cause to remember this White Leghorn hen, which in a class of sixty-seven pullets at Westminster (N.U.P.S.) Show in 1921 I awarded first by score-card with a total of 191 points out of the 200. I think this is the record score by myself on my card. After winning she went on to lay 273 EGGS IN HER FIRST YEAR, and has bred 1922 utility-winning daughters and a Test-winning daughter. Her score-card in 1921 (December) as a pullet was:—

CAPACITY.	CAPABILITY.	BREED, ETC.	TOTAL.
20	12	10	Capacity ... 67
5	3	10	Capability ... 67
10	15	10	Breed, etc. ... 57
9	4	27	191
5	5		
8	9		
10	9		
	10		
67	67	57	

I also awarded the sire 1st by score-card at Tottenham in 1921, so that score-carded birds do not always breed mongrels when mated together as one critic prophesied they would.

THE "RODWELL" HEN.—A White Wyandotte hen I am also proud of has become known as the "Rodwell" hen. Shown under me as a pullet at Westminster in December, 1920, I placed her first in a class of eighty-five pullets, scoring her 175½ points out of my 200. Her score-card then was:—

CAPACITY.	CAPABILITY.	BREED, ETC.	TOTAL.
18	12	10	Capacity ... 57½
5	3	10	Capability ... 63
5	15	8 (small)	Breed, etc. ... 55
10	5	27	175½
2½	4		
10	8		
7	8		
	8		
57½	63	55	

In December, 1921, I received the following letter from Mr. R. Rodwell, whom I had not met or corresponded with before:—

WALVERDEN POULTRY FARM,
NELSON, LANCASHIRE,
December 7th, 1921.

DEAR SIR,
May I congratulate you on your judging at Westminster last year. The pullet which you awarded first and cup for best female in show has created a most remarkable record of 273 eggs in her twelve months, and she never stopped once while she was exhibited nor was she even broody. Her moult was completed in short time and she was at it again. The same bird was awarded second at Tottenham recently by you.

Yours faithfully,
R. RODWELL.

Apart from Shows, then, this pullet laid 273 first-grade eggs in her pullet year after I had judged her as a youngster, and when she was being shown, of course, the eggs laid would not be known. After laying a goodly number of the eggs, I gave her, by score-card, 1st Royal of England at Derby, and when she had completed the year I placed her 2nd in a class of seventy-eight against pullets at Tottenham. I saw nothing of this bird in her second year, but commencing her third year I gave her 1st and "best" Watford and 1st and "best" Wembley for Mr. J. J. Rowsell, the secretary of the Barnet Branch of the N.U.P.S., who had acquired her from Mr. Rodwell in her second year, and confirmed the statement that from April, 1922, to February, 1923, since he had her she had not gone broody or stopped in the moult but just carried on laying large eggs. Her score-card as a three-year-old hen at Watford credited her with 60 odd for capability.

On receipt of Mr. Rodwell's letter I immediately wrote for the laying records of all his birds which I had judged at Westminster in 1920. It was the first Show of the National Utility Poultry Society with my score-card in use, and I had over 700 birds myself to judge the first day, including a class of seventy-three trios. Those who criticised the two pullets already dealt with—especially Mr. Partridge's White Leghorn—as being too small, and so on, may be more than interested to see which was correct—my score-card or their critical opinions, and they were many.

In reply I had the following letter:—

NELSON,
December 23rd, 1921.

DEAR SIR,

My Laying Test birds are closely related to these birds which have scored well under you. The 1920 Westminster results were: 1st and cup best female laid 273 eggs; the two which were in the winning trio laid 248 and 252 large eggs in twelve months.

The best male (1st and cup) bird was from a hen which laid 266 eggs, and he was sired by a cock from a hen which laid 250 large eggs. The male that headed the winning trio pen was from a hen which laid 246 2½-oz. eggs in twelve months, and he is full brother to my 1920-1 winning White Wyandotte pen at Burnley Test.

The 3rd White Leghorn pullet under you in class of eighty-two is full sister to birds which competed at Oldham Laying Trials and championship section at Bentley, where I was second and fourth positions respectively, 1920-1 Tests. I sold this bird, so cannot give her record.

The pullet which won 1st and cup this year [1921, and again under me—Author] was sired by last year's champion Westminster male.

R. RODWELL.

Many will remember me struggling through that final class for trios at Westminster, 1920, when I had sixty-seven trios (201 birds) in the one class to produce layers, both light and heavy breeds. I scored the three birds and the best total won, which happened to be Mr. Rodwell's White Wyandottes referred to above, with Messrs. Abel Latham & Son second with White Wyandottes and third with White Leghorns. I think it was nearly midnight when I put the last award card up. My two pullets in the winning trio laid 248 and 252 eggs, as stated above, and their score-cards were:—

No. 1.—Capacity: 18, 4, 5, 10, 2½, 8, 8, total 55½. Capability: 7 (fat), 3, 10 (fat), 4, 3, 9, 7, 8, total 51. Breed: 10, 10, 25, total 55. Grand total 161½.

No. 2.—Capacity: 20, 5, 10, 10, 2½, 8, 8, total 63½. Capability: 11, 3, 11, 4, 4, 7, 7, 7, total 54. Breed: 10, 10, 10, 25, total 55. Grand total 172½.

The cockerel at head of trio, full brother to the Burnley Test, 1920-21, winners, scored:—

Capacity: 16, 1½, 7, 10, 2½, 8, 10, total 55. Capability: 8, 3, 11, 3, 3, 7, 8, 8, total 51. Breed: 10, 10, 10, 24, total 54. Grand total 160.

At this 1920 Show I placed all cups as well, no judges being available early on the second morning. I placed according to score-card points obtained, and my "best in Show" laid the 273 as stated. If I missed a 274 "egger" my apologies to owner!

You will always find Mr. Rodwell's name at the head of the Laying Tests and also at utility Shows. Critics will tell you that he has the two types for the purpose and that they are not related. I watch all these points and have noted that Mr. Rodwell has never shown any birds under me at the utility Shows during the breeding season. The birds are in the breeding pens, and you have but to see his catalogues to see how the birds are utilised for breeding. Maybe a utility-Show winner is rather different to some past Laying Test winners, but only in their beauty. The utility birds are a combination of laying and beauty, that is all, and that cannot be said of all past Test winning birds. Their photos still adorn the catalogues issued by the owners, and you will be the best judge of the beauty of any individual bird. Then get a catalogue containing photos of my utility winners and you will see the beauty that enters the argument without losing the eggs. And why should we not have eggs and beauty or, in short, a "beautiful layer"? Get busy and breed some of them, but remember egg-production first and beauty second; not all beauty (alleged) and no utility!

Extract from letter in Poultry Press from Mr. R. Rodwell:—

January, 1922.

I am a strong believer in handling system, and when I send my birds to Laying Tests I select the best handlers from a strain, and by practical experience I find the greatest handling birds are generally the most prolific layers.

Extract from letter in Poultry Press from Mr. R. Rodwell:—

January, 1923.

To my mind utility Shows are the best things possible for amateurs to gain knowledge of a good layer. It is only about four or five years since I first heard anything about the handling of birds for ascertaining their laying qualities. And I must admit that it has helped me wonderfully when I have selected birds for competing in the Laying Tests. I can give several facts concerning selected good-handling pullets and their performances at the premier laying competitions. Therefore, if the handling system is a great help to me when selecting birds for laying competitions which you considered practical, is it not a good purpose to have utility Shows?

Extract from letter in Poultry Press by Mr. R. Rodwell:—

February 23rd.

Our trap-nesters are the very class who taught the fancy about handling. One of our most prominent competition winners came to my yards for a Wyandotte cockerel and I refused to sell my 1920 cup winner (then unshown) but offered several brothers. This person, however, stubbornly insisted upon the best handler, but after several hours he was obliged to take the next best handler. The cockerel, he says, has bred the very best layers he has ever had on his farm. Our trap-nesters who are winning the laying competitions when requiring fresh stock always aim to secure the very best handlers possible from the best pedigree trap-nested stock.

230 EGGS—STILL LAYING.—On August 24th, 1922, Messrs. Garrood and Turner wrote:—

You may be interested to know that the Rhode Island Red pullet which you awarded 4th at Westminster, 1921, in open class has been laying continuously since October last, has laid 230 eggs, and is still laying, never having been broody.

241 EGGS—ELEVEN MONTHS.—My 2nd White Leghorn pullet at Chester, 1921, Mr. Dawson, the owner, informed me at Harper Adams Conference had laid 241 eggs with a month to go. Not so bad for out of lay at time of judging. The winning bird was as under:—

LAYING TEST HEN.—In this class at Chester Show the winner was a hen which was shown by Mr. J. H. Mather, Paddock Road, Audlem, Ches., the bird being in the 3rd Burnley winning Test pen and laying at the Test 229 eggs. She scored:—

Capacity: 20, 5, 8, 10, 4½, 10, 9, total 66½—measuring at abdomen 4½″ from thigh to end of abdomen (length), 6½″ in depth from tail bone or socket to end of breast-bone or abdomen (depth), and 4″ from one side to another in flesh (width). Capability (on August 31st): 7½, 3, 11, 5, 5, 8, 9, 10, total 58½.

Mr. Mather also showed under me at Chester his No. 9 White Wyandotte pullet which laid ninety-eight eggs in just under four winter months. She secured first and special best in Show and scored:—

Capacity: 20, 5, 10, 10, 5, 10, 10, total 70 or maximum. At abdomen she measured 4½″ for length; 5¼″ for depth; and 4¼″ for width. Capability at close of laying season: 12, 3, 10, 5, 4, 8 (low eye and thick skull), 8, 10, total 61.

Two cocks were also exhibited by Mr. Mather, and as they have been great stock-getters their score-cards will be of interest. Fifty pounds has been refused for the Wyandotte male. Each secured third prize in its class against cockerels, and of course as adult cocks they were not in best of form in August. The White Leghorn cock, 1918-hatched or in its third year, scored:—

Capacity: 12, 2½, 8, 10, 5, 10, 8, total 55½. Capability: 12, 3, 12, 4, 4, 7, 8, 9, total 59. Abdomen: 4″ for length; 5″ for depth; and 2½″ for width. This cock sired the third Burnley Laying Competition pen (the winning White Leghorn referred to above was one of his daughters), and was bred from the first winning Burnley pen of 1916-17.

The White Wyandotte cock was also 1918-hatched and in its third year when shown. He was the brother of the pen that won the first in American Laying Test, averaging 234 eggs in twelve months. Although a little on the fat side he scored:—

Capacity: 16, 2½, 10, 10, 5, 8, 10, total 61½. Abdomen: 5″ for length; 5″ for depth; and 3″ for width. Capability: 6 (fat), 3, 9 (fat), 4, 4, 8, 8, 8, total 50 (higher for capability when fit).

This was the first time that Mr. Mather had exhibited under me, and it is a noticeable fact that when the big breeders slip in a bird for the first time they invariably get placed. Mr. Rodwell was showing under me for the first time at Westminster in 1920, and won first for White Wyandotte pullets and best female; first and cup for best male with White Wyandotte cockerel; first trio to breed layers, also cup; and third White Leghorn pullet. Mr. Mather secured at his first effort under me two firsts, two thirds, and special best in Show with four entries, and you have above their laying qualities and pedigrees and can best decide whether or not my score-card represents utility merits.

From Mr. Mather's catalogue:—

> My motto is flock-average not abnormal layers. I do not keep any special birds for competition work, as I rely upon selection. Therefore you get exactly what I use myself.

126 Eggs in 155 Days.—Mr. Thos. T. Batt's Columbian Wyandotte pullet first in breeding pen at Yeovil and fourth Westminster, 1921, both under me, laid 126 eggs in 155 winter days.

Bantams and Utility.—The Partridge Wyandotte Bantams which won first and second under me at Winterton averaged over 100 eggs each for the winter months.

Laying of Bantams.—Mr. James E. Cross, of Blackburn, won first and certificate at Westminster, 1921, with a Bantam pullet that laid 110 eggs in the five winter months from October to the February and all eggs 1½ oz. Her weight was 31½ oz. Another Columbian Wyandotte Bantam pullet of Mr. Cross' was second under me at Burton Joyce and laid eighty-one eggs in March, April, and May alone.

202-Egg Bantam.—The White Wyandotte Bantam hen first under me at Northumberland Heath and first at Cr[illegible]ford in

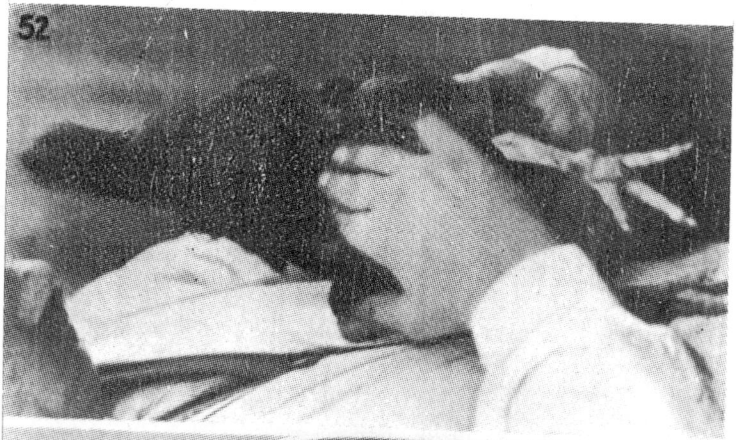
52

53

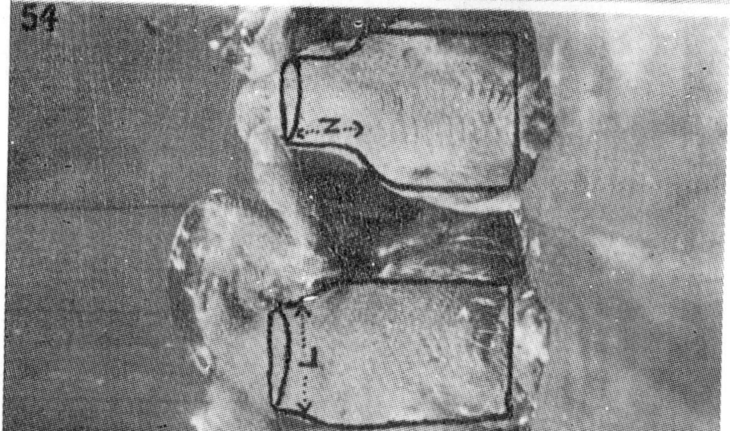
54
Z
L
R

utility laid 202 eggs in twelve months, of which 160 were produced before a broody period.

85, GREEN WALK, CRAYFORD,
November 26th, 1922.

DEAR SIR,

I was pleased to see your remarks on utility Bantams in *Feathered World,* and thought you might be interested to hear that my White Wyandotte Bantam hen which you placed 1st at Northumberland Heath and Crayford has been a grand layer, having laid 202 eggs in twelve months —160 before going broody, when she reared chicks.

Am sorry to note exhibition men are bringing them down too small now to combine exhibition and utility—a great pity, as that was my reason for taking up this breed. I think utility Bantams will go ahead as soon as they are better known, but the public have an idea that they are merely children's pets. They would open their eyes if they knew.

I should like to say how much I enjoy your "Utility Doings" in *Feathered World,* and am getting quite expert in handling through reading same.

Yours sincerely,
W. B. HORSEY.

I judge utility Bantams for utility in exactly the same way as hens or pullets, allowing for the difference in size. It is always difficult to obtain the laying records of Bantams as few trap-nest them, but the above instances have come my way for my winning Bantams.

QUALITY LASTS.—Where one is grading high it is not difficult to pick hens or males which will last for quality. In fact therein lies the key to longevity. The trap-nest may say of a hen 180 for pullet year, and she will handle for a very low number the second and produce eighty. Another pullet with 180 eggs for first year may go on and produce 230 in second year. Things may happen the first year to pull the total down in the latter case, whereas in the former the bird may be low-grade, laying well up to and a little beyond her value.

Longevity of laying is associated with those birds which retain their capability or high quality. You need the four-year-old hen and cock which look like chickens and not the pullet which at nine months old looks like an old hen for facial features and expression. I have referred to the Rodwell hen which won under me in her third year. Here are other cases:—

A Rhode Island Red pullet won first at Ilford and first at Tottenham under me, 1921, for Mr. H. A. Hussey, and cup best in Show (members) at the latter event. After laying hard all the winter she won second under me in the summer at Burton Joyce in a large class and continued to lay on until October, 1922, saw her first again at Tottenham under me against pullets, and reserve for best member's bird in Show. At Burton Joyce she was beaten by Mr. H. Marshall's pullet, which went on to lay 242 eggs in twelve months.

A case of a pullet with lack of longevity and an adult's facial features was this bird's sister, which for Mr. Hussey won under me first at the same Show at Tottenham, 1921, in a "*dud*" *or graded-out class*. Hatched in February, 1921, this pullet did not lay her first egg till March, 1922—thirteen months from hatching out—and after laying a batch of eggs had a broody spell or two and moulted in July.

The White Leghorn cockerel, "Dandie," of Mr. D. C. Gauldie's—the well-known Scottish exhibitor and utility judge—which won under me at the first utility section put on at the Dairy Show, 1920, was third under me in a large class mainly of cockerels at Bradford Show, 1923.

The White Wyandotte pullet which was first and best in Show under me at Northumberland Heath, 1921, was also first under me in 1922 at the same Show, and secured best in Show under all the judges. To show how deceiving high quality can be I stood out for the bird to be a hen when nominating her for the "best," but some of the other judges thought she was a pullet. "Well," said P.-O., "judge her as a pullet; you can't fault her." And she got the verdict. She also won first under me at New Barnet and several medals for "next best," and first Surrey County. She was one of the lasting type, and laid 91 eggs in 120 winter days.

The White Leghorn pullet of Mr. Rodwell's, third in eighty-seven pullets at Westminster, 1920, and afterwards sold, won under me at Burton Joyce Show in 1922 in the new owner's hands second in a large class of pullets and medal for best member's bird.

A White Orpington hen owned by Mr. G. R. Poole (secretary of Bingley Branch of N.U.P.S.) won first at Wembley under me in her fourth year and still laying well in January. She cost five shillings as a pullet.

WINNERS BREED WINNERS.—At the Welsh National Show at Wrexham, 1922, I gave first in a very large class to a White Leghorn pullet which finished the year with 242 eggs, Wrexham being her first time of being shown. She was the mother of the first and special Moore pullet under me by score-card and many other winners. This bird was owned by Mr. S. R. Morgan, of Highfield, Plodder Lane, Farnworth, Bolton, who also won at the same Show under me first, class certificate, cup, and specials with a White Wyandotte cockerel, the son of a 262-egg hen with sire's dam 245 2½-oz. eggs. This cockerel has sired many winning birds in utility also, and his daughters have put up splendid records.

ANOTHER LONGEVITY LAYER.—Mr. Reginald D. Thompson, co-director of Worcestershire Poultry Farm, of Tardebigge, near Bromsgrove, tells me that his Brown Leghorn hen first under me at Gloucester, 1922, is "now in her third year, is in splendid fettle (February 10th, 1923), and laying regularly."

200 EGGS IN SECOND YEAR.—Mrs. E. Sheppard, of Garth, Trevor, Ruabon, writes:—

> Lady Brenda, cup winner under you at Wrexham, has finished her second year with 200 eggs, and last month (January, 1923) in her third year laid 28 eggs. I had an order for 100 day-old chicks from the person who bought the White Leghorn cockerel which you said at Olympia handled so beautifully. He is more than pleased with the cockerel and asked if I had more like him.

ANOTHER WREXHAM 200-EGG HEN.—I must have had a good day at Wrexham, for Mr. William F. Groves, of Belle Vue, Shrewsbury, won first with a Black Leghorn hen which had laid 268 eggs in her pullet twelve months. She was afterwards first Sandy under me and fifth Tenbury against Whites. At Sandy in a hen class she beat Mr. H. Rainbow's White Leghorn hen which had laid over 250 eggs in her pullet year. Another of Mr. Groves' Black Leghorn hens also won second under me at Redruth against Whites of all ages, and laid 259 in her pullet year. The following correspondence is not without interest:—

SHREWSBURY,
August 2nd, 1922.

To the Editress, *Feathered World*.
DEAR MADAM,

Will you kindly send me a list of all the poultry Shows that Mr. Powell-Owen is judging at this year, as I wish to reserve my utility exhibits to be shown under him.

Yours faithfully,
WILLIAM F. GROVES.

The Editress asked me to be kind enough to send the correspondent a list of my engagements at the time. His reply to my letter:—

SHREWSBURY,
August 8th, 1922.

DEAR SIR,

Thanks for your letter regarding Shows at which you are judging. My reason for asking the *Feathered World* for a list of your engagements was that a lot of Show secretaries [secs. please note!—Author] when advertising their Shows do not state the judge's name, which I think is most important when one is studying the ideals of certain judges.

I breed Black Leghorns only at present and intend to mate up my breeding pens next year according to your method of selection as near as I can. I think you will agree that I have learned a bit from your books and writings by being able to win under you each time of showing this year.

Again thanking you.

Yours faithfully,
WILLIAM F. GROVES.

Another interesting item of news is that Mr. Groves is first and second in a Laying Test that is now running, which may prove another case where a broad-minded poultry-keeper can learn something about selection from visits to utility Shows and supporting them with entries.

SHREWSBURY,
August 21st, 1922.

DEAR SIR,

Your letter has taught me more about my birds and their past "relations" than I knew about. I hope by careful breeding and the help of some good "true-to-type" males to have some good or "better poultry" before you in the show-pens in the future. Thanking you for your kindness and all the trouble you have taken.

Yours truly,
WILLIAM F. GROVES.

267-EGG DUCK.—I have only once given a duck the cup for "best fowl" in Show, namely, to Captain G. C. Heseltine's Buff Orpington duck which secured first in the utility duck class at Hants County Show at Winchester, 1922, and against some splendid Buff Orpington ducks. This duck had laid at Bentley Test 267 eggs in twelve months. Capt. Heseltine supplies me with the laying records of his Buff Orpington ducks which have won under me in 1922:—

ROPLEY, HANTS.

DEAR P.-O.,

Very many thanks for your article on Buff Orpington ducks for the Club Year Book.

My Winchester winning duck laid 147 eggs in 147 days at National Duck Test at Bentley and 267 in the twelve months—white eggs and weight 2¾ oz. My 1st Tottenham and special 1st and special Alton and 4th Midland Federation, all under you, laid 231 2⅞-oz. white eggs. My 3rd Tottenham (3rd to above duck) laid 108 in 108 days with 221 for full year. The duck was not in lay at Tottenham and Alton.

Yours sincerely,
(CAPT.) G. C. HESELTINE.

286 EGGS: BEST IN SHOW.—In January, when Sunderland U.P.S. were but a few months old, I judged their first Show and drew 550 entries. When I journeyed to Sunderland in October, 1922, to judge their second Show within the year I was told that my "best bird in Show" at the January fixture had finished with 286 eggs in twelve months, being owned by Mr. J. T. Dawson, of Sacriston. Winning first in a class of sixty pullets she took the cup through scoring the most points in the Show, and one notices how regularly my score-card finds White Wyandottes or White Leghorns scoring the most points. In actual practice they, on the whole, put up best records at the Tests and my card finds them also. This pullet's score-card in January, after starting to lay in October, was:—

CAPACITY.	CAPABILITY.	BREED, ETC.	TOTAL.
20	12	10	Capacity ... 66
5	3	10	Capability ... 66½
10	15	10	Breed, etc. ... 56
8	3½	26	—
5	5	—	188½
9	10	56	—
9	10	—	
—	8		
66	—		
—	66½		
	—		

When I give a utility White Wyandotte 26 for breed characters then it is a real Wyandotte with cobby body and not a Leghorn-Wyandotte cross, or in other words a Leghorn with Wyandotte comb. The second best was a Barred Rock (showing that one finds splendid layers in all breeds to the individual bird) owned by Mr. E. Cawthorn, and he could not show the bird at the October Show because she was moulting and laying, too, but he brought me one of her eggs laid that day. She had laid throughout the year and never broody, scoring in January 183½ points. I am always interested to learn the egg-records of my "best in Show," as I place to the highest points when score-carding.

293 EGGS IN TWELVE MONTHS.—The White Leghorn which I placed third at Moore Show, owned by Mr. W. Strettle, Old Lane, Eccleston Park, Prescot, laid 293 eggs in twelve months. All birds were score-carded at this Show (over 300), and this bird, in a class of thirty-three pullets, etc., scored 164½ points.

Capacity: 16, 5, 7, 8½, 4½, 8, 8, total 57. Capability: 9 (fat), 3, 11 (fat), 4, 4, 7, 8, 7½, total 53½. Breed, etc.: 10, 10, 10, 24, total 54. Grand total 164½.

SHORT-BREASTED MALES.—At Moore Show, Mr. H. Woulds informed me that he had owned a Columbian Wyandotte cockerel which had done much winning but not under me. He was a five-finger bird with very short breast-bone. The bird had never been a good breeder and died suddenly from abdominal dropsy.

SUCCESS FOR A BACKYARDER.—A few seasons back there came to a lecture I gave to a London society a critic who was going to show me up on handling—a heckler! But when he had heard my lecture he found that it had interested him, and that I had given practical reasons for my score-card system of points. From a critic he became a convert, and started grading his birds, and he found they needed severe handling. At the Shows he became always a keen learner and was ever ready to have a chat with me on selection and hand-grading. During the rearing season of 1922 he had become efficient in my system and decided to put on one side eleven Ancona pullets. Of these, nine won him first prizes at utility Shows, and he, as a backyarder, finished the year with first under me at Westminster in open Ancona class and under the cup judges the cup for "best in Show" over 1,000 odd exhibits, the same pullet finishing up with first and special at Olympia in my section. I refer to Mr. C. M. Wortham, of Fulham, S.W. There still remains one critic to convince, namely, the breeder whose strain the bird belonged to. This breeder does not believe in utility Shows but only in Laying Tests, with the result that he does not enjoy the interest attached to a win at a classic for his strain. If he were interested I would inform him also that one of his Anconas in her third year won first and cup for best bird in Show at York Agricultural Show for Mr. Moore. Recently, however, I heard that when a friend called on him a batch of birds had been graded out for market, so he may be getting busy with hand-grading. Many breeders condemn my system without knowing one iota about what it is or is not!

278 EGGS: SECOND PRIZE.—The laying records put up by winning birds under me at Westminster Show—the annual "classic" of the National Utility Poultry Society—always interest me because, firstly, my score-card is the National one and used at this annual event, secondly, there are critics among the breeders who decry the card as a means of selecting the layers. The White Wyandotte pullet which won second under me in a class of eighty pullets at the 1921 fixture, owned by Mr. C. H. C. Partridge, laid 278 eggs in her first year and placed by score-card. I also gave this pullet first and cup "best" at Leicester National. At the December Westminster Show the pullet scored as under, after a month or so of laying, and then went on to do the 278, despite the other outing at Leicester:—

CAPACITY.	CAPABILITY.	BREED, ETC.	TOTAL.
20	12	10	Capacity ... 64
5	3	10	Capability ... 66
8½	15	10	Breed, etc. ... 58
8	4	28	—
4½	5	—	188
8	10	58	—
10	10	—	
—	7		
64	—		
—	66		
	—		

249 EGGS: ELEVEN MONTHS.—The White Wyandotte pullet to which I gave first and special at Handsworth and first and special "best in Show," Tenbury, had laid 249 eggs in eleven months when the owner gave me her record.

270 EGGS: ELEVEN-AND-A-HALF MONTHS.—When judging at Alton Show in November, 1922, I gave the special for best bird in members' section to a White Leghorn cock which had won first under me at the previous Alton Show. My "best opposite sex" was a White Leghorn hen which had laid 270 eggs in eleven-and-a-half months.

240 EGGS: FIRST IN A.O.V.—At Deptford my first White Wyandotte hen in A.O.V. members' class had laid 240 eggs in under the twelve months.

295-EGG HEN.—To my knowledge, I have only placed one bird which exceeded the 300-egg line, but at Bodmin Show my first White Wyandotte hen had laid 295 eggs in her pullet year, being shown in her second season by Mrs. W. G. Tickell. And it is well to remember that I keep a close touch upon type as well as show bloom (allowing for fair wear and tear of production in hens), which shows conclusively that my ideal of utility and beauty can be achieved without loss in eggs, which my critics say is impossible.

285 EGGS: BEST IN SHOW.—When I was at Bodmin Show I met Mr. and Mrs. S. Willis, of Tehidy, and visited their White Wyandotte farm. Tacked on the trap-nests were prize-cards won, and my attention was attracted to a White Wyandotte's score-card carrying my signature as judge. My immediate query was "How many eggs did she lay?" As a pullet I had in 1921 given her first and "best in Show" at St. Columb, and she went on to lay 285 eggs in the twelve months. Is there anything in handling on my score-card? At the Show as a pullet she scored in the 180's with over 60 of the 70 for capability and a true White Wyandotte. It will be seen, then, that both as pullets before they have done the laying, and afterwards as hens when they have attended to the production, my "best in Show" birds put up splendid records.

AT FIRST ATTEMPT.—At Bodmin Show I met Mr. J. Clare Bazeley, of Penzance, who had come through to get a few helpful hints on handling. He is leading the *Feathered World* Laying Trials, which are rather unique for scoring, as the pen turning out the greatest total weight in eggs will win without regard to any lottery of when they start to lay. Showing for the first time, Mr. Bazeley, who is a commercial egg-farmer, just sent hens and pullets on what they had laid by the trap-nest. My judging results saw every bird placed, including fourth with a hen that had laid over 240 eggs in twelve months, together with first and second in White Runners. Four Leghorns and four cards at first attempt and against the best competition shows once again how high-grade stock are awarded prizes by my score-card.

"DAILY MAIL" BLACK LEGHORNS.—I have mentioned previously that many successful breeders and winners in Laying Tests when they commence showing under me secure premier prizes. I could quote dozens of instances. At Watford, Mr. Oliver Hamlin, of Chorley Wood, Herts., won four prizes under me with Black Leghorns in the two strong classes for that variety, including second in cockerels and fourth in pullets. This was, he informed me, his first time of showing, and he had just entered to see what I would do with the birds. In the first place, all the birds were brothers and sisters to his

Daily Mail pen of Blacks, which was the only pen of the variety to gain a prize in that competition, laying 1,669 eggs for the eight pullets, with 196 eggs for the seventh month in twenty-eight days. Another case of sending for quality utility-laying-strain-pedigree birds and me placing them all by score-card system. With a little help and a few demonstrations, backed by a visit to a utility Show or a letter to me to know where birds failed (with stamped envelope, of course), and it would need little more to select from such stock a bird to win under me in strongest competition.

CONSISTENCY IN HANDLING.—I have been told by my critics that it is luck when I am consistent with my previous winners. If they admitted that it was judgment carefully made they would have to credit my score-card system with being a practical proposition, whereas they persist in it being all theory. I always rely upon handling because I insist upon every bird being put down to the moment. I do not want birds following me all over the place unless they hold their condition. I try to penalise mismanagement by deducting points whenever a bird is too fat for the season or has lice at abdomen, contending that good birds should be well looked after and not neglected. I keep always to the practical side in scoring. If the owner has a bird which consistently keeps fit (temperament often decides), then he can show her under me often. But *I am after a graded quality flock per owner and not one particular bird*. That is why I insist upon having every bird shown to the minute. If you overshow a bird you will find that I reduce scoring to a fine art and will deduct points for lack of energy, etc. I want you to have as your ideal a flock of twenty or even half a dozen in a backyard run, any of which, when fit, can win under me. It has been accomplished by very many backyarders in London, and one noted London backyarder, I understand, is proud to show any member of the London societies a pen of about fourteen White Leghorns, every one of which could win under me and twelve of which have already done so up and down the country. A man can take a real interest in such an ideal and such a string of quality birds, I think, and it is the personal side of poultry-keeping which, from the educational side, it is one of my ideals to bring out with my score-card methods.

But one must remember that while males may hold their handling well, not every hen or pullet will stand overshowing. I advise *three outings to Shows per bird per annum*, and always when very fit and up to concert pitch. I demand such a high standard, especially at important or open Shows. I get the "cream" of exhibits, as you will know if you show under me, and my standard has to be very high accordingly. That consistency proves that my score-card is practical and not theory worked out on paper, I can instance hundreds of cases of consistent judging by myself on my card. Am I not judging day in and day out throughout the year and not once in three months as many are? Therefore I have the greater test, and when I am down in Cornwall I get Scottish birds, and when in Northumberland then Somerset exhibits.

But one has also to remember that where quality exhibits clash some must lose, as all cannot win, and often little will divide the birds in the money. Where there is time I score-card in detail and often score every bird in the class, but sometimes just the picked birds in each class. One has to remember that it is all a question of time and number of judges. Soon we shall have more to help us, when the ideal of having every bird scored if so demanded by exhibitors will be possible. One can only score-card in detail at the rate of thirty per hour. Therefore, snowed up with 500 to 700 entries at a single Show as I often am, I have to rely upon hand-grading the exhibits, keeping to my card as my standard without scoring every bird in detail.

At Northumberland Heath, in a very strong class by score-card, I placed Mr. Fenwick's three White Wyandottes first, second, and third. At Surrey County Show, which followed, I placed the three by handling alone in exactly the same positions. A White Wyandotte pullet shown by Mr. J. J. Robinson, of Darlington, commenced under me in December, 1922, with first Newcastle, travelling on to Darlington the same week, where I gave it first and "best." I next found I had given it by score-card first and "best" at Wooler, then first and "best" Bradford; followed by first and "best" Bodmin in February, 1923. At the final Show of the winter season at Wembley, 1923, she got beaten by the "Rodwell" hen. Only one "warm" exhibit can win.

At Alton, for the first time for many years, Mrs. Gooden failed to win the first prize in Buff Orpington pullets under me and my score-card methods. Her pullet came in second. But at Westminster her pullet again started the owner's winning sequence, and I placed her first. Many declare that a "commended" under me is worth a "first" under others, on account of quality and competition, so do give every bird a chance and send her right fit to the minute.

UNIFORMITY OF JUDGING.—Some of the critics desiring to attack utility judging through every conceivable channel declare that there is no uniformity of judging. Among these critics are some exhibition men who go round the other way and, when one is consistent to himself, put it down to coincidence or luck. Why should utility judges be on an equal footing any more than exhibition judges? Let such critics put their own house in order first. It is an everyday experience to see a beginner buy a club Show winner and get out of the money under another judge at the next classic. Again, some exhibition men go for type first and others for colour, while many forget type altogether. I have known a protest lodged against an exhibition Wyandotte for not conforming to standard type and for being more of an Orpington than a Wyandotte while being awarded first prize. But you will be wise to inquire who the writer is when you see him attacking utility birds, utility Shows, or score-carding.

The judging of utility poultry is quite new, whereas that for the exhibition section is as old as the hills. We are young as yet, and time will perfect. But I do not think we shall ever see many men agree as one in anything, because that is contrary to general rules of human nature. Which two men see through the same eyes? The exhibition critics have their club standards for every breed but never practise them in judging the birds. Every breed has its score-card fixed by the club and yet it is never used in judging. Utility judges are but humans, and all will never see alike. I have, for instance, known a utility judge give two birds prizes that should have been marked "in wrong class." I have known another give first prize to a White Leghorn in A.O.V. when there were two mammoth classes for White Leghorns. Again, I have seen the newer breeds passed because the judges were not acquainted with the names of the breeds concerned and took them for crossbreds. One judge will notice a deformity which another will miss. Some will never give satisfaction as judges because they have no patience and make hurried awards. Time and the school of experience must alone be left to settle such problems. Let he who throws mud—if an exhibition judge—cast his mind back to early Shows he adjudicated at, and recollect what mistakes he made.

What I am pleased to note is that utility judges are now getting nearer to each other, *i.e.*, where they are determined to do the judging properly and put their heart and soul into it. Many there are who judge by handling and yet tell all and sundry you cannot pick out a layer. Why judge? Is it for the advertisement offered in exchange? Note the judge who after every Show says it will be his last, as he does not believe in utility Shows; yet he comes up again for the next event! Note the man, too, who attacks utility Shows and declares he is not a bit interested in breeding better poultry because he has made his money out of the business if he makes no more! If the non-progressive poultry-breeders of the old school will not help to keep the utility flag flying and aim at better poultry, then let them make way for the younger generation, who are plenty and moving with the times. Every breeder of utility poultry, if a stockbreeder, carries the onus of keeping up the quality of the utility birds he breeds. That is a duty he owes in all sincerity to the industry.

THE SAME SCHOOL.—I am often reminded that there is a second "P.-O." in Mr. H. A. Hussey, Secretary of Tottenham Utility Poultry Society. We are often branded the "Tottenham Twins," and Mr. Hussey is very near my own decisions when judging utility poultry classes. But being President of his Society and having worked with him at all our Shows, as well as seen him every monthly meeting night for years—not to mention the hundreds of times he has helped me as my steward or score-carder—should we not be of the same school of thought? I remember judging Barnet Show with him in November, 1922, and it came to the question of awarding the cups and medals for the "best birds" in Show. When we had finished judging every bird that had won its class and was brought out for the cups belonged to a member of the Tottenham Society, either a backyarder or a breeder-member. That uniformity comes within the same school of thought I agree.

Here was the list of talent we had out when deciding the winner of the Poultry Club utility cup:—(1) Under Mr. Hussey the first Light Sussex cockerel to which I had given the Poultry Club utility cup and breeder-members' cup at Tottenham Show in October (owned by Mr. H. W. Honey), and which was placed first at Yeovil by Major Potter and awarded Poultry Club cup by Major Potter and P.-O. (2) Under me the first White Wyandotte hen belonging to Mr. Fenwick, and which had won first and Poultry Club utility cup under me at Northumberland Heath, 1921, and the same at Surrey County Show, 1922, together with first under me at Northumberland Heath, 1922, and the Poultry Club utility cup awarded by Major Potter, Mr. Hussey, and P.-O. (3) Lieut.-Col. Tyrell's White Wyandotte cockerel first in my class and first under me at Tottenham and Leicester, with the Poultry Club utility cup at Leicester under Mr. Honey and P.-O. (4) Mr. Partridge's first White Leghorn under me. How would you like to pick the best out of that lot?

For the Middlesex cup Mr. Hussey nominated his first Ancona, and I had already given this pullet first and cup at Tottenham for best backyarder-member's bird in Show.

For the Herts. Poultry Club utility cup I nominated Major Potter's Black Leghorn pullet first under me, and Mr. Hussey brought out his first prize Rhode Island Red cockerel owned by Mr. Aston. The latter bird I placed in the money at

Tottenham in a very large class, and Major Potter gave it first on my score-card at Northumberland Heath, while Mr. Hussey had also given it first at Mill Hill Show.

Here we have winners, then, under those with the same run of thought and views of utility poultry, and shown by exhibitors holding like opinions, and all brought up the Tottenham way. As our school is increased so will we get nearer to uniformity in judging.

HINTS WHEN JUDGING.—The value of a score-card depends upon execution, and some judges abuse my system. It is wrong to judge a class first and then to make the winning birds fit the score-card. Go straight into the class and mark in your judging-book every bird by its grading, writing down a given number against each bird to represent its quality. The first bird may be just an ordinary one and get a figure 3, but the next is a "dud" and receives no mark, while the third is a medium with a figure 6. As we go through the class we find our eighth bird is a really nice pullet and mark her 10. After that one has to bear in mind such a pullet as a key to the class for quality. We may get four above her which may obtain 12 to 15, and three nearly her equal which will get 8 to 9. Then take out the eight birds and score each in detail, remembering that your first bird scored is your key which you must score to. Therefore take out the best bird to score first. You will thus let the total score-card marks decide the placings in the money.

Write in the judging-book, also, notes *re* the failings of each bird, so that any exhibitor writing afterwards for your notes can have them for the benefit of education.

These are some of the duties ahead of the score-card judge, and score-carding will not at first be found an easy task, I can assure you. It needs long practice and experience in grading and handling birds with trap-nest records and those put into houses to be trap-nested and handled at all ages. But remember your "key" bird, as that is imperative if you wish to place the birds in a just order for gradings and all-round merits to my score-card standard. That it is a standard no one can deny. When not scoring in detail, grade every bird with a number as you go through the class, and then place very carefully by again handling those singled out as in the running. It all takes time and thought before the cards can go up, and exhibitors must have patience unless they want hurried decisions—which they will never get from P.-O. When I have placed my cards I can always rest at night knowing I have judged the birds and not rushed them.

Large classes will be more difficult and you must at first, after handling, say, twenty birds, go back to your "key" bird (holding up to then the highest number) and handle her again in order to carry her in your mind.

Get in plenty of practice with your own birds, even investing in a few show-pens, which should be given a place on every utility poultry farm. Score-card plenty of birds of the same breed and sex, and judge them as a class. Practise placing them just on handling, keeping to the score-card, and then after scoring compare the results of each method. I know my card inside and out solely through experience, but anyone can master it by getting down to it and all it stands for. I hope when you start the exhibitors will not flood you as they often do me. I hope you will not get 125 pullets in one class, and you will be well advised to start with small members' Shows and gain confidence.

I remember arranging to go to Derby to judge a small members' Show in the afternoon and to give a lecture in the evening, but word came to hand that there were 400 entries instead of 50. I remember, too, travelling up North by night train expecting (with the first Show and the society only three months old) to have 200 entries, but I had to get my coat off, prepare my own judging-books, help a bit in other ways, and judge nearly 600 birds, despite the breeding season having started. Again, Bodmin Utility Poultry Society was but a few weeks old and ran a Show quickly. A wire came to say there were 600 birds for me to handle in a day, again with the breeding season on. This volume is not dealing with Shows and showing, but I mention these things so that all readers who exhibit utility stock will see the disadvantages under which we judges work. It is nice for exhibitors to come along to the Show at 1 o'clock and expect all cards to be up. One could do that if the birds were judged by the judging-stick and not handled. I have been to a Show to judge where the public were in at 1 o'clock and the pens and birds were not ready for me till 12.45, the whole of the county having to be scoured for the necessary pens owing to a mishap.

MAGPIE DUCKS.—At Winchester Show apart from giving first to Capt. Heseltine's Buff Orpington duck which at Bentley laid so well, I placed a* Magpie duck owned by Field-Marshal Sir Arthur Barrett, of Sharnbrook, Beds., third in a mixed class, Mr. J. S. Parkin has informed me that she has laid eighty-nine eggs in three winter months. This would be the first Magpie duck to be placed in a utility class, and naturally she must have handled well to have been placed by me against the popular breeds.

* *From October 2nd to April 12th, this duck laid* 167 *eggs, more than* 60 *above any other Magpie on the farm.*—Author.

226 Eggs: V.H.C. Bird.—I have judged Heanor for two years running, every bird being score-carded throughout. The winner of the class in which the bird referred to below competed scored 191, which is the same total as my highest-scoring White Leghorn pullet at Westminster, owned by Mr. Partridge, and which laid 273 eggs in her pullet year. Letter reads:—

Wood Lane, Horsley Woodhouse,
Derbyshire,
December 1st, 1922.

Dear Sir,

I am writing to tell you that my two pullets—winners of 2nd and 3rd under you at Derby last week—were bred from a hen which got very highly commended at Heanor under you with a score of 180½ points; the winner of this class, closely related, scored 190—almost a record, I believe.

I thought you would be interested to know the year's record of this hen. From October 2nd to June 21st, 1922, she laid 191 eggs and went broody. I allowed her to sit to give her a rest, and she resumed laying on August 3rd, laying 35 eggs up to September 24th, when she went into moult, thus making a total of 226 eggs for the year. She commenced to lay again on November 2nd, and is still laying.

Yours sincerely,
Wm. Hinds.

Way to be Consistent.—Mrs. M. A. Adams, of Sunnybank Poultry Farm, near Usk, Mon., who judges many utility Shows in Wales, has had experience with my score-card, and her opinion will be quite unbiassed. I am at all times pleased to meet any exhibitors or poultry-keepers at the Shows and to help them with my handling methods if they are interested and intend to get down to actual practice and testing. If they condemn handling and then ask what it means or how exactly to practise it—plenty do that—I am not interested in helping to convert them, because no destructive critic can see good in anything. It happened that Mrs. Adams met me at a Show and displayed interest, and I arranged for her to act as my steward for notes at several Shows. That is the best way to get an insight into the work if interested in judging utility poultry. And I am pleased to see so many well-known breeders desirous of stewarding me when I judge their local events. In due course Mrs. Adams judged her Shows by my score-card system, and here is an unsolicited letter:—

Sunnybank Poultry Farm,
Near Usk,
December 30th, 1922.

Dear Mr. Powell-Owen,

I shall certainly stick to your score-card, as I feel certain it is the only way to be consistent. At three Shows I have given the same Wyandotte pullet 1st, also the cockerel and pullet which I gave 1st and 2nd to, you gave them two 2nds at Worcester Show afterwards. The White cockerel you gave 3rd to at Tenbury Wells I gave 3rd to at the same Show as the cockerel which came 1st and you awarded 2nd to at Worcester.

I thought perhaps the details would interest you, the owner telling me of the birds at Newport Show.

Yours sincerely,
(Mrs.) M. A. Adams.

To counterbalance this proof of consistency the critic will tell you that the judge recognises the bird. Ask him if a pullet which, like the Rodwell hen, wins in her first year and does not come out again till the third year, to commence winning once more first and "best" can be remembered all that time! In exhibition birds one recognises certain of the best by the small defects they possess. You look at a cock which stands out for colour and type but has a small defect, and it is the latter which hits you every time. Again, exhibition birds are judged for external points, which means they can be often recognised again within a reasonable time. Fanciers, too, have long discussions over the winning exhibits at the large Shows, and certain birds are recognised afterwards at classics that follow. Now at very few Shows do I have a chance to get round and see my winning birds, and when I place by score-card just the adding up of the points awarded decides the placing. Yet, again, the good utility judge will back his sense of touch against his eye every time, for the hand tells of utility type and condition. Once one concentrates on observation alone the whole "touch" has gone from the class and you just place four good birds, while at the next Show, with the same birds competing, you will put another four in the money. I have seen that happen often where a bird is good enough to win, on observation, a mammoth class and to secure "best in the Show," while at another event within a very short time and under the same judge it has obtained nothing. I have trained my hands, and am not prepared to lose such an advantage, so that my stewards hand me every bird in a particular way so that it always faces me, and I can take it by the shoulders in the two hands. I have no time to be looking out for certain birds, and it would take me all my time, seeing that when I am in Northumberland I would have to be thinking of all the winning White Leghorn cockerels which had won under me of late in Somerset and Devon. I have always made up my mind to go solely by the bird and how she handles, cutting very severely any out of tone or condition, "handling light," or overshown.

THREE TIMES CUP-WINNER.—It interests me to note the catalogues of breeders as they come out because from them one gets an idea as to the number of eggs winning birds under me have laid by trap-nest. The White Wyandotte which won under me as a pullet first and Poultry Club cup, Northumberland Heath, in 1921, as well as in 1922 as a hen, and the same at Surrey County Show, 1922, with first, also three medals (reserve for Poultry Club utility cup to another bird that had won it three times) at Barnet, in 1922, for its three outings for the year for Mr. Fenwick, laid 91 eggs in 120 winter days in her pullet year. His other White Wyandotte hen first and special under me at Melksham (Wilts. County) and the same at Redruth laid 70 eggs in 120 days at Wye Winter Test, being in the second-prize pen. After leading for two of the four months one of the pullets moulted, the cup being missed by one point only.

237 EGGS FROM COLUMBIAN.—In his advertisement in the Columbian Club Year Book, Mr. William Hamnett, of Myrtle Poultry Farm, Blackpool, quotes No. 60 hen as winning first Trowbridge (under me) and laying 237 eggs in the twelve months.

THE COCK THAT LAID.—As already mentioned, the bird which at the N.U.P.S. Show at Westminster, 1921, I declared to be a cock-hen "IN LAY," and which Mr. Tom Barron betted me £500 was a cockerel and, mated to a few small hens, would prove fertile, caused a stir in the daily as well as the poultry Press.

Extract from *Daily Mail*, December 9th, 1921:—

> PUZZLED SCIENTISTS.—The cock bird justified its fame at the National Utility Society Show in the Royal Horticultural Hall, Westminster, where for three days it has been a centre of the greatest interest and expectation. At three o'clock one of the Show authorities took a warm brown egg from the pen. Opinion is about equally divided as to whether this egg was genuine. But there is no doubt that it was warm from the body and was too big to have been placed through the wires of the fast-closed pen [and the egg was the right colour for the breed—Author]. Four scientists are eager to have the bird.

As I was the culprit and judge to say the bird was in full lay any doubt that existed in the minds of the critics was removed when the bird lived on to find its way to Dr. Crew's Experimental Station, to lay eggs, go broody and sit on them, as reported elsewhere in the book. One triumph for handling if we never get another.

ADVANTAGES OF SCORE-CARDING.—After I judged Bradford Show I paid a visit to Mr. G. R. Poole's farm at Bingley (of which poultry society he is the secretary), Yorks. One has but to handle a noted White Orpington hen he has which, although ancient, is like a chicken for quality and won under me in 1923 at Wembley for the first time of showing, although she has won under nearly all judges in her time. When going the rounds of farms with him a breeder pointed to a solitary White Wyandotte pullet which was nothing but a Bantam. It appears that this owner had a White Wyandotte cockerel which he was proud of, and in 1921 he exhibited the bird under me and got nothing, after it had won so many prizes. He had written me at the time for my reasons, and I had told

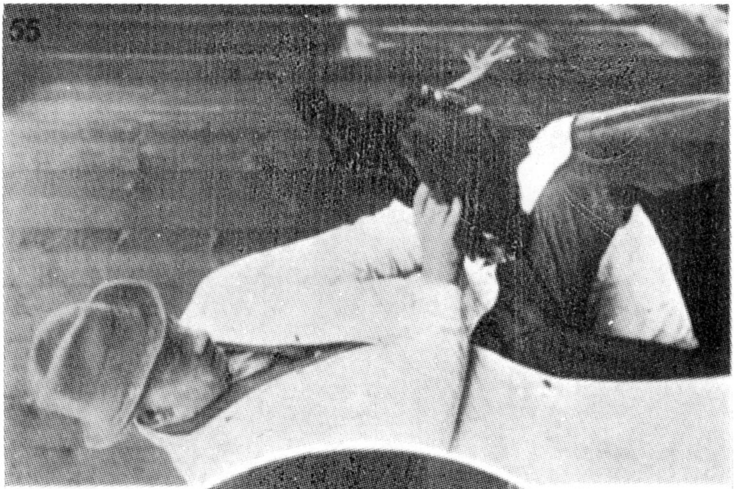

55

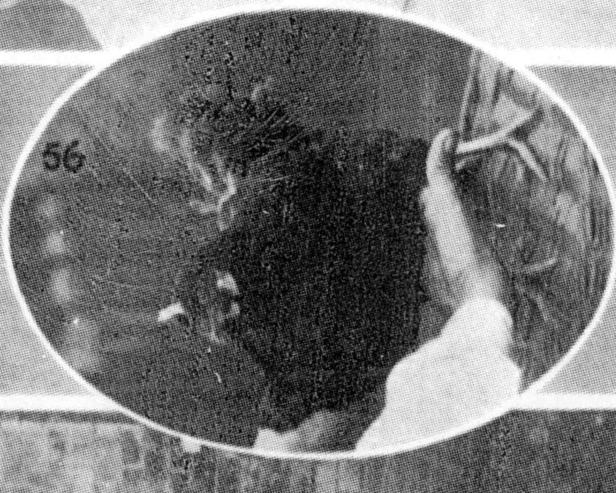

56

57

him that he would breed superfine Bantams from him. Practice proved me to be right as he used the cockerel for breeding and this daughter is kept on to show him his folly, so to speak.

BINGLEY,
March 4th, 1922.

DEAR MR. POWELL-OWEN,

I am doing what I can to push forward the P.-O. score-carding and methods, not because of any personal feeling but because I can see through it and see its advantages. There are many who cannot, they say! I think the truth is "I didn't see it first and won't now." I think it is a case of "won't" not "can't," as some people are pig-headed.

Faithfully yours,
G. R. POOLE.

IN THE FACTORY.—As stated elsewhere, many cannot trap-nest, as my correspondent points out:—

COLNE, LANCS.,
October 22nd, 1921.

DEAR SIR,

I am very pleased to say that your notes on handling have given me some information on a few points I was not sure of. I have lent your notes to a few poultry-keepers, all of whom seem very interested therein. In fact you pointed out one or two things they did not have much idea about.

Most of us here in Lancashire have got to use the handling system, having no time to trap-nest with working in the factory.

Yours truly,
A. PETTY.

AUSTRALORPS.—The Australorps Farms, Ltd., of Street Court, Kingsland, Herefordshire, report laying as follows for their winners under me: My first and special club Show winner at Leicester laid 221 eggs in England from May, 1921 (imported), to 1922, with hen's winter record of 48 eggs in three winter months. Their hens first and special Market Drayton, first Winchester, and first Tottenham, were in the second prize Bendigo Laying Test team (4th against all breeds), which laid 1,505 eggs in twelve months for six birds. My second Leicester hen laid 207 eggs despite odd season of importation and Shows. My second Westminster hen laid 104 eggs in 157 days from May 7th to September 31st; score-card 175 ex 200.

262 EGGS FOR LIGHT SUSSEX.—Mr. M. W. Slade's Light Sussex hen first under me at Stokenchurch, 1921, laid 262 eggs in under eleven months, size $2\frac{1}{4}$ to $2\frac{1}{2}$ oz. Her score-card at the Show was: Capacity: 20, 5, 10, 9, 5, 9, 9, total 67. Capability: 10, 3, 10, $3\frac{1}{2}$, $4\frac{1}{2}$, 9, 10, $8\frac{1}{2}$, total $58\frac{1}{2}$. Breed: 8, 10, 9, 24, total 51. Grand total $176\frac{1}{2}$. Once broody.

315 EGGS: TWELVE MONTHS.—I have only found one hen in the money under me exceeding the 300 eggs, although if any not trapped have done this I would not know. I have been very often in the 280's and 290's. The record is held by a Light Sussex hen of Mrs. Goodden's, West Coker, which

laid 315 eggs in twelve months, and which secured third prize under me at Melksham, in June, 1922, in her second year, being but one point behind the second prize bird, and losing for weak hackle under breed characters.

West Coker,
June 6th, 1922.

Dear Mr. Powell-Owen,

Sorry to bother you, but I particularly want to know your criticism of my Light Sussex which I showed under you at Melksham. I should not have picked her out for show in the ordinary way as she is rather big and hardly fine enough about the head, though she scores well enough. I sent her for this reason: that she laid 315 eggs in her first twelve months (325 in twelve-and-a-half months), when she moulted and rested for six weeks and started off again, never missing more than two days in each month, and eggs averaging well over 2 oz.

She is quite the best layer I have ever trap-nested—never broody or off colour—and I was anxious to see how you would place her.

Yours very truly,
(Mrs.) J. M. Goodden.

This was the exception we so often meet, where only a detailed score-card which she got at the Show under me (I scored all birds) would find her by giving her credit marks where she scored them. Her score-card was:—Capacity: 20, 5, 8½, 9, 5, 10, 9, total 66½. Capability: 10, 3, 14, 4, 4, 8, 8½, 5, total 56½. Breed: 8, 10, 9, 23, total 50. Grand total 173.

Twenty-one Eggs per Month.—At Westminster, December, 1920, Mr. Rodwell won the White Wyandotte pullet class with a bird that went on to lay 273 eggs, as mentioned. The third pullet in the same class owned by Mr. F. J. Marston, of Biddenden, laid 172 eggs from September 22nd, 1920, to May 31st, 1921, when the traps were stopped owing to the hot weather. I happened to handle this pullet on the farm on September 23rd, 1921, along with many other birds, and give her card. She was still in lay and had not been broody: Capacity: 20, 5, 8½, 7, 4½, 8, 8½, total 61½. Capability: 8½, 3, 12, 4, 5, 8, 8, 7, total 55½.

250 Eggs per Annum.—The White Wyandotte hen I placed seventh at Yeovil in a class of eighty-seven, mainly pullets, laid 250 eggs in her pullet year for Mr. Wm. Runciman, Lapford, North Devon.

A 242-Egg Red.—Mr. Harold Marshall, of Bramshott Manor, Hants, in his 1923 catalogue adds:—

We cannot let the season close without remarking how accurately such judges as Mr. Powell-Owen are in the awards in utility classes where they have only handling to justify their choice.

One of Mr. Marshall's Rhode Island Red pullets won under me in 1922 first and special Burton Joyce, first and best female in Show Redruth, and first and special York, where she got something sharp into her foot, causing poisoning and necessitating removal of her toes on one foot to save her.

Despite this she finished the year with 242 eggs. Her score-card at Redruth Show, July, 1922, was:—Capacity: 20, 5, 8, 8½, 5, 10, 10, total 66½. Capability: 9, 3, 12, 3½, 4½, 8, 8½, 8½, total 57. Breed: 10, 10, 10, 26, total 56. Grand total 179½.

One of this utility-exhibition breeder's Red Sussex pullets won first under me at Burton Joyce, also first at York, and finished the year with 276 eggs. With Minorca pullets he won first and second under me at York, and they laid 223 and 211 in the year despite exhibitions.

TWENTY-ONE EGGS MONTHLY.—My v.h.c. White Wyandotte pullet at Wrexham, for Mr. T. Moreland, laid 211 eggs in ten months to date of Show (July 25th, 1922). His v.h.c. White Leghorn laid 164 eggs in ten months despite a partial moult in January and February.

Moving about as I do among poultry-farmers I am constantly tested for hand-grading; surely if, as my critics assert, I am just a "bluffer," and my handling system is "all theory," both myself and my system would have been graded out long since.

A HANDLING TEST.—A friendly critic at Burton Joyce Conference ridiculed handling, so a party of us on our way back called upon him and ran through his birds. I think he had better views of hand-grading when I located one bird in the flock with diphtheritic roup, some cripples, and another that had not laid for ages suffering from scrofulous liver which filled the abdomen.

PLACING TEST DUCKS.—When judging at Winterton Show in June, 1922, I spent some enjoyable time with Mr. Oscar C. Brown, who is so well known for Khaki Campbell ducks. In the traps for the night were his winning Khaki Campbell ducks which had been at Bentley Test, so I suggested that I should judge them and place them as if at a Show. I placed them on handling exactly as they had laid to number of eggs at the Test. There were only a few eggs difference between the first two of the pen, but I take no credit for placing the one above the other except that I liked the two very much better than the rest, and in my final placings after careful consideration of the two nominated the highest for premier place.

TEST AT GLOUCESTER SHOW.—At Gloucester Show in June, 1922, Mr. L. G. Price—a good progressive county poultry expert—had an educational stand and set me the following tests with bad, good, and super layers. I had to pick out a " dud " Rhode Island Red which had laid but twenty eggs in nine months; a Black Leghorn with fifty to her credit in the same period; name the better of two good Barnevelders after saying they were good layers, and the same with two White Leghorns; the Barnevelders having done twenty eggs per month since September and the two White Leghorns being over 200 each in ten months. The Red pullet had to be picked from others by observation and the rest by handling. In all ten handling tests—as Mr. Arkell, my steward, will bear me out—I was correct. Mr. Price has selected many winning birds in Laying Tests and always puts up a good educational display at the Shows for his county, teaching poultry-keepers how to grade, as does Mr. Cockeram, the Wilts. County instructor.

EGGS IN ABDOMEN.—While at Gloucester Show, judging the Australorps, I visited Mr. Arkell's farm and went with him to a friend's farm for some grading. Aided by a lamp I located four flighty pullets of the "squeaker" type and handled them. Two post-mortems followed the next morning, the one bird having three shelled eggs in abdomen, and the other a shelled egg in abdomen together with a broken egg and putrid contents (hard boiled) in the shelling department. All through an argument as to temperaments in birds and a remark from me that "the docile bird is the best layer."

TEN MONTHS: NO EGGS.—When judging at York Show in July, 1922, I lectured to the society members on the show-ground the night before. A member put up a bird for me to tell him how many she had laid, although I was grading young pullets as a guidance. I do not stand for numbers of eggs but mentioned that "I should not be surprised if the owner told me the pullet had not laid an egg for the last five months or so." The owner then said the bird had yet to lay her first egg since the previous October. It reminds me of a similar case when lecturing at Nottingham. My summing up in this case was that I would not give fourpence for the bird, thereby avoiding guessing at numbers. At the York Show the next day I found exhibited a hen with spurs, and a member asked what was the matter with the bird, seeing that I had dealt with cock-hens and spurred hens at the evening lecture. So we handled this exhibition hen and found a shelled egg in abdomen; later we saw the bird lay a soft egg in the pen.

DEFORMED BACKS.—When grading a flock of White Leghorns in the Midlands I found every male with a badly "roached" or rounded back, and traced this defect back to the original male, which had cost a large sum. Every cockerel had a deformity, and many had badly wry and squirrel tails.

When grading down the West I met a similar case, with all the Rhode Island Red cockerels from one mating deformed in back, all traceable to the original sire, which could have been avoided by handling before he was accepted from the pedigree-breeder at a big price. When such deformities arise there are many young birds which are completely deformed. Mr. Bird found forty such pullets when grading a farm flock in Denmark.

Another grading at Wrexham in 1922, also of Rhode Island Reds, found all cockerels deformed, while many pullets (sisters) were humpty-backed and useless. When you buy a cockerel get a perfectly-formed one, which is what I hope you pay a good price for—a graded article.

At London and District Show I pointed out such a failing in an exhibitor's Ancona cockerel, and he afterwards reported that twenty-seven cockerels (all sons) out of twenty-seven had severe deformities.

TEST FOR BREAST-BONES.—When visiting one of our best breeder's farms after judging at Market Drayton, we had a friendly argument as to short or long breast-bones. I suggested that after supper we should visit any house where trapped pullets were finishing their laying period. This we did, and the first bird I tested for excessively short breast-bone and sagging abdomen while on the perch had abdominal dropsy, the abdomen being full of fluid. Another had abdomen full of eggs, and I think I convinced my friend with so brief a grading; the trap-records spoke of no eggs from these birds for months. People talk of my theories, but I am always putting them to practical test with men that matter. Time will see things viewed in a broader light, believe me!

TEST FOR DOCILITY.—When I was on my way to judge Trowbridge Show in October, 1922, I called at the Worcestershire Poultry Farm at Tardebigge, near Bromsgrove, where there has only been one case of roup, and that twenty years ago. I was much impressed by the stories of their early systems of recording. A White Wyandotte pullet had to be dropped into a pen of White Leghorns so that her brown eggs could be distinguished, and vice versa. The first 200-egg hen was bred on this farm, laying 220 odd eggs all on her own in a dog-kennel, including two shelled eggs in a day. Having discussions with Mr. Dudley Thompson, a co-director with Mr. Dixon, we leaned against a poultry house until it was quite dark. Discussions wandered to docility, and I tested my handling for that part by going into the house we were leaning against and bringing out a hen from the perch which I said was the best-tempered bird I could find there. Striking a match, Mr. Thompson took her ring-number, and when we got into the house we compared her record and she had laid 256 eggs in the twelve months. A tap on the back as the birds perch will often help the hand-grader.

COMB IN EYE.—When judging Australorps at Bristol, a friendly discussion arose regarding the hanging of a comb in a Black Leghorn pullet. As is well known I have always stood out for a comb in Leghorn, Ancona, Minorca, etc., pullets that falls completely clear of the eye. An owner of such a bird, when I pointed out this failing (in my view) with his bird, informed me that during the summer the pullet, although a good layer, had several times been down on her back (brain trouble) and been saved by prompt treatment.

MATING-UP BY HAND-GRADING.—When judging at Newcastle in December, 1922, my host was again Mr. George Simpson, of Cleadon—the popular chairman of the Northumberland and Durham branch of the N.U.P.S. Owner of the premier restaurant in Newcastle he specialises at Cleadon in new-laid eggs and roses, the former for use at the restaurant and the latter to give a breath of Nature to its surroundings. Keeping each year about 160 head of laying pullets he set out in the winter of 1922-3 to obtain from them 100 eggs daily for the full week, and at the time of my visit he had reached his 700 for the week, the top weight for any one day having been till then 112 eggs. Later when I judged Wooler I again stopped with him and mated up the breeding pens for 1923. I have just as a friend mated up his pens by hand-grading for several years without any knowledge of their laying-records, relying just on handling with both hens and males. In the 1921-2 season he averaged daily throughout the year 50 per cent. of the stock kept, *i.e.*, 50 eggs from every 100 head of birds. I am quite aware that the egg-getting depends upon his personal care and clever feeding, but for my handling I claim the type of bird bred—especially as a few years back his Leghorns were very small (over which we had long discussions), and he as a broad-minded poultry-keeper saw my views. I had to smile when the three-year-old cock again graded in for active service, and nearly all the three-year-old hens, showing that they had answered for longevity to my grading the year before. It is quality we want in our breeding stock, because through them we obtain our progeny-pullets—the next winter's layers.

EGG-PRODUCTION IMPROVED.—After judging Winterton Show, Mr. A. Thompson wrote:—

WINTERTON,
June 29th, 1922.

DEAR MR. POWELL-OWEN,

I am pleased to hear you arrived home safely. First one and then another member stops me and says he hopes we shall be able to get you here again.

I keep on picking up a bird to practise your handling method. I learned more from you than . . .

Yours sincerely,
A. THOMPSON.

A later report dated February, 1923, reads:—

> Last year my poultry cost 25s. to 30s. per week in food, with egg-average fair. This year the same number has cost me 12s. 6d. to 15s. a week, with egg-record very much better. They came into lay in mid-October, and seventy-two birds have exceeded 2,000 eggs in just over two months. I did a bit of hand-grading and got rid of those I thought useless.
>
> A. THOMPSON.

CAPABILITY MEANS PEDIGREE.—When judging at Bodmin Show, 1923, I visited Admiral Cayley's well-arranged farm at St. Kew Highway, Cornwall, where over 1,000 layers are kept commercially in White and Exchequer Leghorns. A splendid recording system is practised here, and all his stock cocks in White Leghorns handled splendidly to my card and they had proved valuable stock-getters. What I liked about the birds was their good frames, with plenty of quality attached. While grading a few cockerels to demonstrate my handling system I hit upon one cockerel and suggested that his dam was a tip-top layer. The egg-record book was fetched, and my judgment was correct and for longevity of production as well. If I had had no trap-nest records and had graded this male in for use, I would then have employed the right class of bird. What is that to a man who cannot trap-nest when the choice of the male means so much, seeing that every chick produced will carry his good or bad qualities from the utility view-point? Fancy picking a cockerel haphazardly whose dam laid 120 in pullet year and 100 in second, with six broody periods thrown in, and breeding from him all for the sake of not knowing something about my handling system.

EGGS IN ABDOMEN.—One soon convinces a critic regarding my score-card when he attends one of my lectures on "The Hen from Within," so that I can explain by practical demonstrations how I arrived at my score-card. At the Shows I find all manner of birds that are wrong internally, and when I judged Bodmin in February, 1923, the officials wished me to give them this lecture. As I was due at Redruth the next evening for the same lecture I suggested I should give their members a "general." In the end I promised that if I found an interesting bird exhibited under me at Bodmin Show I would, if they could buy it, give the needful demonstration. When judging, therefore, I made a note against several likely birds and at the close nominated them. We were able to buy a Rhode Island Red pullet which was in one of the classes, and when I cut the body up to demonstrate upon I withdrew from the bird two putrid fully-shelled eggs, a hard yolk, and two soft yolks. At Bradford Show a poultry-keeper brought me a hen which had seemed queer, and he thought it had crop disorder. It had died, and he brought it up for me to suggest the cause of death. Handling her I could feel an egg inside the abdomen so we had this bird for the lecture. Putrid eggs were found in the abdomen, and the bird had died just from internal poisoning through the gases.

"NEVER-LAYERS."—At Tottenham lecture in January, 1923, "The Hen from Within," planned for the first monthly meeting each year for "new members," I had two birds to demonstrate upon. Again I had an audience of over 250. The one bird had not laid for months through a very large egg getting into the abdomen, while the second had been "laying" for eighteen months without dropping an egg in the nest-box. She had always thought she was laying, and handled in lay, but in this case she had been making the yolks and absorbing them back again. She had spent twelve months in a Laying Test. The original cause of the trouble was traced to a watery cyst in the opening to the oviduct near the vent, which may have been caused originally by the breaking within the shelling department of the first egg to come down. No egg was ever recorded to her for twelve months at a Test and five months of the second year. It was a case not of the yolks bursting and being poured into the abdomen to be absorbed or to melt away. They arrived at their ripe size when the yolk matter was just sucked back into the system for the yolk casings to look saggy.

A TEST AT NEWBURN.—When I was judging at Newcastle Show in December, 1922, I visited Newburn, accompanied by Mr. George Simpson and Mr. Dale, of "Spratt's," to give "The Hen from Within" to members of the Newburn society. I had asked the secretary to get a hen and have it killed and plucked so that I could cut it open and demonstrate with it. The secretary asked members if anyone would supply the bird, and the selected member visited his flock and picked up the first hen that he came to haphazardly. As I had been told she was killed for the purpose of my lecture I was caught napping, and did not notice even the spurs she carried. After cutting up the bird I found she had not laid for many a day, and in the abdomen I found a "football" of hard-boiled yolks numbering some forty, I should say, in concentrated form. Even then I did not notice the spurs, but having told members about cock-hens and warned them to suspect and test all hens with spurs, a member at the back of the audience asked if I could explain why this hen had not started to grow spurs when the ovary went wrong. For the first time I glanced at the shanks, and there I found spurs over an inch long! A minute spent on handling this bird when it was alive would have located the "football" of hard yolks and saved many months if not years of feeding for no return. Such hens, through going dormant in laying organs, put on fat, and it is surprising how well this fat acts as a "preservative" and seals the disorder or obstacle.

LAYING WITHIN.—At Bethnal Green, 1922, the hen demonstrated on had in its abdomen a similar ball of hard yolks weighing 1¼ lb., and the cause was a watery cyst rather large in size and hanging from the ovary on its cord. With the ball cut through only yolk-matter was noticed, showing that the trouble started from the ovary and not from piercing the oviduct walls, when albumen layers would have been present. Where such yolks drop and form into a ball they harden with the heat of the body and often become coated with a skin when the cavity is filled, the bird then drying up in ovary and oviduct, which remain dormant. No poisonous gases arise, and the bird lives on until the weight of ball exhausts the bird and plays on the heart, seeing that internal fat comes along through non-laying. But if eggs become shelled and drop through the oviduct walls into the abdomen, the contents often go putrid and the poisonous gases which arise kill the bird.

PULLETS GO WRONG.—Pullets when starting to lay should be watched, because disorders of this kind are common then. I have very many pullets shown under me in September to November with shelled eggs in abdomen. At Ilford Show in 1922 the first pullet I handled had several eggs in abdomen. It may arise from the pullet holding an egg too long in the shelling department through being nervous over the new duty of egg-laying.

WINNING A SUSSEX SITTING. Many years ago I had a friendly argument with the late Revd. George Crawshay, who was very interested in all animal subjects. He wrote to say that one of his hens had died, and he had withdrawn a large tumour from the bird which he would be pleased to send me if I wished him to. I replied that if he would cut it through he would find the ball a mass of hard yolks. In reply he sent the ball to me and said that if it was not a tumour, but what I so judged it to be, he would present me with a sitting of eggs from his best club Show, Dairy and Palace winning Light Sussex hens. Upon cutting it through I found I was right and returned him the two halves so that he could see the layers of yolks, and he sent me the eggs on the condition I did not exhibit the progeny. I handed the eggs to one of my friends who was then editor of Nash's Magazine, and some grand progeny and foundation stock for him resulted. When the ball of yolks has filled the cavity Nature often puts on a kind of pretty covering or skin around the whole, which so many mistake for a tumour.

IN HOLLAND.—When visiting Holland, Mrs. J. M. Walker, of Chatteris, handled on my system some Barnevelders at a trap-nested farm. Her report:—

CHATTERIS, CAMBS.,
April 24th, 1922.

MR. POWELL-OWEN,

When in Holland I spent a morning with the breeder, and he was very astonished that I told him his good and bad layers by handling. I was right every time, so you see you have taught me something I am very grateful for. He begged for your score-card, which I promised to send him, and details as to its use.

(MRS.) J. M. WALKER.

A BREEDER'S VIEW.—Here is an extract from a current letter, and—while my plan in handling is not to dispense with the trap-nest—I have the idea of my handling system helping all who trap-nest to save labour by trapping only "trappable" stock of real future value, and to judge likely second and third year laying possibilities of hens trapped and untrapped. Also to give to those who cannot trap-nest a sound system as against a haphazard one.

I was discussing a few days back the question of trap-nests with a very well-known man in the industry, and he remarked what a nuisance trap-nesting was, but unfortunately at the moment it was the only possible method. He regretted the fact that this country did not possess a few more Powell-Owens, which would do away with the necessity of using trap-nests at all.

SYMPATHIES.—In looking over my letters I came across the following:—

TOTTENHAM,
January 27th, 1921.

The committee of the Tottenham Branch of N.U.P.S. having heard with great satisfaction that your score-card system has been adopted for Liége Show wish me to convey to you their hearty congratulations.

They also express the hope that you will continue to take a wide view of all poultry matters and so include the education of the small poultry-keeper, for whom you have already done so much.

Also that you will not in any way be influenced or discouraged by the destructive criticism by those who, by their writings, show that they know little or nothing about your system.

Signed on behalf of the committee,
H. A. HUSSEY,
Hon. Sec.

It has been the constant receipt of such letters of appreciation that has helped me along, single-handed, to fight the destructive criticism that has been so rampant.

THE WESTMINSTER COCK-HEN.—The life of the Westminster or *Daily Mail* cock-hen was disclosed through an article I wrote which Dr. Crew happened to read:—

THE UNIVERSITY, EDINBURGH,
ANIMAL RESEARCH DEPARTMENT,
June 23rd, 1922.

DEAR MR. POWELL-OWEN,

I was greatly interested in your article in *Feathered World* in which you refer to the condition of the ovary in these so-called cock-hens or hen-cocks. I have been collecting such birds for two years or more and now have over 20 under observation. The *Daily Mail* bird, by the way, is now sitting on nine of her own eggs. I doubt they are fertile, since the male birds were all inclined to regard her as a male.

My observations are in line with yours.

Yours truly,
F. A. E. CREW.

But why I write is to ask that if you encounter any of these birds and do not wish to keep them yourself, you may remember that I want to collect as many as I can.

EDINBURGH,
June 29th.

I am very glad indeed I wrote you. Your letter is indeed most helpful. I am particularly interested in what you say about the *Daily Mail* bird, which is news to me. I got the bird on February 2nd, 1922, and in the basket in which she came was a brown egg. You will remember the bird—the head and voice of a female; male feathers on hackle and tail; spurs as of the male but in general build far more like a hen than a cock.

At first she was mobbed by the hens and not regarded as a hen by the cocks. Then she began to moult (February 14th to 27th), lost her tail-feathers, and then the cocks regarded and treated her as a hen. The bird itself behaved at all occasions as a hen. Her eggs following this were fertile.

Thanks to a protracted moult, her egg-yield has been low. I have had twenty-seven eggs, averaging in weight 56-7 grammes. On June 16th she, being broody, was set on nine of her own eggs.

With many thanks.

Yours truly,
F. A. E. CREW.

SCORE-CARD OF HEN.—As already mentioned, Miss Barbara Raye's White Wyandotte pullet which was picked on my system and secured bronze medal at Harper Adams in single-pen section, 1920-1, laid 235 eggs at the Test, and 249 for the full twelve months. In October, 1921, after she had done the laying, when on Miss Raye's farm I score-carded her among many birds as follows:—Capacity: 20, 5, 10, 8½, 5, 10, 9, total 67½. Capability: 10, 3, 15, 5, 4, 10, 9, 8, total 64. Weight 5¼ lb.

Among other birds score-carded was her White Leghorn hen which, picked on my system and her first entry in any Test, won a medal in the single-pen section at Harper Adams the year before. She had laid 260 eggs, 1919-20, and 203 (still in lay) in 1920-1 when I scored her in October, 1921:—Capacity: 16, 5, 8½, 8, 4½, 9, 8, total 59. Capability: 10, 3, 15, 5, 4, 8½, 9, 8½, total 63. Weight 4¼ lb.

It may be of interest to give the score-card of an old Rhode Island Red hen taken at the same time. Hatched in 1917, she was not trapped the first year but laid 230 second year, 198 third year, and 149 (still in lay) up to October, 1921, when I scored her:—Capacity: 20, 5, 10, 10, 5, 8, 9, total 67.

Capability: 8, 3, 12, 5, 5, 8½, 10, 6, total 57½. Weight 5 lb. 6 oz. Note size and texture of vent, also fine pliability at abdomen for such an old hen, while her eggs had taken second that year at Tottenham Show, showing that texture of egg remained with the handling. Had she scored very low at abdomen, pelvis, etc., the eggs would have been deficient in shape and texture, also size.

LONGEVITY OF PRODUCTION.—Longevity of production, season after season, depends upon the *individual bird keeping high in capability*. There is great variation in the total number of eggs put up by birds year after year, which would not be credited by the average poultry-keeper. Most breeders only trap-nest their birds for the first year of lay and then breed from the highest producers, discarding the lowest performers. But should the first-year total stand for the lifetime quality of the particular bird? Many of those who win the Laying Tests trap-nest such winning birds for many years, although not a few fail to carry it out. Being able to obtain the trap-nest records of birds I handle, this variation, I can assure you, is very marked. For instance, one pullet lays extra well for the first year and then is all to pieces for numbers ever after. Another starts just medium and, the second season, surprises the owner with sixty more eggs than in the first year. Other birds lay on, year after year, for three and four years, like Miss Raye's Rhode Island Red hen.

The question arises as to how far this trap-nesting of pullets for their first-year lay helps us. If the record put up by the bird in her pullet year is true to her quality, any low output in future seasons or in the second year may be due to mishap or internal disorder such as weakened or fatty heart, or less virile muscles and oviduct. But with trap-nesting I advocate the use always of those birds which stand out for longevity of production, knowing that they exist and should be given preference in the breeding season. Supposing we put into a house a breeding pen of hens each of which has laid 250 eggs in the twelve pullet-year months. You may take it for granted, providing you have taken these birds solely by their records, each will yield a different total of eggs the second year. Which, then, should you breed your males from? I know that many save the chicks from the highest hens for numbers and toe-punch chicks from them to find later that their best hen for longevity is one which laid 220 the first year and 200 the second, but they have no chicks toe-punched from her because they backed the hen laying 270 the first year but only putting up 120 the second!

A lot of hens lay badly the second year because of being fattened during the first season and getting fatty hearts, while not a few fall by the way in longevity through saggy abdomens (excessively short breast-bones) and through being bronchial, not possessing the ideal type of backs. But should we take it for granted that all hens laying well the first year are up to quality whether they produce many eggs afterwards or not? If that is to be the case what does one do with the hen which lays 200 the first year and 230 the second, nowadays discarded as regards having her chicks toe-punched or marked because her first-year record is not as good as that of other preferred hens?

Where trap-nesting is practised this will say what the hen has laid the first year, but it will only tell the owner what she has done the second season after the close, when one has already bred from her and obtained a host of progeny. It is useless to attempt to reason out likely second-year production as being solely covered by the number of eggs laid the first year, *i.e.*, if very high the first then low the second year, and vice versa. One can, however, after trap-nesting the pullets for the first season place a dozen selected on their trap-numbers into a breeding house to be handled later on before the breeding season commences, so that any not handling well can be withdrawn. *The number of eggs likely for the second year will be decided as stated by the high or low capability of the bird when she starts her second effort.* That is easily understood seeing that the number of eggs to be laid the second (and even third) year will be governed by the moult. If a hen is soon through the moult after her first year of lay she will, as a general rule, put up a good total of eggs the second. If she is months in the moult she will be low for total the next year. Taking months over the moult she will, through not laying, get excessively fat at pelvis and abdomen. Handle a hen in her fourth season where she stops production in July of her third year and restarts in March of her fourth season, and she will have pelvis bones an inch thick with fat and gristle, and abdomen like a bladder of lard. That is where a hen loses her capability. After each year's production I want you to keep to those hens which handle well for production in the coming year—January and February being the best months for the handling of such hens. Where you do not trap-nest you will rely upon those hens which handle well at the beginning of each year of lay, so that every year they have to run the gauntlet of being checked up for future quality or capability held or lost by them. You need only check up, then, the capability as being high or low as the individual bird may show.

If only we use hens which are good layers for two years at least that will be better than high yield the first year and low the second. A hen may, as stated, not lay many eggs the first year yet looks and handles too good for it to be true, whereas another lays plenty but does not handle well for second-year output, giving one the impression she had plenty of luck with the first season's total and will not be equal to the task another twelve months. In breeding I always advocate the use of males bred from hens which have graded in high for capability for two or more years and laid well for the same period, if trap-nesting is employed. To breed cockerels I always go for males and hens which at the second year look fresh and young, retaining their quality. Thus a hen looking like a pullet at four years would be useful for breeding cockerels, just as a three-season cock looking like a cockerel would be ideal for longevity of quality or laying.

When it comes to egg-production and not breeding, it is just as important to know which birds handle well for second-year laying and longevity, because we do not want to keep for a whole second year a hen which will not pay for her keep from the new-laids she will lay. Nor do we want to keep any bird the moment she drops below the paying line. Readers will learn, too, the lesson of longevity as applied to management, because hens carefully managed through the moult will be better layers than those which are allowed to get as fat as butter and to become fatty at heart. Naturally, the coarser the bird or the lower she is in capability the lower will be her ability to lay. Do not, therefore, expect a cart-horse, beef type of hen well below the 50 capability line to lay "like smoke" for three or four years.

I have score-cards of thousands of birds which I have taken from time to time with their past and subsequent production, and they provide an interesting study. I have not given them publicity in this volume, wishing to keep to birds which have been shown under me at public Shows or entered in official Laying Tests. I will, however, dive into a few score-cards made by me on poultry farms of birds which come into line with the matter under discussion, viz., longevity of production.

I will deal only with the capability scores, as they are what concern us with longevity of laying or maintaining high capability over many years. Bear in mind that my ideal is over 50 for capability and as much over as possible.

On one farm I score-carded a pen which won at Harper Adams, scoring them between the close of their first laying year and the commencement of their second. Individual results of the seven birds, including the reserve, are as follows (scores taken by me in February):—

No. 1.—Laid 137 eggs at Test. Scored 44 capability (4, 3, 10, 3½, 4, 6, 7, 7). Last egg, September. First egg on restart, February. Graded out as well below 50 capability. Laid 3 eggs in February and 8 in March, and then discarded by breeder.

No. 2.—Laid 130 eggs at Test. Scored 46 capability (4, 3, 10, 3½, 3½, 7, 8, 7). Laid 116 eggs second year from October to July.

No. 3.—Laid 181 eggs at Test. Capability 48 (5, 3, 9, 4, 4, 7½, 8, 7½). Laid 122 eggs second year from October to July.

No. 4.—Laid 212 eggs at Test. Capability 50 (6, 3, 12, 4, 4, 5, 8, 8). Laid 139 eggs second season from October to July.

No. 5.—Laid 126 eggs at Test (no eggs first three months). Capability score 61. Laid 165 eggs second year from October to July. A hen my handling spoke up for as regards likely good production second season. Capability score 61 (10½, 3, 15, 4, 3½, 9, 8, 8).

No. 6.—Laid 157 eggs at Test. Capability score 53½ (7½, 3, 12, 4, 4, 8, 7, 8). Card marked "dropped abdomen; grade out." Laid 52 eggs second year in following up-and-down sequence: 7, 5, 2, 14, 3, 21. Then discarded. Note that I have always referred to this up-and-down monthly laying.

No. 7.—Laid 193 eggs at Test. Capability 51 (6, 3, 12, 5, 3½, 6, 7, 8½). Also marked "abdomen low." Laid 119 eggs second year, October to July, as follows: 0, 16, 20, 6, 12, 21, 13, 18, 13, and 0 for July.

On another farm I was able to score-card some birds which had won a Laying Test and been kept trap-nested. Again the grading was done in February between the laying seasons.

No. 1.—Laid 280 eggs first year, 140 second, 92 third (restarted March). Scored capability 46½ (5, 3, 8, 2½, 5, 7½, 8, 7½). Restarted in March and laid, fourth year, 60 eggs.

No. 2.—Laid 292 eggs first year, 180 second, 32 third. Scored capability (February) 49½ (5, 3, 10, 2½, 3½, 9, 9, 7½). Laid 90 eggs fourth year (restarted in March).

No. 3.—Laid 200 eggs first year, 171 second. Scored capability (February) 51 (6, 3, 12, 4, 4, 7, 8, 7). Laid 116 eggs third year; moulted from October of second year to January of third year before restarting in February.

No. 4.—Laid 160 eggs first year, 210 second. Scored capability 51½ (7, 3, 11, 4, 4, 7, 8, 7½). Laid 160 eggs third year.

Now for two hens with dropped abdomens, one scoring high for capability less the serious fault (for which I grade out a bird as undesirable without hesitation). I take no risks of 270 eggs or 27 for the twelve months, and I have proof of breeders in ignorance selecting solely for excessively short breast-bones and breeding the same stamp, only to get into difficulties and having to rebuild their strains.

No. 5.—Laid 182 eggs first year. Capability (February) 40, and marked "dropped abdomen probable; short breast-bone." Laid 68 eggs second year in following sequence: 9, 13, 7, 15, 5, 12, 7, for the months. Up one month and down the next. Capability 42 (5, 3, 6, 2, 4, 8, 7½, 6½).

No. 6.—Laid 210 eggs first year. Capability score (February) 53½ (6, 3, 9, 3, 3, 10, 9, 8½). A good handling type of hen but marked "abdomen down." Laid 61 eggs, laying right through the winter months as follows: November, 2 eggs; December, 8; January, 12; February, 4; March, 17; April, 9; May, 14; June, 5; July, 0; August, 0; September, 0; and October, 0. One can see from the above that the hen was in lay from November to June, and yet could only produce in "parts" showing the dropped abdomen was affecting it.

My matings have been responsible for only two 300-egg hens, and the White Leghorn hen of Miss Nancy Clayton's is by far the best bird for longevity I have known among my students. This hen laid 307 eggs in her first year (winter, 112), and I scored her as follows in August of her second year, which she finished with 173 eggs:—Capacity: 16, 5, 5, 8, 4, 7½, total 53½. Capability: 10, 3, 13, 4, 5, 8, 9, 8, total 60. She finished her third year of production with 177 eggs. Another White Leghorn laid 286 eggs first year (winter, 87), and was scored in August of second season as under, finishing with 203 eggs:—Capacity: 16, 5, 10, 8½, 4½, 8, 9, total 61. Capability: 7 (fat), 3, 10 (fat), 5, 4, 8½, 9, 8, total 54½. She laid 177 eggs in her third year, ending 1922, and is rather a better hen for capacity (more my ideal).

A TEST WITH DUCKS.—The importance of management cannot be too severely pressed upon readers, even if you have the right type of bird or duck. I remember being called in to a farm at Walton-on-Thames in April, 1922, because a flock of 180 ducks were not laying. Stagnant water was suspected as the likely cause, but I thought it was a case of low feeding which was compelling the ducks to take up anything whether fit to eat or not. Putting them on a good feeding menu saw the eggs start in May with 38 eggs daily from the 180 flock, reaching 130 by the end of May, with a grand total of 3,170 eggs for the month of June. Such should have been found out in the winter and not as late as May when the moult was coming along. Do not blame the stock until you have tested personal management.

"DUDS" BY MISMANAGEMENT.—Another case I took up for a poultry-farmer who engaged me to visit his farm, grade out a few of the best and send the rest to market, because he was sure the birds were "duds." Handling soon told me what was wrong and that he had good birds being spoilt. I located the cause, as he had misread a formula containing Sussex ground oats as a base. The formula said, dry off with Sussex ground oats, and not double it up by drying off with another lot of Sussex ground oats. Hence a fattening mash in use for months. A proper conditioning system was evolved by me—and drastic at that—and eggs ran up from 30 per day or 200 weekly from 150 fowls in April to 986 eggs for May and 1,706 for June, etc. We often smile over the "duds," seeing that from some of their progeny we picked a winning pen of Laying Test pullets in the autumn. *Again a case of handling telling accurately the condition of the flock.*

MATING BY HANDLING ALONE.—In August, 1922, I received a letter from Major Robert C. B. Lethbridge asking if I would visit his farm and do some grading. You will see from his letter that *his interest started with a win in utility* at the local Agricultural Show under Messrs. Jones Robins and C. H. Toy. They had suggested such a visit from me.

BROOMHILL,
TIVERTON, NORTH DEVON,
July 30th, 1922.

DEAR SIR,
My quite unexpected success at our utility Show, at which one of my pens won, among other specials, that for best pen in the Show, coupled with the judges' complimentary remarks, have encouraged me to ask if you could inspect the plant and stock and give me your opinion, which I should value highly whether favourable or otherwise.

Yours faithfully,
ROBERT C. B. LETHBRIDGE.

Early in September, 1922, I spent a day with Major Lethbridge at his farm and handled every hen, male, and pullet on the place. I planned the matings of three pens, ringing all hens and pullets so selected with distinguishing coloured rings. I set aside two breeding pens of early-hatched pullets for use in 1923 and a third pen of hens, and rung sixty to seventy pullets for trap-nesting and for future breeding as hens in 1924. *I had my pick from* 500 *pullets* and did not take into account pedigrees but went entirely by my handling. What is more, I studied closely colour (which we were out to improve), breeding value, etc. I mated up exactly as if the stock were my own and I were going to rely upon handling and not trap-nests. All the pullets so selected were trap-nested from my grading, so that records are available once again to check my methods. It was the first letter I had received from Major Lethbridge, so that I did not know him; yet if I were a "bluffer," as the critics suggest, would I have faced such a test? Here was I, in September, picking the future breeders which so far were maturing and would go ahead to be trapped for actual records individually.

Here is a breeder who is a member of the Poultry Club, Scientific Poultry Breeders' Association, and British Rhode Island Red Club. He specialises in Reds only for profit for stockbreeding and commercial egg-production. His farm is an approved station holding for the county under the egg-distribution scheme of the Ministry of Agriculture. He enters a pen of Reds in the utility section at his local Agricultural Show and secures a first prize and "best pen in Show." Receiving appreciation of his stock from the judges concerned, he writes seeking a visit from me. You can see how progressive such a breeder is, and even if he believed the critics that there was nothing in handling he went ahead on the test. He has always trap-nested his birds, so that he could have continued as before had he not been broad-minded and wished to know more about hand-grading.

I had not intended referring to my autumn, 1922, selections, but seeing Major Lethbridge's catalogue for 1923 I thought no harm would accrue from writing and getting the winter records of the pullets I selected, whether good or bad. In his catalogue, for instance, appears a photo of his *first Barnstaple Red pullet* which won under Mr. Wm. Runciman and against all breeds except White Wyandottes. The inscription stated that the pullet had laid *seventy-one eggs in three winter months from October 1st to December 31st despite four days at Show.* Naturally I was interested to know if that pullet was picked by me when grading in the September, and what kind of ring I had put on her. As you will see from his letter I had rung her blue for pen 1, for 1923 breeding. Mr. Runciman is a commercial egg-farmer himself who regularly shows utility birds and eggs under me with success.

Reading on, I found in pen 1 (1922 pullets) twelve pullets, six of which had records for three winter months of 80 eggs, 78, 66, 62, and 64; rest incompleted. Pen 2, also of February, 1922, pullets, consisted of thirteen, and seven had laid as follows for three winter months: 77 eggs, 75, 75, 72, 67, 66, 64; rest incomplete. Pen 3 of second-year hens had averaged 219 eggs in pullet year.

In the catalogue I also noticed a reference to Major Lethbridge's five Red pullets in *Feathered World* Test (his first entry in a laying competition and made on my advice at time of visit) which have led for three winter months all pens of the breed, and have held a leading place in its section against other breeds, always mentioned monthly in the official reports of leading pens.

November 1st, 1922.

DEAR MR. POWELL-OWEN,

The pullets that you graded for me are laying very well, and I expect to have some nice coloured birds in 1923. I picked out my Laying Trial birds on the lines you taught me but I doubt if I am yet experienced enough to pick out the best. I tried to send birds that had not laid, of this I was certain by trap-nest, but I fear that my learning is not yet advanced enough to ensure that my birds will all start at dawn on November 1st. All I can hope is that they will not disgrace me.

I won 1st utility cockerels and 1st utility pullets yesterday (all heavy breeds except White Wyandottes) at Barnstaple under Mr. Runciman, much to my delight, so I think that some of your lessons have borne fruit.

Yours truly,

ROBERT C. B. LETHBRIDGE.

I only have the first monthly report to hand from Harper Adams but I notice that four of the pullets in Major Lethbridge's pen laid 62 eggs for the month, or an average of 15½ eggs per bird, so some started at dawn! The only other entry by him was the *Feathered World* pen.

In reply to my request for detailed egg-records I have received the following letter:—

BROOMHILL,
TIVERTON, NORTH DEVON,
February 22nd, 1923.

DEAR MR. POWELL-OWEN,

Am very glad to be able to give you some figures concerning the birds which you graded for me last September, but you may remember that you departed in a measure from your usual grading because I asked you (1) to pick out two breeding pens of February pullets for 1923 breeding; (2) to pick out sixty or seventy good pullets to keep entirely for trap-nesting and for 1924 breeders as hens; (3) to keep for general egg-production the birds of inferior colour and type which nevertheless gave promise of heavy egg laying powers.

You put a blue ring on the Barnstaple winner [laid 71 eggs from October 1st to December 31st, less four days at show; won 1st under Mr. Wm. Runciman—Author], but the pullet that laid 80 eggs in three months was a bird that you chose for trap-nesting and subsequently for a 1924 breeder.

Including my breeding pullets you selected, from upwards of 500, 89 pullets, many of which I managed to keep from laying until they were fully grown, so my egg-records might easily have been better than they are. I kept moving the birds from house to house because I wanted big frames and no early precocity. Up to to-day these 89 pullets have laid 4,839 eggs since October 1st, an average per bird of 54 eggs each, though many of them did not start till well in December.

Of these birds 74 *have completed their three months since their first egg, and the average of their three months' records* is 55·3 eggs per bird, which does not take into consideration at all 215 eggs which were laid in the pen but not in the trap-nests; if these are divided up it means an additional 3 eggs each (nearly) to their records, or 58 *eggs each* approximately. One bird (pullet No. 4) laid 1 egg only, only ten pullets laid less than 40 eggs in three months, fifteen laid 70 or over, nineteen laid between 60 and 70, sixteen laid between 50 and 60, fourteen laid between 40 and 50. One pullet laid 70, and has three days to complete her three months. Another has laid 69, and has two full weeks to complete her three months.

As regards my highest records to date since October 1st I have 105, 98, 93, 90, 98, 91, 90, 99, 95, and a number of 80's.

All the pullets referred to above were selected by you, either for trap-nesting or as breeders, the former you rung white and the breeders blue, of the breeders there were twenty-four in all, but I cut out the one which has proved no good.

The pullets selected by you for breeding pen 1 have averaged, since October 1st to date, 68·2 eggs each, and those you selected for breeding pen 2 have averaged exactly 83 *each,* but these figures are not quite fair to you, for the pullets you chose for pen 1 had in many cases laid in August and September before I began to trap-nest and *quite half of them did a pretty thorough moult in November.* Most of them are now laying much larger eggs in consequence of the rest.

As regards the pullets which you advised me to keep for laying they have also done extremely well, considering they have all been together in a large flock, out in all weathers on absolutely sodden ground and fed on a non-stimulating mash. Since February 1st the number in the house has been 185, and the lowest daily total of eggs from them since that day was 104. I usually get from 110 to 120, and the weather has been consistently appalling.

I know that the foregoing figures are not record-breaking, but I do not force my birds but go for stamina and vigour. *I have had a perfect bill of health* since you were here, two adults only have died, one of them being caught in a rabbit wire and the other had a broken egg inside her.

I have a few very nice chicks out—between 600 and 700—and though they are always up to their hocks in mud and rain they are very strong and hearty. I think I have made some strides forward in the matter of colour.

I am most grateful to you for your selection of my birds and for your advice. *Your Score-card System must be sound if such good results can be obtained as I have experienced*—the trouble is, how many people can handle a bird properly? I'm sure I can't, but I hope to some day.

With many thanks.

Yours very truly,

R. C. B. LETHBRIDGE.

You will see from the letter that many of the early-hatched pullets moulted, as they usually do, which upsets their winter records, but at the same time if I am pleased with a bird I utilise her as a breeder. All were fed for quality of egg and not quantity, as should be the case with breeders. *Note, apart from production, how things panned out for reproduction* in that 600 to 700 chicks are reported out already by February 22nd. *A perfect bill of health* with only two deaths (accidental) among the adults. *Seventy-four pullets of the eighty-nine have finished their three winter months of production since the first egg, and the average is* 58 *eggs per bird, or nearly* 20 *per month per bird for three winter months.* This despite all moulting, etc. One bird laid only one egg, and naturally the critics or inexperienced would say that had damned the hand-grading. Naturally that first egg she laid sent her wrong internally, and she would have been discovered later on when refusing to come up to the food-trough at feeding-time.

I advocate, as I have so often declared, the use of my hand-grading system, first, to pick out "trappable" pullets and thus save labour, etc. What is more, I study reproduction from the start before I know a pullet's record, and trap-nest only those that will be of use as breeding hens and strain-builders when they have done their laying.

Secondly, I advocate the use of my handling system for all who cannot trap-nest, and here is a case where in one day I picked out the 1923 and likely 1924 breeders, and you can see what they have laid by trap-nest. Had they not been trapped, how far would I have been from sound matings made solely by experienced hand-grading? The hatching results you may judge by their reproductive returns, because only one other mated pen is catalogued, and that of untrapped pullets picked for frames and size of egg. *Am I not conscientious in advocating and teaching my handling system through my score-card?*

As I go to press, Major Lethbridge sends me a full report of the *six winter months' laying* of the pullets I selected by handling:—

TIVERTON,
April 9th, 1923.

. . . . The 28 breeding pullets for 1923 were housed in separate breeding houses, and the 60 pullets for 1923 trapping and 1924 breeding in a flock with free ranges and house 20 ft. by 16 ft. Trapping started on October 1st, and the records go to March 31st. Most of the February-hatched pullets were in lay in August and September, and in consequence moulted in November and December, reducing the average considerably.

The average of the whole 88 pullets by actual trap-nest for the six winter months is 91·1 eggs per bird. Taking the number of days in lay the total is 12,789 (since first egg in each case), with number of eggs 8,022, or a daily percentage of production in each case of 62·7 per cent.

Of 28 birds, which laid 100 eggs or over in the six winter months, only 6 have been in lay for the whole period.

28	pullets	laid	over	100 eggs within the period.			
10	,,	,,	,,	90	since	first	egg.
15	,,	,,	,,	80	,,	,,	,,
15	,,	,,	,,	70	,,	,,	,,
9	,,	,,	,,	60	,,	,,	,,
11	,,	,,	under	60	,,	,,	,,

Only 25 have laid less than 50 per cent. production since the first egg.

As regards those pullets picked for laying, the average was 17 eggs in January, 17 in February, and 19·2 in March, or 53 eggs each for the three months in a flock of over a hundred. Only one bird lost (other than by accident), viz., No. 8, which laid 65 eggs in three days less than three months and then broke an egg inside her.

I have over 1,500 chicks now and have finished hatching for the year—all splendidly healthy, and some weigh 3 lb. already.

R. C. B. LETHBRIDGE.

Had all the 88 pullets I picked out of the 500 been due to lay from October the average would have been higher, but, as it was, these 88 *pullets averaged* 91 *eggs for the six winter months.* Naturally, for breeders in 1923 I made full use of February pullets, knowing they would moult later, but sound reproduction was obtained, as witness 1,500 vigorous youngsters out as a season's operation from only three small breeding pens. They included, too, 300 April chickens hatched out especially for Laying Tests starting next autumn, showing that reproduction was maintained in addition to laying. All birds were especially picked as handling to my ideal as breeder-layers. Forty cockerels averaged 4 lb. at 16 weeks.

VISITS TO NESTS: NO EGGS.—Elsewhere in this volume I have referred to hens that visit the nest-box and never lay. Just as I close for press I have received the following letter from Mrs. B. Buckle—the successful breeder of Reds—which is the first occasion I have had confirmation of my own discoveries. I publish her letter, and you will see that one of her hens laid only one or two eggs in 1922, although constantly visiting the nest-box. Starting her 1923 laying season this hen went off with five eggs weekly. Being an exhibition Red she happened to be kept on, for which I am very pleased, as so many critics are apt to fall heavily on any new discoveries I make. During 1922 this hen was either making giant yolks and absorbing the fluid yolk-matter as it dropped into the abdomen, or she was just making yolks and absorbing them back again without the yolk-casings bursting and letting out the contents. The moult gave a resting-period to ovary, and after the moult she came into lay again as a normal hen. One can be pardoned for pointing out how little is known of the "Hen from Within," but I hope in time to fathom ALL its interesting problems.

68, Hazlehurst Brow,
Daisy Mill,
Bradford,
February 18th, 1923.

Dear Mr. Powell-Owen,

You will remember me telling you the other week at Bradford Show about a pullet which only laid one or two eggs last year and never deposited the goods again, though frequenting the nests—capacity but not capability.

I thought you would be interested to hear she is still laying five a week, and we have two batches of chicks from her; eight from eleven—the hen crushing one a fortnight ago—and to-day ten out of eleven—the eleventh being crushed with the broody sitting too tightly. In fact I had to change her, as she had two others with all the shell off them and I was afraid she would crush the lot.

We have four under another hen and nearly another sitting ready, so if she laid at the same rate last year what a number of eggs she must have consumed. No wonder she had such a sheen on her!

It was a good thing we did not kill her, as she is bred from a hen that won the "Cadbury Brown Bowl" at the club Show, and reported by the *Feathered World* as being the best pullet exhibited.

We are hoping great things from her, as the first sitting was sired by the 2nd club Show B.B. and the second sitting by the 5th Dairy cockerel.

We have chosen a couple of utility pens, all good coloured birds as near like our Bradford 3rd pullet as we could get them, and mated to cockerels from a 250-large-egg hen, so are hoping to build up a strain of good layers and winners.

Yours sincerely,
(Mrs.) B. Buckle.

Production and Reproduction.—The Guernsey cow is often quoted as proof that in-breeding or line-breeding does not mean a loss of stamina. But when recently visiting Guernsey I was able to get first-hand information concerning this beautiful "utility" cow. How well she answers to utility and beauty, to be sure. For over a hundred years the law has forbidden the importation of any live cattle into the island, so that the Guernsey has been bred within the existing strains there. Note, however, how well breeders keep to the law of the survival of the fittest. No bull can be used for service until he is over fifteen months old, and then he has to be passed by the Stud Book committee, and his dam must be present at such inspection. Again, the dam's and sire's dam's production must be registered, so that reproduction as well as production are concentrated upon in the official testing-out. I commend the same ideals to poultry-breeders, because too many put into pen 1 the layers of the most eggs (producers) as if they were the best birds. What of reproduction? Keep strictly to the pullet that lays well for longevity and is yet good enough in frame and construction to make a nice breeding hen after the production. Keep up vigour at all costs; breed well and kill well; and never trap-nest with your eyes shut. The eye and the hand are the brains of the trap-nest, and all three are essential to stockbreeding.

Guernsey is now taking a live interest in utility poultry, and in February, 1923, when I visited the island to give a poultry lecture, I was the guest of the Utility Poultry Council of the Royal Guernsey Agricultural and Horticultural Society, consisting of Miss R. G. Peek, Messrs. C. B. Blampied, F. F. Peek, H. W. Barnsley, F. W. Bichard, W. J. Dorey, H. Le Page, H. B. T. Boucher, and F. Brooke. I was motored to very many farms during my three days' stay, and, like Denmark, roup was unknown, and, again like Denmark, one mammoth incubator was installed. With a population of around 40,000, and an island ten miles by five miles, one will be able to watch the new industry develop. The members run a Laying Test and have utility poultry classes, and it may be decided to have so many stockbreeders who will trap-nest and a given number of commercial egg-farmers, and to encourage the small poultry-keepers to go for utility and beauty, which would mean an extra channel of demand for utility birds for showing as well as laying. I visited Mr. H. W. Barnsley—who is a dairy farmer and tomato grower—and he had 500 layers, reporting excellent profits. He is relying on hand-selection for his breeding hens, buying in reliable stock cockerels. Messrs. Ogier Brothers—also successful farmers—are adding a commercial egg side. Mr. Barnsley won the Laying Test with White Wyandottes, and in showing him the ideal type to aim at I picked what happened to be the best layer in the Test. Mr. F. Brooke—the secretary of the Utility Poultry Club—is adding poultry in a large way, while Mr. Boucher—who won an English Laying Test—specialises in Reds and Khaki Campbell ducks. Mr. Dorey specialises in White Wyandottes of Mr. Leslie Williams' strain, and Mr. Le Page is also increasing his poultry stock on his farm. Mr. Bichard—the Test manager—is often over in this country for poultry events.

Already one trap-nesting station is established, and here I saw some splendid utility White Wyandottes. I refer to Miss R. G. Peek's Delancey Poultry Farm. While staying as the guest of Mr. Gervase F. Peek—who has planned a clever poultry plant on labour-saving lines for his daughter—I hand-graded all the hens and pullets and marked the best for 1923 and 1924 breeding seasons. After the grading, the trap-records of the birds were examined and I found I had placed in pen 1 hens with the records of 464 (in just over two years)—which would be the best trap-record hens in the island; and others with records for pullet year and up to February 26th of second season of 285 eggs, 314, 332, 273, and 333. And to make up pullets with high winter records to date. Two pullets I graded out in Reds laid seventeen and twenty-three eggs respectively for the winter months to February 26th. I endeavoured while making my tour to leave a " key " bird for each poultry-keeper to study closely and to keep to as an ideal. While visiting the Vauxbelets Agricultural College with Mr. Granier (the director) and Brother Firme (the vice-president) I noticed vines in the poultry runs for shade, and the fowls were enjoying the insides of the stems of moellier—a French cabbage which might interest poultry-keepers here. The capacious stem is full of succulent matter. A visit was also paid to Mr. E. de Garis' farm, where Guernseys were the main line, he being responsible for all registrations, etc. And at Mr. William Mauger's we had a change to orchids, bulbs, and flowers. Guernsey is a land of agricultural activity, the thousands of glass-houses in the island being in ideal preparation for the season's crops of tomatoes, figs, grapes, etc. Fields upon fields of bulbs and flowers were a pleasant set-off to the number of glass-houses which dotted every estate. The largest glass-house I saw was 500 ft. long and 30 ft. wide at Mr. Gervase F. Peek's vineries.

ANOTHER PROBLEM SOLVED.—As I go to press I am able to report in brief the solution to another problem. The hen which constantly lays half-ounce eggs minus yolks, but containing albumen, has a functioning ovary, the yolks dropping into the body and being absorbed as in the " never-layer." Albumen is secreted in the oviduct and shelled.

Recently a hen demonstrated upon at a lecture possessed two ovaries and two oviducts. Another hen had infantile ovary and oviduct and masculine headgear.

61

62

63

65

64

INDEX.